AFLATOXINS
Progress and Challenges

AFLATOXINS
Progress and Challenges

Roland Kuruville
Carl Bialek

Kruger Brentt
Publishers
2023

Kruger Brentt Publishers UK. LTD.
Company Number 9728962

Regd. Office: 68 St Margarets Road, Edgware, Middlesex HA8 9UU

ISBN: 978-1-78715-036-2

For information on all our publications visit our website at http://krugerbrentt.com/

PREFACE

Aflatoxins are a family of toxins produced by certain fungi that are found on agricultural crops such as maize (corn), peanuts, cottonseed, and tree nuts. The main fungi that produce aflatoxins are *Aspergillusflavus* and *Aspergillusparasiticus*, which are abundant in warm and humid regions of the world. Aflatoxin-producing fungi can contaminate crops in the field, at harvest, and during storage.

People can be exposed to aflatoxins by eating contaminated plant products (such as peanuts) or by consuming meat or dairy products from animals that ate contaminated feed. Farmers and other agricultural workers may be exposed by inhaling dust generated during the handling and processing of contaminated crops and feeds. Aflatoxins are a major source of disease outbreaks due to a lack of knowledge and consumption of contaminated food and feed worldwide. Excessive levels of aflatoxins in food of non-industrialized countries are of major concern. Several effective physical, chemical, biological, and genetic engineering techniques have been employed for the mitigation, effective control and management of aflatoxins in food. However, developing fungal resistant and insect resistant hybrids/crops to combat pre-harvest infections and their outcome is a major issue of concern. Post-harvest treatments to remove aflatoxins such as alkalization, ammonization, and heat or gamma radiation are not generally used by farmers. However, some of the microorganisms naturally present in soil have the ability to degrade and reduce the aflatoxin contamination in different types of agricultural produce. Therefore, methods of using these organisms to reduce aflatoxin are currently being focused on.

The present book entitled ***Aflatoxins: Progress and Challenges*** contains 20 chapters covering all related disciplines. These chapters include Aflatoxins: An Introduction, Aflatoxins Issues, Food Safety, Health and their Management, Aflatoxins: Hazards in Grain, Aflatoxicosis and Livestock, Biological and Health Effects of Aflatoxins on Body Organs and Body Systems, Aflatoxins in Milk, Application of Mycotoxin Binders for

Control of Aflatoxins in Feed, Poultry Mycotoxicosis and its Control, Approaches for Reducing Aflatoxin M1 Content in Milk and Dairy Products, Reduction of Aflatoxin Contamination in Groundnuts and Maize, Transformation and Channeling of Aflatoxins into Food Raw Materials in Livestock, Aflatoxins in Food and Feed, Mycotoxins and their Consequences in Aquaculture, Innovative Technologies to Manage Aflatoxins in Foods and Feeds , Interventions to Prevent Aflatoxin-Induced Carcinogenesis in Developing Countries, Aflatoxins Occurrence in Spices, Mechanisms of Mycotoxicity, Fungal Growth and Mycotoxin Development, Mycotoxins and Human Disease, Mycotoxicoses of Animals, Risk Assessment and Regulations for Mycotoxins and Related terminology is given at the end for ready reference. This book would be an invaluable asset for the scientists, students, teachers, and manufacturers working in the field of human and animal nutrition, processing of food and feeds, animal and public health. This book is also a valuable reference source for anyone interested in knowing more about worldwide impact of aflatoxins and describes recent advances in research on aflatoxin production, analysis, and toxicology. We are grateful to all those persons as well as various books, manuals, periodicals, magazines, journals etc. that helped in the preparation of this book. In spite of the best efforts, it is possible that some errors may have occurred into the compilation and editing of the book. Further queries, constructive suggestions and criticisms for the improvement of the book are always welcome and shall be thankfully acknowledged.

Roland Kuruville

Carl Bialek

CONTENTS

1

AFLATOXINS: AN INTRODUCTION

1.1. INTRODUCTION

Mycotoxin (*Gk. mykes*, mushrooms-fungus) is a toxin produced by an organism of the fungus includes mushrooms, moulds and yeasts. Aflatoxins are naturally occurring mycotoxins that are produced as secondary metabolites by some strains of *Aspergillus flavus* and *A. parasiticus*, plus related species *A. nomius* and *A. niger* on a variety of food stuff causing health hazards to animals consuming them.

1.2. OCCURRENCE AND DISTRIBUTION OF AFLATOXINS

- Aflatoxin in swine has been reported in Georgia USA in 1940. The death of swine was traced as a result of feeding mould maize. Similar incident occurred in 1950 at Alabama. In 1960 more than 100,000 young turkeys on poultry farms in England died in the course of a few months from an apparently new disease that was termed Turkey X disease. It was soon found that the difficulty was not limited to turkeys. Ducklings and young pheasants were also affected and heavy mortality was experienced. A careful survey of the early outbreaks showed that they were all associated with feeds, namely Brazilian peanut meal.
- An intensive investigation of the suspect peanut meal was undertaken and it was quickly found that this peanut meal was highly toxic to poultry and ducklings with symptoms typical of Turkey X disease.
- Speculations made during 1960 regarding the nature of the toxin suggested that it might be of fungal origin. In fact, the toxin-producing fungus was identified as *Aspergillus flavus* in 1961 and the toxin was given the name Aflatoxin by virtue of its origin (*A. flavus* > Afla). This discovery has led to a growing awareness of the potential hazards of these substances as contaminants of food and feed causing illness and even death in humans and other mammals.

- In India Aflatoxin levels were found in range of 1000 to 5000 ppb (parts per billion) in 12% of groundnut samples in 1965 during a study conducted in Andhra Pradesh, 50% of groundnut cake samples were positive for aflatoxin in 1976 in Madhya Pradesh, similarly a study conducted by National Institute of Nutrition (N.I.N) (Hyderabad) recorded aflatoxin contamination in maize grain up to 1560 ppb which resulted death of 100 people by acute hepatitis in Tribal belts in western India in 1986.

1.3 BIOLOGY OF *ASPERGILLUS FLAVUS* AND *ASPERGILLUS PARASITICUS*

The two fungi *A. flavus* Link ex Fr. and *A. parasiticus* Spear, are closely related and grow as saprophytes on organic material left on and in the soil. They are distributed worldwide, with a tendency to be more common in countries with tropical climates that have extreme ranges of rainfall, temperature and humidity.

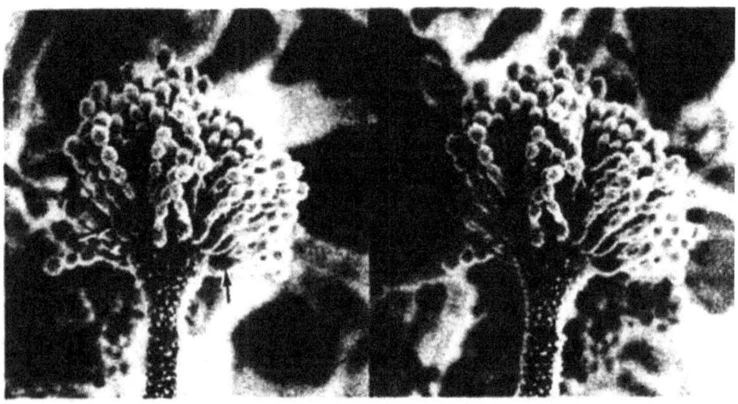

Fig. 1. Terminal portion of a conidiophore of *A. flavus* showing the basal portion of the vesicle and distribution of radiation phialides (arrows) 1000X.

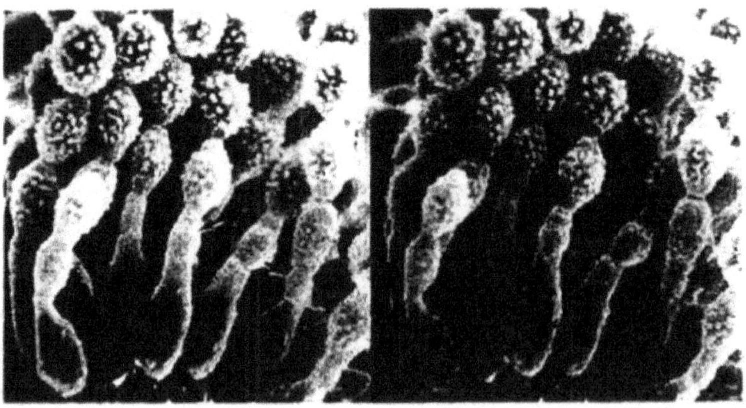

Fig. 2. Phialides and chains of conidia of *A. parasiticus* illustrating basipetal development of conidia. Those at the base of the chains (arrows) are least mature 3000X

Members of the genus *Aspergillus* are characterized by the production of non-septate conidiophores, which are quite distinct from hyphae and which are swollen at the top to form a vesicle on which numerous specialized spore-producing cells, known as phialides or sterigmata are borne either directly (uniseriate) or on short outgrowths known as metulae (biseriate).

Some time difficulty may arise especially to determine because the primary sterigmata are tiny and are easily obscured by spores or other sterigmata. Colonies of *A. flavus* are green- yellow to yellow-green or green on Czapek's agar. They usually have biseriate sterigmata; reddish-brown sclerotia are often present, conidia are finely roughened, variable in size and oval to spherical in shape.

Colonies of *A. parasiticus* dark green on Czepak's agar, remain green with age. Sterigmata are uniseriate, sclerotia are usually absent; conidia are coarsely echinulate, uniform in shape, size and echinulation. There are about 18 different types of aflatoxins identified major members are four plus two additional metabolic products, designated as $B_{1;}$ B_2, $G_{1;}$ G_2, and M_2.

1.4 . FLUORESCENCE PRODUCTION OF AFLATOXINS

- The aflatoxins fluorescence strongly in ultraviolet light (365 nm). Designation of aflatoxins B_1 and B_2 resulted from the exhibition of blue fluorescence of the relevant structures under ultraviolet light, G_1 and G_2 form yellow green fluorescence under ultraviolet light.
- While M_1 and M_2 were first isolated from milk of lactating animals fed aflatoxin preparations. *A. flavus* typically produces B_1 and B_2, whereas *A. parasiticus* produce G_1 and G_2 as well as B_1 and B_2. Four other aflatoxins M_1, M_2, B_{2A} and G_{2A} were isolated from cultures of *A. flavus* and *A. parasiticus.*

1.5. PROPERTIES OF AFLATOXINS

Studies revealed that aflatoxins are produced primarily by some strains of *A. flavus* and by most, if not all, strains of *A. parasiticus*, plus related species, *A. nomius* and *A. niger*. Moreover, these studies also revealed that there are four major aflatoxins: B_1, B_2, G_1, G_2 plus two additional metabolic products, M_1 and M_2, that are of significance as direct contaminants of foods and feeds.

The aflatoxins M_1 and M_2 were first isolated from milk of lactating animals fed aflatoxin preparations; hence, the M designation. Whereas the B designation of aflatoxins B_1 and B_2 resulted from the exhibition of blue fluorescence under UV-light, while the G designation refers to the yellow- green fluorescence of the relevant structures under UV-light.

Table 1. Properties of Aflatoxins

Aflatoxin	Molecular formula	Molecular weight
B_1	$C_{17}H_{12}O_6$	312
B_2	$C_{17}H_{14}O_6$	314
G_1	$C_{17}H_{12}O_7$	328
G_2	$C_{17}H_{14}O_7$	330
M_1	$C_{17}H_{12}O_7$	328
M_2	$C_{17}H_{14}O_7$	330
B_2A	$C_{17}H_{14}O_7$	330
G_2A	$C_{17}H_{14}O_8$	346

These toxins have closely similar structures and form a unique group of highly oxygenated, naturally occurring heterocyclic compounds. Food products contaminated with aflatoxins include cereal (maize, sorghum, pearl millet, rice, wheat etc.), oilseeds (groundnut, soybean, sunflower, cotton), spices (chilies, black pepper, coriander, turmeric, zinger), tree nuts (almonds, pistachio, walnuts, coconuts) and milk.

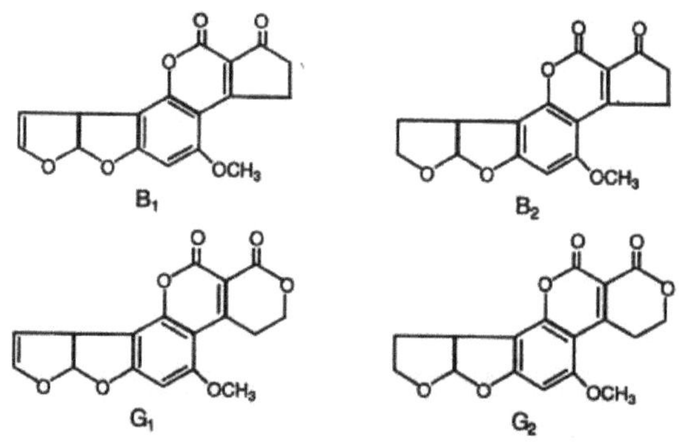

Fig. 3. Structures of aflatoxins B_1, B_2 and G_2

Aflatoxins are potent toxic, carcinogenic, mutagenic, immunosuppressive agents, produced as secondary metabolites by the fungus Aspergillus flavus and *A. parasiticus* on variety of food products. Among 18 different types of aflatoxins identified, major members are aflatoxin B_1; B_2, G_1 and G_2 (Fig.4). Aflatoxin B_1 (AFB_1) is normally predominant in amount in cultures as well as in food products.

Pure AFB1 is pale-white to yellow crystalline, odorless solid. Aflatoxins are soluble in methanol, chloroform, acetone, acetonitrile. Aflatoxin M_1 and M_2 are major metabolites of aflatoxin B_1 and B_2 respectively, found in milk of animals that have consumed feed contaminated with aflatoxins.

Aflatoxins are normally refers to the group of difuranocoumarins and classified in two broad groups according to their chemical structure; the difurocoumarocyclopentenone series (AFB_1, AFB_2, AFB_2A, AFM_1, AFM_2, AFM_2A and aflatoxicol) and the difurocoumarolactone series (AFG_1, AFG_2, AFG_2A, $AFGM_1$, $AFGM_2$, $AFGM_2A$ and AFB_3). The aflatoxins display potency of toxicity, carcinogenicity, mutagenicity in the order of $AFB_1 > AFG_1 > AFB_2 > AFG_2$ as illustrated by their LD_{50} values for day-old ducklings.

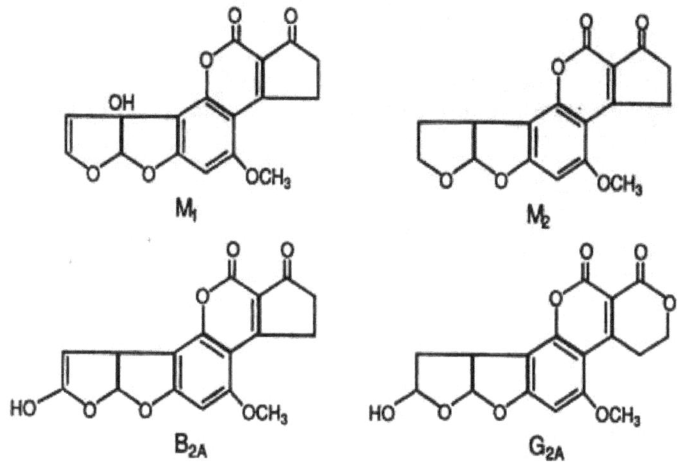

Fig. 4. Structures of aflatoxins M_1, M_2, B_{2A} and Q_{2A}.

Structurally the dihydrofuran moiety, containing double bond, and the constituents liked to the coumarin moiety are of importance in producing biological effects. The aflatoxins fluorescence strongly in ultraviolet light (ca. 365 nm); B_1 and B_2 produce a blue fluorescence where as G_1 and G_2 produce green fluorescence.

1.6. HEALTH HAZARDS CAUSED BY AFLATOXINS

In 1967, twenty people of Taiwan became ill with apparent food poisoning due to moldy rice that was later confirmed to contain Aflatoxin B_1. In 1978 Significant levels of aflatoxin were found in the livers 23 of children who had died of Rye's syndrome in Thailand. Children who died in Czechoslovakia and New Zealand in 1977 have also been found to have aflatoxin in their liver at autopsy.

The outbreak of hepatitis that effected 400 people in western states of India in 1974 of whom 100 died by consuming corn contaminated with *A. flavus* containing about 15 mg/kg aflatoxin. Aflatoxicosis is primarily a hepatic disease. The susceptibility of individual animals to aflatoxins varies considerably depending on species, age, sex, and nutrition. In fact, aflatoxins cause liver damage, decreased milk and egg production, recurrent infection as a result of immunity suppression (e.g., salmonellosis), in addition to embryo toxicity in animals consuming low dietary concentrations. While the young of a species are most susceptible, all ages are affected but in different degrees for different species.

Clinical signs of aflatoxicosis in animals include gastrointestinal dysfunction, reduced re-productivity, reduced feed utilization and efficiency, anemia, and jaundice. Nursing animals may be affected as a result of the conversion of aflatoxin B_1 to the metabolite aflatoxin M_1 excreted in milk of dairy cattle.

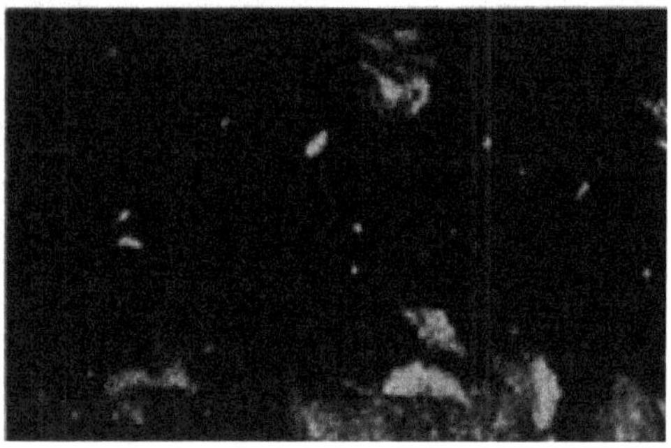

Fig. 5 Six rat livers injected with increasing doses of aflatoxin B_1. The liver in the upper left corner received no aflatoxin (control), while the one at the lower right corner received the highest dose. Observe the color difference in these livers.

Fig. 6. A rat live fed with high doses of aflatoxins B_1. Notice the induced tumors in the liver.

Table 2. Acute toxocity of aflatoxin B_1 expressed as a single oral dose LD_{50} (Anonymous, 2008)

Species	LD_{50} mg Kg^{-1} body weight
Rabbit	0.30
Duckling (11 day old)	0.43
Cat	0.65
Pig	0.60
Rainbow trout	0.80
Dog	0.50-1.00
Sheep	1.00-2.00
Guinea pig	1.40-2.00
Baboon	2.00
Chicken	6.30
Rat (male)	5.50-7.20
Rat (female)	17.90
Macaque (female)	7.80
Mouse	9.00
Hamster	10.20

The induction of cancer by aflatoxins has been extensively studied. Aflatoxin B_1, aflatoxin M_1 and aflatoxin G_1 have been shown to cause various types of cancer in different animal species. However, only aflatoxin B_1 is considered by the International Agency for Research on Cancer (IARC) as having produced sufficient evidence of carcinogenicity in experimental animals to be identified as a carcinogen (Figs. 5-6). Thus aflatoxins were proved to be potent toxic carcinogenic, mutagenic, immunosuppressive agents. Their toxicity is illustrated by their LD_{50} values for day old ducklings (Table 2).

1.7 AFLATOXIN: PRODUCING POTENTIAL OF FUNGI

Toxin producing potentials vary with the nature of the fungal organism and it may remain confined to specific or even at strain level. More than 150 species of fungi are known to produce different types of mycotoxins in laboratory as well as under natural conditions.

The concern about mycotoxins producing potentials of molds has increased since 1960 i.e. after the discovery of 'Turkey-K-disease' which was attributed to aflatoxin elaborated by *Aspergillus flavus*, since then several hundred strains belonging to this group have been isolated from various food and feed materials. All the strains of *A. flavus*, however, do not possess the capacity of elaborating aflatoxins. Screening of *A. flavus* isolates for aflatoxin production also received considerable importance in India.

Aflatoxins elaboration by *A. flavus* on various food substrates largely depends upon the nature of the substrates and the environmental conditions. Moisture and temperature are the two most important factors which exert decisive influence on mycotoxin elaboration. Besides environmental factors, aflatoxin elaboration also varies with the nature. A particular danger from consumption of aflatoxin contaminated food and feed lies in the possibility that some may damage the hereditary material of man. If this is so, what is the magnitude of the effect, are regrettably unknown. Surely, one of the greatest responsibilities of our generation is our temporary custody of the genetic heritage received from our ancestors. We must make every responsible effort to ensure that this heritage is passed on to future generation undamaged.

Fig. 7. Consumption of aflatoxin contaminated food may damage the hereditary material of humans.

Socially responsible scientists should be concerned about the potential hazards of such mycotoxin-induced mutations for two more reasons *i.e.* the genetic impairments, and somatic hazards. The former may enhance our existing burden of disease and premature death because it is a well realised fact that about 40% of our health problem is of genetic origin.

Somatic hazards include the manifestations of the action of mycotoxin on DNA and other cellular constituents. Needless to say that our existing genetic load is a summation of these three categories of genetic damage (replication, translation and transcription) as well as the availability of surface areas of the substrates. Among all the mycotoxins, aflatoxins occupy key position with regards to carcinogenic effects on animals and human systems. These are one of the potent hepatocarcinogens known so far and can induce carcinoma (Table 3).

Table 3: Mycotoxins present in different commodities and associated mycotoxicoses

Mycotoxins	Mycotoxicoses	Commodities
(A) *Aspergillus* toxin		
(i) Allatoxin B_1, B_2, G_1, G_2	Liver cancer	Corn, peanuls, milk
(ii) Sterigmatocystin	Carcinogenesis	Green coffee, grains
(iii) Ochratoxin	Renal tumour	Corn, coffee, wheat flour bread
(B) *Fusarium* toxins		
(i) Moniliformin	Onlyai disease	Rice
(ii) Fumomisins	Leukoencephalomalacia promote cancer	Corn, wheat flower Corn flakes
(iii) Trichothecenes (T-2 toxin. Deoxyninalend),	Dermatitis.oesophageal digestive disorder	Corn, wheat, wine, cancer, Commercial cattle feed.
(iv) Zearelenone	Cervical cancer abonion	Corn meal, corn flakes. walnut
(C) *Penicillium* toxins		
i) Citrovirdin	Cardiac damage	Mouldy peanut,
ii) Citrinin	Kidney damage	dry fruit, nice, corn
(iii) Cyclopiazonic acid	Kodua poisoning	Cheese crust. corn
(iv) Patulin	Capillary damage in vital organs	Cider. apple juice, jam, scented beetle nut
(v) *Penicillinis* acid	Edema carcinogenesis	Corn. bean, apple,
(vi) Penitrem A	Bloody diarrhoea death	Mouldy cream cheese
(viii) Rubraloxin	Liver disease	Mouldy grains
(D) Other mycotoxins		
(i) *Agaricus* toxins	Kwashiorkor	*Agaricus bisporus.* frozen mushrooms
(ii) *Amanitins*	Mushroom poisoning	Mushrooms
(E) Ergot alkaloides		
(i) Ergosine, Ergometrin. Ergocristine	Ergotism	Flour of rye, wheat, triticalc. baby cereals

1.8 EFFECT OF *A. FLAVUS* & AFLATOXINS CONTAMINATION

Deteriorate in grain quality due to *A. flavus* growth and become unfit for marketing and consumption. In groundnut, seed and non-emerged seedling decay and aflaroot disease was observed due to fungus attack. Aflatoxins contamination in grain poses a great threat to human and livestock health as well as international trade. According to FAO estimates, 25% of the world food crops are affected by mycotoxins every year. And also crop loss due to aflatoxins contamination costs US producers more than $100 million per year on average including $ 26 millions to peanuts ($ 69.34/ha).

1.9 FACTORS FAVOURING AFLATOXIN PRODUCTION

- Fungal growth and aflatoxin contamination are the consequence of interactions among the fungus, the host and the environment. The appropriate combinations of these factors determine the infestation and colonization of the substrate, and the type and amount of aflatoxin produced. However, a suitable substrate is required for fungal growth and subsequent toxin production, although the precise factor(s) that initiates toxin formation is not well understood.
- Water stress, high-temperature stress, and insect damage of the host plant are major determining factors in mold infestation and toxin production. Similarly, specific crop growth stages, poor fertility, high crop densities, and weed competition have been associated with increased mold growth and toxin production.
- Aflatoxin formation is also affected by associated growth of other molds or microbes. For example, pre-harvest aflatoxin contamination of peanuts and corn is favored by high temperatures, prolonged drought conditions, and high insect activity; while post harvest production of aflatoxins on corn and peanuts is favored by warm temperatures and high humidity.

1.10 ANALYSIS FOR AFLATOXINS IN FOODS AND FEEDS:

1.10.1 Sampling and Sample Preparation

Sampling and sample preparation remain a considerable source of error in the analytical identification of aflatoxins. Thus, systematic approaches to sampling, sample preparation, and analysis are absolutely necessary to determine aflatoxins at the parts-per-billion level. In this regard, specific plans have been developed and tested rigorously for some commodities such as corn, peanuts, and tree nuts; sampling plans for some other commodities have been modeled after them. A common feature of all sampling plans is that the entire primary sample must be ground and mixed so that the analytical test portion has the same concentration of toxin as the original sample.

1.10.2 Thin-Layer Chromatography

Thin layer chromatography (TLC), also known as flat bed chromatography or planar chromatography is one of the most widely used separation techniques in aflatoxin analysis. Since 1990, it has been considered the AOAC official method and the method of choice to identify and quantitative aflatoxins at levels as low as 1 rg/g . The TLC method is also used to verify findings by newer, more rapid techniques.

1.10.3 Liquid Chromatography

Liquid chromatography (LC) is similar to TLC in many respects, including analytic application, stationary phase, and mobile phase. Liquid chromatography and TLC complement each other. For an analyst to use TLC for preliminary work to optimize LC separation conditions is not unusual. Liquid chromatography methods for the determination of aflatoxins in foods include normal-phase LC (NPLC), reversed-phase LC (RPLC) with pre- or before-column derivatization (BCD), RPLC followed by post column derivatization (PCD), and RPLC with electrochemical detection.

1.10.4 Immunochemical Methods

Thin layer chromatography and LC methods for determining aflatoxins in food are laborious and time consuming. Often, these techniques require knowledge and experience of chromatographic techniques to solve separation and interference problems. Through advances in biotechnology, highly specific antibody-based tests are now commercially available that can identify and measure aflatoxins in food in less than 10 minutes. These tests are based on the affinities of the monoclonal or polyclonal antibodies for aflatoxins. The three types of immunochemical methods are radioimmunoassay (RIA), enzyme-linked immunosorbent assay (ELISA), and immunoaffinity column assay (ICA).

1.10.5 PCR-Method

Manonmani (2006) developed a rapid method for assessment of aflatoxigenic fungi in food using an indigenously designed primers for aflatoxin regulatory gene aflR in PCR shows positive amplification with DNA of aflatoxigenic *A. flavus* and *A. parasiticus* can detect up to 100 cfu of target toxigenic fungi.

1.11 CONTROL AND MANAGEMENT OF AFLATOXINS

1.11.1 Regulatory Control

- Aflatoxins are considered unavoidable contaminants of food and feed, even where good manufacturing practices have been followed. The FDA has established specific guidelines on acceptable levels of aflatoxins in human food and animal feed by establishing action levels that allow for the removal of volatile lots from commerce.
- The action level for human food is 20 ppb total aflatoxins, with the exception of milk which has an action level of 0.5 ppb for aflatoxin M_1. The action level for most feeds is also 20 ppb. However, it is very difficult to accurately estimate aflatoxins concentration in a large quantity of material because of the variability

associated with testing procedures. No presentable data is available to form an experimentally valid basis for regulation of aflatoxin levels in food and feed hence, the true aflatoxin concentration in a lot cannot be determined with 100% certainty.

- In US and Europe aflatoxins are only mycotoxins jurisdictionally regulated. In Asian countries regulation is primarily introduced to protect export market. On the other hand domestic regulatory measures on aflatoxin received little attention. In India mycotoxin legislation have been introduced but its implementation is inadequate.

1.11.2 Detoxification Strategies

Because aflatoxin contamination is unavoidable, numerous strategies for their detoxification have been proposed. These include physical methods of separation, thermal inactivation, irradiation, solvent extraction, adsorption from solution, microbial inactivation, and fermentation. Chemical methods of detoxification are also practiced as a major strategy for effective detoxification.

Structural Degradation Following Chemical Treatment: A diverse group of chemicals has been tested for the ability to degrade and inactivate aflatoxins. A number of these chemicals can react to destroy (or degrade) aflatoxins effectively but most are impractical or potentially unsafe because of the formation of toxic residues or the perturbation of nutrient content and the organoleptic properties of the product.

Two chemical approaches to the detoxification of aflatoxins that have received considerable attention are ammoniation and reaction with sodium bisulfite. Many studies provide evidence that chemical treatment via ammoniation may provide an effective method to detoxify aflatoxin-contaminated corn and other commodities. The mechanism for this action appears to involve hydrolysis of the lactone ring and chemical conversion of the parent compound aflatoxin to numerous products that exhibit greatly decreased toxicity. On the other hand, sodium bisulfite has been shown to react with aflatoxins (B_1, G_1 and M_1) under various conditions of temperature, concentration, and time to form water-soluble products.

Modification of Toxicity by Dietary Chemicals: The toxicity of mycotoxins may be strongly influenced by dietary chemicals that alter the normal responses of mammalian systems to these substances. A variable array of chemical factors, including nutritional components (*e.g.*, dietary protein and fat, vitamins, and trace elements), food and feed additives (*e.g.*, antibiotics and preservatives), as well as other chemical factors may interact with the effects of aflatoxins in animals.

Alteration of Bioavailability by Aflatoxin Chemisorbents: A new approach to the detoxification of aflatoxins is the addition of inorganic sorbent materials, known as chemisorbents, such as hydrated sodium calcium aluminosilicate (HSCAS) to the diet of animals. HSCAS possesses the ability to tightly bind and immobilize aflatoxins in the gastrointestinal tract of animals, resulting in a major reduction in aflatoxin bioavailability.

1.11.3. Biological Control of Aflatoxins

- Ciegler (1966) after screening about 1000 microorganisms found Flavobacterium aurantiacum NRRL B-164 ability to remove aflatoxin B_1 form solution. Milk corn oil and peanut butter were artificially contaminated with 600, 700 and 700 micro gram of aflatoxin B_1 respectively when viable cellos of Flavobacterium aurantiacum were added to each of these foodstuffs aflatoxin concentration were reduced to about 0. When viable cells were mixed with soybean, corn and peanut seeds contaminated with aflatoxin were completely removed from corn and peanut and about 86% from soybean. Displacement of toxigenic strains of *A. flavus* with a toxigenic strains is one of the effective strategy.
- This strategy is possible because of great variability of phenotypes of *A. flavus* in agricultural fields and common occurrence of a toxigenic strains. Use of bio-fungicides can be a better choice for controlling aflatoxin fungi. Neem extracts have been reported to block aflatoxin biosynthesis up to 98% (Bhatnagar and McCormic, 1988). Essential oils of Thymus *eriocalyx* and *Gracinia* indica (extract) have been reported inhibitory against aflatoxin producing fungi and also inhibit aflatoxin production.

1.1.4 Miscellaneous Control Measures of Aflatoxins

- Elimination of these compounds from the food and animal feed is necessary. Three basic approaches *i.e.* prevention, inactivation and detoxification have been proposed to control the mycotoxins.
- A large number of chemical fungicides and fumigants like propionic acid, acetic acid, ethylene bromide, sulfur dioxide, luprosil *etc.* have been found to be very effective in preventing mould growth and aflatoxin elaboration.
- Propionic acid and crystal violet were found to exhibit greatest antifungal activity in liquid medium against toxigenic strains of *A. parasiticus.* These compounds also significantly prevented the aflatoxin elaboration by *A. parasiticus.* Prevention of alfatoxm production by the use of plant product have achieved aflatoxin inhibition.
- Several attempts have been made to detoxify aflatoxin contaminated food materials by physical chemical and biological means. However, physical separation procedures have been found to be most successful. An alternative method would be to use UV light to detect potentially contaminated lots and segregate them out but use of X-rays and UV rays have been found to be ineffective in minimizing aflatoxin level in food stuffs.

- Detoxification of aflatoxin have been achieved through heat treatments. Destruction of aflatoxin in food materials has also been suggested through roasting, frying in oils, spray drying, baking and autoclaving. Sunlight reduces alfatoxm level in peanut oil.

2

AFLATOXINS ISSUES, FOOD SAFETY, HEALTH AND THEIR MANAGEMENT

2.1 INTRODUCTION

Aflatoxins are a major source of disease outbreaks due to a lack of knowledge and consumption of contaminated food and feed worldwide. Excessive levels of aflatoxins in food of non-industrialized countries are of major concern. Several effective physical, chemical, biological, and genetic engineering techniques have been employed for the mitigation, effective control and management of aflatoxins in food.

The aflatoxin producing fungi, *Aspergillus* spp., are widely spread in nature and have severely contaminated food supplies of humans and animals, resulting in health hazards and even death. Therefore, there is great demand for aflatoxins research to develop suitable methods for their quantification, precise detection and control to ensure the safety of consumers' health. Aflatoxins are one of the highly toxic secondary metabolites derived from polyketides produced by fungal species such as *Aspergillus flavus, A. parasiticus*, and *A. nomius* .These fungi usually infect cereal crops including wheat, walnut, corn, cotton, peanuts and tree nuts and can lead to serious threats to human and animal health by causing various complications such as hepatotoxicity, teratogenicity, and immunotoxicity (Fig. 1). The major aflatoxins are B_1, B_2, G_1, and G_2, which can poison the body through respiratory, mucous or cutaneous routes, resulting in overactivation of the inflammatory response.

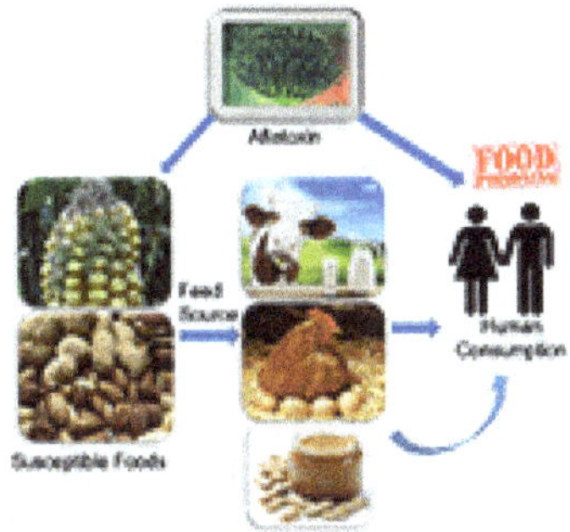

Fig. 1 Overview of aflatoxin effects on humans.

Food safety is one of the major problems currently facing the world; accordingly, a variety of studies have been conducted to discuss methods of addressing consumer concerns with various aspects of food safety. Since 1985, the United States Food and Drug Administration (USFDA) has restricted the amount of mycotoxins permitted in food products. The USDA Grain and Plant Inspection Service (GPIS) have implemented a service laboratory for inspection of mycotoxins in grains. Additionally, the Food and Agricultural Organization (FAO) and World Health Organization (WHO) have recognized many toxins present in agricultural products. When mycotoxins are contaminated into foods, they cannot be destroyed by normal cooking processes. However, there have been many recent advances in food processing developed to keep final food products safe and healthy, such as hazard analysis of critical control points (HACCP) and good manufacturing practices (GMP;. Moreover, several physical, chemical and biological methods can be applied to partially or completely eliminate these toxins from food and guarantee the food safety and health concerns of consumers. This review provides an overview of aflatoxigenic fungi, chemistry and biosynthesis of aflatoxins, along with their diversity in occurrence, and their health related risks to humans and livestock.

2.2 MAJOR OUTBREAKS DUE TO AFLATOXINS

In 1974, a major outbreak of hepatitis due to aflatoxin was reported in the states of Gujarat and Rajasthan in India, resulting in an estimated 106 deaths. The outbreak lasted for 2 months and was confined to tribal people whose main staple food, maize, was later confirmed to contain aflatoxin. The preliminary analysis confirmed that consumption of *A. flavus* had occurred. Another outbreak of aflatoxin affecting both humans and dogs was reported in northwest India in 1974. A major aflatoxin exposure outbreak was subsequently documented in Kenya in 1981. Since 2004, multiple aflatoxicosis outbreaks have been reported worldwide, resulting in 500 acute illness and 200 deaths. Most outbreaks have been reported from rural areas of the East Province of Kenya in 2004 and occurred because of consumption of home grown maize contaminated with molds. Preliminary testing of food from affected areas revealed the presence of aflatoxin as reported in 1981. In 2013, countries in Europe including Romania, Serbia, and Croatia reported the nationwide contamination of milk with aflatoxin.

2.3 MAJOR SOURCE OF AFLATOXIN

The major sources of aflatoxins are fungi such as *A. flavus, A. parasiticus*, and *A. nomius* , although they are also produced by other species of *Aspergillus* as well as by *Emericella* spp.. There are more than 20 known aflatoxins, but the four main ones are aflatoxin B_1 (AFB_1), aflatoxin B_2 (AFB_2), aflatoxin G_1 (AFG_1), and aflatoxin G_2 (AFG_2;), while aflatoxin M_1 (AFM_1) and M_2 (AFM_2) are the hydroxylated metabolites of AFB_1 and AFB_2 .

2.3.1 Aspergillus spp.

The *Aspergillus* species are an industrially important group of microorganisms distributed worldwide. *A. niger* has been given Generally Recognized as Safe (GRAS) status by the USFDA. However, some species have negative impacts and cause diseases in grape, onion, garlic, peanut, maize, coffee, and other fruits and vegetables. Moreover, *Aspergillus* section *nigri* produces mycotoxins such as ochratoxins and fumonisins in peanut, maize, and grape. Plant–pathogen interactions have been studied using molecular markers such as green fluorescent protein (GFP) isolated from *Aequorea Victoria.* The GFP gene has been successfully inserted into *Undifilum oxytropis* , *Fusarium equiseti* , and *Muscodor albus* and utilized to study the expression of different proteins and production of mycotoxins. *A. flavus* and *A. parasiticus* infect many crops in the field, during harvest, in storage, and during processing. *A. flavus* is dominant in corn, cottonseed, and tree nuts, whereas *A. parasiticus* is dominant in peanuts. *A. flavus* consists of mycelium, conidia, or sclerotia and can grow at temperatures ranging between 12 and 48°C. *A. flavus* produces AFBI and AFB۲, whereas *A. parasiticus* isolates produce AFGI, AFG۲, AFM۱, AFBI, and AFB۲. *A. flavus* produces a number of airborne conidia and propagules that infect plants such as cotton. A high number of propagules was reported in soil, air, and on cotton leaves during mid- to late August, while soilborne inoculum increased drastically between April and December in cotton fields in Arizona. This fungus can even colonize moribund rye cover crop and peanut fruit debris.

2.3.2 Aflatoxin (AFT)

Among the mycotoxins affecting food and feed, aflatoxin is the major one in food that ultimately harms human and animal health. The level of toxicity associated with aflatoxin varies with the types present, with the order of toxicity being AFTs-B_1 > AFTs-G_1 > AFTs-B_2 > AFTs-G_2 .

2.4 CHEMISTRY AND BIOSYNTHESIS OF AFLATOXINS

Chemically, aflatoxins (AFTs) are difuranocoumarin derivatives in which a bifuran group is attached at one side of the coumarin nucleus, while a pentanone ring is attached to the other side in the case of the AFTs and AFTs-B series, or a six-membered lactone ring is attached in the AFTs-G series. The physical, biological and chemical conditions of *Aspergillus* influence the production of toxins. Among the 20 identified AFTs, AFT-B_1, and AFT-B_2 are produced by *A. flavus,* while AFT-G_1 and AFT-G_2 along with AFT-B_1 and AFT-B_2 are produced by *A. parasiticus* . AFT-B_1, AFT-B_2, AFT-G_1, and AFT-G_2 are the four major naturally produced aflatoxins. AFTs-M_1 and AFTs-M_2 are derived from aflatoxin B types through different metabolic processes and expressed in animals and animal products. AFT-B_1 is highly carcinogenic, as well as heat resistant over a wide range of temperatures, including those reached during commercial processing conditions.

The biosynthetic pathway of aflatoxins consists of 18 enzymatic steps for conversion from acetyl-CoA, and at least 25 genes encoding the enzymes and regulatory pathways have been cloned and characterized. The gene comprises 70 kb of the fungal genome and is regulated by the regulatory gene, *aflR* . The metabolic grid involved in the aflatoxin biosynthesis. Hydroxyversicolorone (HVN) is converted to versiconal hemiacetal acetate

(VHA) by a cytosol monooxygenase, in which NADPH is a cofactor. Monooxygenase is encoded by the *moxY* gene, which catalyzes the conversion of HVN to VHA and the accumulation of HVN and versicolorone (VONE) occurs in the absence of the *moxY* gene.

2.4.2 Gene Responsible for Aflatoxin Production

Various genes and their enzymes are involved in the production of sterigmatocystin (ST) dihydrosterigmatocystin (DHST), which are the penultimate precursors of aflatoxins. The aflatoxin biosynthesis gene *nor-1*, which was first cloned in *A. Parasiticus*, is named after the product formed by the gene during biosynthesis. These genes named according to substrate and the product formed *nor-1* (norsolorinic acid [NOR]), *norA*, *norB*, *avnA* (averanti [AVN]), *avfA* (averufin [AVF]), *ver 1* (versicolorin A [VERA]), *verA* and *verB* while those based on enzyme functions *fas 2* (FAS alpha subunit), *fas-1* (FAS beta subunit), *pksA* (PKS), *adhA* (alcohol dehydrogenase), *estA* (esterase), *vbs* (VERB synthase), *dmtA* (mt-I; *O*-methyltransferase I), *omtA* (*O*-methyltransferase A), *ordA* (oxidoreductase A), *cypA* (cytochrome P450 monooxygenase), *cypX* (cytochrome P450 monooxygenase), and *moxY* (monooxygenase). Initially, the aflatoxin regulatory gene was named *afl-2* in *A. flavus* and apa^{-2} in *A. parasiticus*). However, it was subsequently referred to as *aflR* in *A. flavus, A. parasiticus*, and *A. nidulans* because of its role as a transcriptional activator. Previous studies have shown that *aflA* (fas^{2-}), *aflB* (fas^{-1}), and *aflC* (*pksA*) are responsible for the conversion of acetate to NOR. Moreover, the *uvm8* gene was shown to be essential for NOR biosynthesis as well as aflatoxin production in *A. parasiticus*. The amino acid of sequence of the gene is similar to that of the beta subunit of FASs (FAS_1) from *Saccharomyces cerevisiae* . FAS forms the polyketide backbone during aflatoxin synthesis; hence, the *uvm8* gene was named fas^{-1} . Fatty acid syntheses (FASs) is responsible for sterigmatocystin (ST) biosynthesis in *A. nidulans* and further identified two genes *viz.*, *stcJ* and *stcK* that encode FAS and FAS subunits (FAS^{2-} and FAS^{1-};).

2.4.3 Occurrence in Various Food

Aflatoxins are found in various cereals, oilseeds, spices, and nuts. These *Aspergillus* colonize among themselves and produce aflatoxins, which contaminate grains and cereals at various steps during harvesting or storage. Fungal contamination can occur in the field, or during harvest, transport and storage. Aflatoxins contamination of wheat or barley is commonly happen by the result of inappropriate storage. In milk, aflatoxins is generally at 1–6% of the total content in the feedstuff. AFTs infect humans following consumption of aflatoxins contaminated foods such as eggs, meat and meat products, milk and milk products.

2.4.4 Effects on Agriculture and Food

Mycotoxins, including aflatoxin, have affected most crops grown worldwide; however, the extent of aflatoxin toxicity varies according to the commodities. Aflatoxin can infect crops during growth phases or even after harvesting. Exposure to this toxin poses serious hazards to human health. Commodities such as corn, peanuts, pistachio, Brazil nuts, copra, and coconut are highly prone to contamination by aflatoxin, whereas wheat, oats, millet, barley, rice, cassava, soybeans, beans, pulses, and sorghum are usually resistant to aflatoxin contamination. However, agricultural products such as cocoa beans, linseeds, melon seeds and sunflower seeds are seldom contaminated. Aflatoxin was on the Rapid Alert System for Food and Feed (RASFF) of the European Union in 2008 because of its severe effects, and the International Agency for Research on Cancer (IARC) later categorized AFB1 as a group I carcinogen for humans. Despite several research and control measures, aflatoxin is still a major threat to food and agricultural commodities.

2.5 MECHANISM OF TOXICITY AND HEALTH EFFECTS

Aflatoxin are specifically target the liver organ. Early symptoms of hepatotoxicity of liver caused by aflatoxins comprise fever, malaise and anorexia followed with abdominal pain, vomiting, and hepatitis; however, cases of acute poisoning are exceptional and rare. Chronic toxicity by aflatoxins comprises immunosuppressive and carcinogenic effects. Evaluation of the effects of AFT-B_1 on splenic lymphocyte phenotypes and inflammatory cytokine expression in male F344 rats have been studied. AFT-B_1 reduced anti-inflammatory cytokine IL4- expression, but increased the pro-inflammatory cytokine IFN-γ and TNF-α expression by NK cells. These findings indicate that frequent AFT-B_1 exposure accelerates inflammatory responses via regulation of cytokine gene expression. Furthermore, it is observed that AFT-B_1 interrupts the process of antigen-presenting capacity of porcine dendritic cells, suggested this perhaps one of mechanism of immunotoxicity by AFT-B_1.

Aflatoxins cause reduced efficiency of immunization in children that lead to enhanced risk of infections. The hepatocarcinogenicity of aflatoxins is mainly due to the lipid peroxidation and oxidative damage to DNA. AFTs-B_1 in the liver is activated by cytochrome p450 enzymes, which are converted to AFTs-B1-8, 9-epoxide, which is responsible for carcinogenic effects in the kidney. Among all major mycotoxins, aflatoxins create a high risk in dairy because of the presence of their derivative, AFTs-M_1, in milk, posing a potential health hazard for human consumption. AFTs-B_1 is rapidly absorbed in the digestive tract and metabolized by the liver, which converts it to AFT-M_1 for subsequent secretion in milk and urine. Although AFTs-M_1 is less mutagenic and carcinogenic than AFTs-B_1, it exhibits high genotoxic activity. The other effects of AFTs-M_1 include liver damage, decreased milk production, immunity suppression and reduced oxygen supply to tissues due to anemia, which reduces appetite and growth in dairy cattle. Several studies have shown the detrimental effects of aflatoxins exposure on the liver, epididymis, testis, kidney and heart . It has been found that aflatoxin presences in post-mortem brain tissue, suggested that its ability to cross the blood brain barrier. AFTs also cause abnormalities in the structure and functioning of mitochondrial DNA and brain cells. The effects of aflatoxin on brain chemistry have been reviewed in details. Furthermore, few reports have described the effects of AFTs-B_1 administration on the structure of the rodent central nervous system liver toxicology of aflatoxin is also a critical issue. Limited doses are not harmful to humans or animals; however, the doses that do cause-effects diverse among Aflatoxin groups. The expression of aflatoxin toxicity is regulated by factors such as age, sex, species, and status of nutrition of infected animals. The symptoms of acute aflatoxicosis include oedema, haemorrhagic necrosis of the liver and profound lethargy, while the chronic effects are immune suppression, growth retardation, and cancer .

2.5.1 Effects of Processing on Aflatoxin

Techniques to eliminate aflatoxin may be either physical or chemical methods. Removing mold-damaged kernels, seeds or nuts physically from commodities has been observed to reduce aflatoxins by 40–80%. The fate of aflatoxin varies with type of heat treatment (*e.g.*, cooking, drying, pasteurization, sterilization, and spray drying . Aflatoxins decompose at temperatures of 237–306°C; therefore, pasteurization of milk cannot protect against AFM1 contamination. It is reported that neither pasteurization nor boiling influenced the level of AFM1 in bovine milk. However, boiling corn grits reduced aflatoxins by 28% and frying after boiling reduced their levels by 34–53%). Roasting pistachio nuts at 90°C, 120°C, and 150°C for 30, 60 and 120 min was found to reduce aflatoxin levels by 17–63%. The decrease in aflatoxin content depends on the time and temperature combination. Moreover, alkaline cooking and steeping of corn for the production of tortillas reduces aflatoxin by 52%). It also reported reductions in aflatoxin content of 50–80% after extrusion alone. When hydroxide (0.7 and 1.0%) or bicarbonate (0.4%) was added, the reduction was enhanced to 95%. The highest aflatoxin reduction was found to be 59% with a moisture content of 35% in peanut meal, and the extrusion variables non-significantly affected its nutritional composition reported an 84% reduction in aflatoxin of peanut meal when cooked in the presence of calcium chloride.

2.5.2 Impact of Environmental Temperature on Aflatoxin Production

Climate change plays a major role in production of aflatoxin from *Aspergillus* in food crops. Climate change affects the interactions between different mycotoxigenic species and the toxins produced by them in foods and feeds. Changes in environmental temperature influence the expression levels of regulatory genes (*aflR* and *aflS*) and aflatoxin production in *A. flavus* and *A. parasiticus* . Temperature interacts with water activity (a_w) and influences the ratio of regulatory genes (*aflR*/*aflS*), which is directly proportional to the production of AFB1. The interactions between water activity and temperature have prominent effect on *Aspergillus* spp. and aflatoxin production. Increasing the temperature to 37°C and water stress significantly reduces the production of AFB1 produced, despite the growth of *A. flavus* under these conditions. The addition of CO_2 under the same temperature and water activity enhances AFB_1 production. fungal biomass and AFB_1 production were reported to be highest at 28°C and 0.96 a_w, while no fungal growth or AFB_1 production was seen at 20°C with a_w values of 0.90 and 0.93. There was also no

AFB_1 production observed at 37 °C. Reverse transcriptase quantitative PCR also revealed that the regulatory genes *aflR* and *aflS* were highly expressed at °28C, while the lowest expression was observed at 20 and °37C, suggesting that temperature plays a significant role in gene expression and aflatoxin production.

2.6 MAJOR DETECTION AND IDENTIFICATION TECHNIQUES

The detection and quantification of aflatoxin in food and feed is a very important aspect for the safety concerns. Aflatoxins are usually detected and identified according to their absorption and emission spectra, with peak absorbance occurring at 360 nm. B toxins exhibit blue fluorescence at 425 nm, while G toxins show green fluorescence at 540 nm under UV irradiation. This florescence phenomenon is widely accepted for aflatoxins. Thin layer chromatography (TLC) is among one of the oldest techniques used for aflatoxin detection, while high performance liquid chromatography (HPLC), liquid chromatography mass spectroscopy (LCMS), and enzyme linked immune-sorbent assay (ELISA) are the methods most frequently used for its detection. ELISA can be used to identify aflatoxins based on estimation of AfB_1-lysine (metabolite of AFB_1 toxin) concentration in the blood. Specifically, the test detects levels of AfB_1 in blood as low as 5 pg/mg albumin, making it a cost effective method for routine monitoring that can also be utilized for the detection of hepatitis B virus. Room temperature phosphorescence (RTP) in aflatoxigenic strains grown on media is commonly used in food mycology. Aflatoxins immobilized on resin beads can induce RTP in the presence or absence of oxygen and heavy atoms and also have high sensitivity and specificity. Moreover, several biosensors and immunoassays have been developed to detect ultra-traces of aflatoxins to ensure the food safety.

2.7 DEGRADATION KINETICS

Various treatments including chemical, physical, and biological methods are routinely utilized for effective degradation, mitigation and management of aflatoxin. The aflatoxins AFB_1 and AFG_1 are completely removed by ozone treatment at 8.5–40 ppm at different temperatures, but AFB_2 and AFG_2 are not affected by this method. The degradation of aflatoxin followed first order kinetic equation. However, microbial and enzymatic degradation is preferred for the biodegradation of aflatoxin due to its eco-friendly nature. The bacterium *Flavobacterium aurantiacum* reportedly removes AFM_1 from milk and *Nocardia asteroides* transforms AFB_1 to fluorescent product. *Rhodococcus* species are able to degrade aflatoxins and their ability to degrade AFB_1 occurs in the following order: *R. ruber* < *R. globerulus* < *R. coprophilus* < *R. gordoniae* < *R. pyridinivorans* and < *R. erythropolis*. Fungi such as *Pleurotus ostreatus, Trametes versicolor, Trichosporon mycotoxinivorans, S. cerevisiae, Trichoderma* strains, and *Armillariella tabescens* are known to transform AFB1 into less toxic forms reported purification of extracellular enzymes from the bacterium *Myxococcus fulvus* ANSM068 with a final specific activity of 103 × 569.44 U/mg. The pure enzyme (100 U/mL) had a degradation ability of 96.96% for

AFG_1 and 95.80% for AFM_1 after 48 h of incubation. Moreover, the recombinant laccase produced by *A. niger* D-15Lcc118) 3#2 U/L) was found to lead to a decrease in AFB_1 of %55 within 72 h.

2.7.1 Management and Control Strategies

The biocontrol principle of competitive exclusion of toxigenic strains of *A. flavus* involves the use of non-toxigenic strains to reduce aflatoxin contamination in maize. The use of biocontrol agents such as *Bacillus subtilis*, *Lactobacillus* spp., *Pseudomonas* spp., *Ralstonia* spp., and *Burkholderia* spp. are effective at control and management of aflatoxins. Several strains of *B. subtilis* and *P. solanacearum* isolated from the non-rhizosphere of maize soil have been reported to eliminate aflatoxin. Biological control of aflatoxin production in crops in the US has been approved by the Environmental Protection Agency and two commercial products based on atoxigenic *A. flavus* strains are being used (Afla-guard® and AF36®) for the prevention of aflatoxin in peanuts, corn, and cotton seed. Good agricultural practices (GAPs) also help control the toxins to a larger extent, such as timely planting, providing adequate plant nutrition, controlling weeds, and crop rotation, which effectively control *A. flavus* infection in the field.

Biological control is emerging as a promising approach for aflatoxin management in groundnuts using *Trichoderma* spp, and significant reductions of 20–90% infection of aflatoxin have been recorded. Use of inbred maize lines resistant to aflatoxin has also been employed. Potential biochemical markers and genes for resistance in maize against *Aspergillus* could also be utilized. Additionally, biotechnological approaches have been reviewed for aflatoxin management strategies. Advances in genomic technology based research and decoding of the *A. flavus* genome have supported identification of the genes responsible for production and modification of the aflatoxin biosynthesis process.

2.8 AFLATOXINS POSE A HEALTH RISK TO HUMANS AND LIVESTOCK

Aflatoxins are poisonous substances produced by certain kinds of fungi (moulds) that are found naturally all over the world; they can contaminate food crops and pose a serious health threat to humans and livestock. Aflatoxins also pose a significant economic burden, causing an estimated 25% or more of the world's food crops to be destroyed annually.

2.8.1 Human Exposure Comes From Nuts and Grains

- Two closely related species of fungi are mainly responsible for producing the aflatoxins of public health significance: *Aspergillus flavus* and *A. parasiticus*. Under favourable conditions typically found in tropical and subtropical regions, including high temperatures and high humidity, these moulds, normally found on dead and decaying vegetation, can invade food crops.

- Drought stress, insect damage and poor storage can also contribute to higher occurrence of the moulds including in more temperate regions. Several types of aflatoxin (14 or more) occur in nature, but four – aflatoxins B_1 , B_2 , G_1 and G_2 are particularly dangerous to humans and animals as they have been found in all major food crops; but most human exposure comes from contaminated nuts, grains and their derived products.
- Additionally, aflatoxin M_1 (AFM_1), a product of aflatoxin B_1 (AFB_1) metabolism, can be found in milk in areas of high aflatoxin exposure. Subsequently humans may be exposed to this aflatoxin through milk and milk products, including breast milk, especially in areas where the poorest quality grain is used for animal feed. Food crops can become contaminated both before and after harvesting.
- Pre-harvest contamination with aflatoxins is mainly limited to maize, cottonseed, peanuts and tree nuts. Post-harvest contamination can be found in a variety of other crops such as coffee, rice and spices. Improper storage under conditions that favour mould growth (warm and humid storage environments) can typically lead to levels of contamination much higher than those found in the field.

2.8.2 Dietary Exposure to Aflatoxins

National estimates of dietary exposure to aflatoxins indicate differences between developed and developing countries. In developed countries, mean aflatoxin dietary exposures are generally less than 1 ng/kg body weight (bw) per day (a nanogram is one billionth [1×10^{-9}] of a gram), whereas estimates for some sub-Saharan African countries exceed 100 ng/kg bw per day, although these latter estimates are often based on very few data. Estimates of dietary exposure to AFM_1 have rarely exceeded 1 ng/kg bw per day in any country (although up to 6.5 and 8.8 ng/kg bw per day for young children and breastfed infants have been reported).

2.8.3 Chronic Exposure Can Have Serious Health Consequences:

Chronic exposure to aflatoxins has several health consequences including:

- Aflatoxins are potent carcinogens and may affect all organ systems, especially the liver and kidneys; they cause liver cancer, and have been linked to other types of cancer –AFB_1 is known to be carcinogenic in humans; the potency of aflatoxin to cause liver cancer is significantly enhanced in the presence of infection with hepatitis B virus (HBV);
- Aflatoxins are mutagenic in bacteria (affect the DNA), genotoxic, and have the potential to cause birth defects in children;
- Children may become stunted, although these data have yet to be confirmed because other factors also contribute to growth faltering *e.g.*, low socioeconomic status, chronic diarrhoea, infectious diseases, malnutrition;

- Aflatoxins cause immunosuppression, therefore may decrease resistance to infectious agents (*e.g.*, HIV, tuberculosis);

2.8.4 Large Doses of Aflatoxins Lead to Acute Poisoning:

- Large doses of aflatoxins lead to acute poisoning (aflatoxicosis) that can be life threatening, usually through damage to the liver. Outbreaks of acute liver failure (jaundice, lethargy, nausea, death), identified as aflatoxicosis, have been observed in human populations since the 1960s. Most recently deaths attributed to aflatoxins were reported during the summer of 2016 in the United Republic of Tanzania. Adults are more tolerant to acute exposure than children.
- The consumption of food containing aflatoxin concentrations of 1 mg/kg or higher has been suspected to cause aflatoxicosis. Based on past outbreaks it has been estimated that, when consumed over a period of 1–3 weeks, an AFB_1 dose of 20–120 μg/kg bw per day (μgram is one billionth [1×10^{-9}] of a kilogram) is acutely toxic and potentially lethal.

2.8.5 Aflatoxins Cause a Variety of Adverse Effects In Animals

- In chickens, the effects of aflatoxins include liver damage, impaired productivity and reproductive efficiency, decreased egg production, inferior eggshell quality, inferior carcass quality and increased susceptibility to disease. Pigs are also highly affected by aflatoxins, with the chronic effects largely apparent as liver damage.
- In cattle, the primary symptoms are reduced weight gain as well as liver and kidney damage; where milk production is also reduced. Different forms of the enzymes that metabolize aflatoxins (*e.g.*, cytochrome P450s, glutathione S-transferases) are considered responsible for the different susceptibilities of different animals to the toxic effects of aflatoxins.

2.8.6 Aflatoxicosis Detection in Humans and Animals Is Difficult

- Detecting aflatoxicosis in humans and animals is difficult due to variations in clinical signs and the presence of other factors such as suppression of the immune system caused by an infectious disease.
- Of the two techniques most often used to detect levels of aflatoxins in humans, one measures a breakdown product in urine (which however is only present for 24 hours after exposure), and the other measures the level of an AFB–albumin compound in the blood serum, providing information on exposure over weeks or months. These biomarker measurements are important in investigating outbreaks where aflatoxin contamination is suspected.

2.9 A VARIETY OF METHODS TO DETECT AFLATOXINS IN FOOD AND FEED

- Aflatoxins are of major importance and techniques for their detection and analysis have been extensively researched to develop those that are highly specific, useful and practical. A plethora of methods are available for different needs, ranging from techniques/methods for regulatory control in official laboratories (such as HPLC-MS [high-performance liquid chromatography-mass spectrometry]) to rapid test kits for factories and grain silos (such as ELISA [enzyme-linked immunosorbent assay]).
- Potential novel aflatoxin-detection systems, based on emerging technologies, include dip-stick kits, hyperspectral imaging, electronic noses, molecularly imprinted polymers, and aptamer-based biosensors (small organic molecules that can bind specific target molecules). The latter technologies may have relevance in remote areas because of their stability, ease of production and use.

2.10 SAMPLING PROCEDURES ARE PROBLEMATIC

- Because moulds and aflatoxins are not evenly distributed throughout bulk shipments and batches of stored grain, appropriate sampling is critical to get a representative result. Protocols for sampling procedures have been developed, in particular in the context of regulatory control. For instance, in setting maximum levels for aflatoxins, the Codex Alimentarius Commission has specified the protocols to be used for peanuts, almonds, Brazil nuts, hazelnuts and pistachios intended for further processing and for ready-to eat almonds, Brazil nuts, hazelnuts, pistachios and dried figs. The Food and Agriculture Organization of the United Nations (FAO) has developed a mycotoxin sampling tool which is available on-line1 .
- Recommended sampling methods are a problem especially for subsistence farmers in rural areas who do not produce enough grain to spare the quantities needed for accurate testing. Thus there is a need to develop rapid, low-cost, low technology, accurate detection methods for aflatoxins to improve surveillance and control in rural areas. Organizations such as the Partnership for Aflatoxin Control in Africa and the World Food Programme are addressing these issues, *e.g.*, the World Food Programme has instituted the Purchase for Progress Program to ensure grain quality by creating the Blue Box, which contains test kits for grain quality, including aflatoxins.

2.11 AFLATOXINS CAN BE CONTROLLED BOTH PRE- AND POST-HARVEST

- Control measures are required both pre- and post-harvest. The most long-term, stable solution to controlling pre-harvest aflatoxin contamination is through enhancing the ability of the crop to resist fungal infection and/or prevent production of aflatoxins by the invading fungus. This can be achieved through plant breeding or through genetic engineering of crops of interest. However, these processes are laborious and time consuming. Effective, sustainable and universally applicable pre-harvest intervention strategies are needed.
- One strategy that has received significant attention for reduction of aflatoxins prior to harvest has been biological control using non-toxigenic *A. flavus* isolates. The non-toxigenic strains occupy the same niches as the naturally occurring toxigenic strains, and are capable of competing and displacing the toxigenic strains. This approach has been deployed on crops such as cotton, maize, peanuts, figs and pistachios in the USA, maize in Africa, and peanuts in Australia, Argentina and China. The strategy has also been used for maize in Thailand to measure the effectiveness of this treatment pre-harvest and post-harvest; the results were promising, but inconsistent.
- Post-harvest interventions include preventive measures to address adequate storage conditions (moisture, temperature, mechanical or insect damage, and aeration), which influence contamination and toxin production by mould. Other measures, such as chemical decontamination or use of enterosorbents, can be used to remove aflatoxins from already contaminated foodstuffs.

2.11.1 Integrated Approach for Control of Aflatoxins

An integrated approach, whereby aflatoxins are controlled at all stages from the field to the table, is required for reduction in risk. Such an approach includes targeted plant breeding practices, enhancement of host plant resistance, and biological control methods, coupled with post-harvest technologies such as proper drying and storage of potentially affected crop products, as well as development of appropriate alternative uses to retain at least some economic return on value of damaged crop. Therefore by removing the sources of contamination, promoting better agricultural and storage techniques, ensuring adequate resources are available for testing and early diagnosis, enforcing strict food safety standards, informing and educating consumers and (small/subsistence) farmers, promoting better livestock feeding and management, and creating general awareness about personal protection, are some of the ways in which national authorities can help to control aflatoxins.

2.11.2 WHO in Collaboration of FAO Supports Countries in Controlling Aflatoxins

WHO, in collaboration with FAO, evaluates the science and develops risk assessments to define safe exposure levels. Based on these risk assessments, maximum levels for aflatoxins in different foods are recommended. These form the basis for national regulations to limit contamination. Since first being noted in the 1960s, aflatoxins have several times been the subject of toxicological evaluation and dietary exposure assessment by the Joint FAO/WHO Expert Committee on Food Additives (JECFA). These evaluations inform the Codex Alimentarius Commission which has worked, since 1963, to create harmonized international food standards to protect the health of consumers and ensure fair trade practices. Codex standards set the maximum levels for contaminants and natural toxins such as aflatoxins in food and are the reference for the international trade in food, so that consumers everywhere can be confident that the food they purchase meets agreed standards for safety and quality, no matter where it was produced. The maximum levels for aflatoxins in various nuts, grains, dried figs and milk are in the range of 0.5 to 15 μg/kg (a μgram is one millionths [1×10^{-6}] of a gram). To prevent and reduce the risk of aflatoxins in food and feed, Codex has also developed codes of practice, which detail appropriate preventive measures. In support of this work, WHO collects food contamination data from nationally recognized institutions through the Food Contamination Monitoring and Assessment Programme of the WHO Global Environment Monitoring System, commonly known as GEMS/Food. The GEMS/Food contaminants database informs governments, the Codex Alimentarius Commission and other relevant institutions, as well as the public, on levels and trends of contaminants in food. The GEMS/Food programme has also developed a Consumption Cluster Diets database, which provides an overview of food consumption patterns worldwide, through 17 dietary patterns (based on population food choices) covering more than 180 countries. These estimates, together with reported contamination levels, allow assessment of the potential exposure of populations to contaminants such as aflatoxins in food. The consumption diets are based on Food Balance Sheet (FBS) data collected by FAO, and are routinely used by international risk assessment bodies.

2.11.3 National Authorities are Developing Regulations to Limit Contamination

- Exposure to aflatoxins needs to be kept as low as possible to protect the consumer. Many countries have regulations governing aflatoxins in food with prescribed acceptable limits, and most have maximum permitted or acceptable levels for different foodstuffs.
- Aflatoxins damage health and business opportunities, and importing countries are imposing increasingly more stringent regulations.

- The Codex recommendations, maximum levels, and codes of practices mentioned above, serve as guidance for national authorities.

CONSUMERS CAN DO

Mouldy foods are potentially contaminated with aflatoxins and therefore are possibly harmful when consumed. The moulds do not just grow on the surface but penetrate deep into food. To reduce exposure to aflatoxins, the consumer is advised to:

- carefully inspect whole grains and nuts for evidence of mould, and discard any that look mouldy, discoloured, or shriveled;
- buy grains and nuts as fresh as possible; that have been grown as close to home as possible, and which have not been transported over a long time;
- buy only reputable brands of nuts and nut butters – aflatoxin moulds are not entirely killed by processing or roasting, so can show up in products e.g. peanut butter;
- make sure that foods are stored properly and are not kept for extended periods of time before being used; and
- try to ensure his/her diet is diverse; this not only helps to mitigate aflatoxin exposure, but also improves health and nutrition. Consumers who lack dietary diversity need to pay extra attention to minimize the risk of high exposure to aflatoxins. For example, extensive aflatoxin exposure has been reported from areas where people get a major part of their daily calorie intake from maize; this foodstuff is commonly contaminated with aflatoxins and needs to be handled properly both before and after harvest.

3

AFLATOXINS: HAZARDS IN GRAIN, AFLATOXICOSIS AND LIVESTOCK

3.1 INTRODUCTION

Many different fungi may grow as molds on stored grains. Fusarium and Aspergillus fungi are among the most common grain molds. Not all fungi produce toxins, but Aspergillus, which produces aflatoxin, is among the most common grain mold fungi. Aflatoxins are poisonous, carcinogenic byproducts produced during the growth of several species of the mold fungus Aspergillus. These byproducts are produced as the fungi grow in feed grains, processed feed, and food products.

Aflatoxins are highly toxic to livestock, poultry, and people. Even when fed at nonfatal levels, aflatoxin can seriously impair animal health and productivity. For lactating dairy cattle, do not exceed 20 ppb aflatoxin in rations to avoid exceeding the Food and Drug Administration level of 0.5 ppb in milk. Aflatoxin is just one of many mycotoxins that can adversely affect animal health and productivity.

Aflatoxins are primarily a problem in corn but can also occur in other grain crops. Aflatoxins are highly toxic to livestock, poultry, and people. Consumption of low concentrations by animals sensitive to aflatoxins can lead to death in 72 hours. In general, at nonfatal levels, the health and productivity of animals fed contaminated feed are seriously impaired. As a result, the Food and Drug Administration (FDA) has set an action level for aflatoxins in corn at 20 parts per billion (ppb). Corn containing aflatoxin levels of 20 ppb or more cannot be sold in interstate commerce, and, in general, should not be fed to young poultry, swine, and livestock, or to lactating animals, and must not be milled for human consumption.

3.2 *ASPERGILLUS* AND AFLATOXIN CONTAMINATION

The development of aflatoxins depends on the infestation and growth of the *Aspergillus* mold in grain. High temperatures and high humidity favor the infection of corn kernels through the silks by the Aspergillus fungi. In the southeastern United States and the eastern Corn Belt this environmental condition occurs more frequently and is the main reason Aspergillus infections and aflatoxin contamination occur more frequently. Below-normal soil moisture (drought stress) has also been found to increase the number of Aspergillus spores in the air. Therefore, when drought stress occurs during pollination, the increased inoculum load (spores in the air) greatly increases the chances of infection. Furthermore, drought stress, nitrogen stress, and other stresses that affect plant growth during pollination can increase the level of aflatoxins produced by *Aspergillus* fungi. Often, Aspergillus will grow on the unfilled portions of the ear . In the past, insect injury to the maturing ear of corn was considered a requirement for infection to occur. This is now known to be false. However, insect damage to ears provides wounds that allow Aspergillus to more readily infect the kernels. Insects also transport Aspergillus spores to the silks and the kernels. Therefore, insect damage, especially during pollination in drought-stressed corn, can increase the occurrence of Aspergillus and the levels of aflatoxins.

Several other factors play a role in the development of Aspergillus mold and aflatoxin production. Because drought stress plays such an important role, practices that reduce drought stress in plants should reduce the levels of infection and aflatoxin production.

Irrigation has been shown to be very effective in reducing Aspergillus infection and aflatoxin development, even if done only during pollination. Tillage practices have not been as effective and have only been demonstrated to reduce aflatoxin by subsoiling in areas with hardpans. Occasionally during droughty periods, hybrids of differing maturities or those planted early will pollinate during periods when drought stress is less often observed in South Dakota. Escaping drought with planting dates and hybrid maturity may differ from one year to the next.

Time of harvest has also been shown to be important in influencing the occurrence and levels of aflatoxin because Aspergillus does not compete well with other molds when corn is above the 20 percent moisture content. Harvesting corn when moisture content is above 20 percent followed by rapid drying to at least a moisture content of 14 percent within 24 to 48 hours of harvest keeps further Aspergillus growth and toxin production at a minimum.

3.2.1 Aflatoxin Development in Storage

Mature corn that remains in the field or corn that is stored without adequate drying can be subject to Aspergillus growth and aflatoxin production. Temperatures between 80F and 100F and relative humidity of 85 percent (corresponding to 18 percent grain moisture) are optimum for growth of Aspergillus. Growth of the fungus is poor below 55F, but if the grain is moist enough, toxins can still be produced. However, simply reducing the moisture content to as low as 12 percent does not kill the fungus and does not reduce the levels of toxins that have already been produced. If moisture levels rise again above 12 percent anytime during storage, and temperatures are high enough, then mold growth and toxin production will resume. It is important to note that conditions favoring the growth of Aspergillus also favor the growth of other fungi that can have harmful effects on humans if they are inhaled or ingested while working in grain handling facilities. Always wear a dust respirator when working in grain or feed storage and handling areas.

3.2.2 Detecting Aflatoxin Contamination

Once aflatoxin is produced, it is stable. Heat, cold and light do not affect it. It is also colorless, odorless and tasteless, and because of the low concentrations involved and the uneven distribution in grain bins, aflatoxins are difficult to detect.

In the past, elevator operators and buyers used the blacklight test, but this test simply detects compounds that fluoresce (aflatoxins and others) and should only be used to select samples that require further testing. Similarly, minicolumn tests are no longer recommended, as they were prone to give false positive results if used improperly.

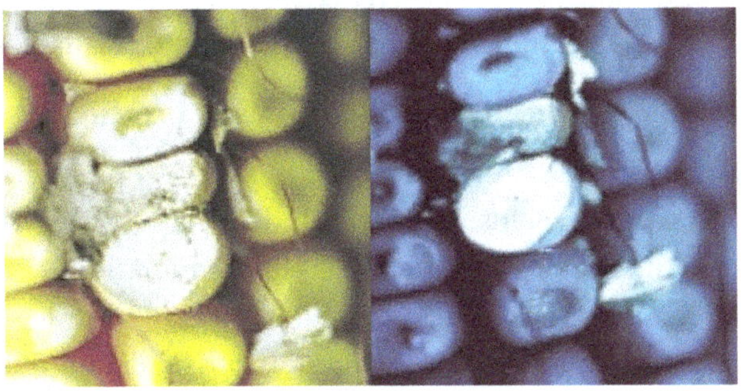

Fig.1: Fluorescence of grain under ultraviolet light is an indicator of possible aflatoxin contamination, but does not confirm the presence of aflatoxin.

Serological tests are now considered to be more reliable and their accuracy has been validated by comparison to more costly and time-consuming analytical procedures. Serological test kits using such methods as ELISA(Fig. 2) do not require specialized labs, equipment, or training and when conducted according to manufacturer's instructions can give accurate results for the presence (qualification) and amount (quantification) of aflatoxin in grain samples.

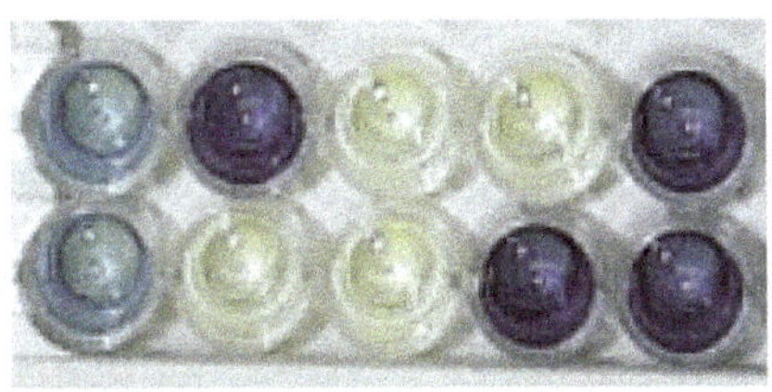

Fig. 2: An ELISA test can detect very small quantities of specific contaminants in grain. The development of color (blue in this test) indicates the presence of the toxin. Intensity of the color can indicate the amount of toxin present.

3.2.3 Sampling for Aflatoxins

Regardless of the test procedure used, the single most important factor for reliable and accurate testing of grain for aflatoxins is obtaining a representative sample. The ideal sample size should be at least 10 pounds of corn. The sample should consist of several smaller subsamples (10 or more 1-pound samples) that have been taken from different spots and then mixed together. Handle each bin or truck separately, and take a 10-pound sample from each source.

Place samples in a cloth or paper container that allows air exchange. Air-tight containers or plastic bags allow condensation, which raises the moisture content, resulting in the possibility of continued growth and toxin production of the fungus.

Send or take samples to a testing lab as quickly as possible. Monday through Wednesday is the preferred time of week to ship samples for testing. Samples mailed later in the week are more likely to be delayed in transit. This gives the fungus a chance to continue to grow and produce toxins. Samples should be properly identified and include: the source of the sample (truck or bin), the sender of the sample, full address, and telephone number.

3.3 PREVENTION OF AFLATOXIN CONTAMINATION

Resistance to aflatoxin accumulation in corn kernels has been recently identified. Hybrids resistant to aflatoxin and other fungal toxins should become available in the near future.

3.3.1 To minimize aflatoxin contamination of corn products:

- **Reduce plant stress.** Use recommended production practices to minimize plant stress and maximize yields. These include insect, weed, and disease control practices, and the use of recommended plant populations and fertility practices. When possible, irrigation during pollination can decrease predisposition of the crop to aflatoxin. Plant corn as early as possible, and plant several different hybrids of different pollination periods to reduce the chance of environmental stress at pollination in at least part of the crop. Care should be taken to store uncontaminated hybrids away from those that are contaminated.
- **Harvest corn early and dry it immediately:** Harvesting corn when it is above the 20 percent moisture content and drying it within 24 to 48 hours to a moisture content no greater than 14 percent greatly reduces the infection, growth, and toxin production by Aspergillus.
- **Avoid damaged kernels:** Damaged kernels are more likely to become infected with molds both in the field and in storage. Corn hybrid selection and insect control can play a role in reducing kernel damage. Corn hybrids with good husk coverage of the ear have been shown to have less infection and aflatoxin development. Also, *Bt* corn hybrids derived from transformational events that express the *Bt* trait in the ears and the silks are less likely to become infected with toxin-producing fungi as a result of reduced kernel damage. Adjustment of the combine to reduce mechanical damage of the kernels at harvest is a very important means of reducing contamination in storage.
- **Store corn at 12 percent moisture content:** Maximum moisture content for corn storage should be 14 percent. Moisture content at or below 12 percent is

ideal for storage of corn, because growth and toxin production by Aspergillus cannot occur.

- **Keep storage and feeding facilities clean:** *Aspergillus* fungi can survive on residues left in storage areas. When environmental conditions become favorable, infection and toxin production can reoccur year after yearin storage systems that are not properly cleaned and disinfected.

3.3.2 Using Contaminated Corn

Recommended levels are 0 ppb aflatoxins in feed: However, aflatoxin-contaminated feed can be tolerated by some livestock, particularly older animals. Obviously, the higher the level of contamination, the greater the risk in feeding contaminated corn to animals. Furthermore, continued proper storage is essential so that aflatoxin levels do not continue to increase in the corn or feed before use. Detoxification of feed continues to be an elusive goal. However, certain feed additives have been successfully used to inhibit mold growth and to reduce the incidence of aflatoxicosis in animals. Organic acids such as propionic, sorbic, and benzoic acids as well as their salts such as calcium propionate and potassium sorbate, and copper sulfate can be used to inhibit mold growth in feed. Mineral clays such as zeolite and bentonite as well as hydrated sodium calcium aluminosilicate (HSCAS) can bind to aflatoxin, protecting animals from absorbing the toxin that may be in the feed. These products, according to FDA rules, cannot as yet be labeled as mycotoxin binders, and are sold as anti-caking and free-flow feed additives. There are no clear-cut safe levels for different animal species regarding their resistance or tolerance to aflatoxins. The following section on aflatoxicosis and ruminants and the general guidelines for dealing with aflatoxin-contaminated feed may assist you in deciding whether to assume the risk.

3.4 AFLATOXICOSIS AND LIVESTOCK

Aflatoxicosis is a disease caused by the consumption of aflatoxins. Aflatoxins are secondary mold metabolites produced by some strains of Aspergillus flavus and other related species of Aspergillus fungi. The four most common aflatoxins are B_1, B_2, G_1, and G_2. Contaminated grains and grain byproducts are the most common sources of aflatoxin. Corn silage may also be a source of aflatoxins, because the ensiling process does not destroy toxins already present in silage. Aflatoxins are metabolized in ruminants by the liver and are excreted in the bile. Aflatoxin B_1 is the most potent mycotoxin (toxic substance produced by a mold) to affect cattle. B_1 increases the apparent protein requirement of cattle and is a potent carcinogen (cancer causing agent). When significant quantities of B_1 are consumed, the metabolite M_1 appears in milk within 12 hours. Research suggests M_1 is not as carcinogenic or mutagenic as B_1, but it does appear to be as toxic as its parent compound.

3.4.1 Symptoms

Beef and dairy cattle are more susceptible to aflatoxicosis than sheep or horses, although other mycotoxicoses occur in these species, such as facial eczema in sheep and leukoencephalornalacia in horses. Young animals of all species are more susceptible than mature animals to the effects of aflatoxin. Pregnant and growing animals are less susceptible than young animals, but more susceptible than mature animals. Feed refusal, reduced growth rate and decreased feed efficiency are the predominant signs of chronic aflatoxin poisoning. In addition, listlessness, weight loss, rough hair coat and mild diarrhea may occur. Anemia along with bruises and subcutaneous hemorrhage are also symptoms of aflatoxicosis. The disease may also impair reproductive efficiency, including abnormal estrous cycles (too short and too long) and abortions. Other symptoms include impaired immune system response, increased susceptibility to disease, and rectal prolapsed.

3.4.2 Pathology

Clinical laboratory findings vary with the animal species, level of aflatoxin in the ration, and the duration of feeding. There are no consistent diagnostic changes in hematocrit, hemoglobin, and differential cell counts in animals fed aflatoxin. Leukocytosis may occur in animals with secondary bacterial infections. Serum bilirubin levels may be elevated and typically serum protein levels are decreased. Lesions observed at necropsy related to either acute or chronic liver disease are dependent upon the level of aflatoxin and the duration of feeding. A majority of acute liver damage observed has been the result of experimentally high doses, while chronic liver damage is a more common field observation. The liver is usually pale tan, yellow or orange. Hepatic fibrosis and edema of the gallbladder may also be observed.

3.4.3 Diagnosis

The diagnosis of aflatoxicosis is often difficult because of the variation in clinical signs, gross pathological conditions and the presence of infectious diseases due to the suppression of the immune system. On the farm, more than one mold or toxin may be present in the contaminated feed, which often makes definitive diagnosis of aflatoxicosis difficult. The prognosis of aflatoxicosis depends upon the severity of liver damage. Once overt symptoms are noticed the prognosis is poor. Treatment should be directed at the severely affected animals in the herd and further poisoning prevented.

3.4.4 Treatment

Aflatoxicosis is typically a herd rather than an individual cow problem. If aflatoxicosis is suspected, the ration should be analyzed immediately. If aflatoxins are present, the source should be eliminated immediately. Levels of protein in the ration and vitamins A,D,E,K and B should be increased as the toxin binds vitamins and affects protein synthesis. Good management practices to alleviate stress are essential to reduce the risk of secondary infections. Secondary infections must receive immediate attention and treatment.

4

BIOLOGICAL AND HEALTH EFFECTS OF AFLATOXINS ON BODY ORGANS AND BODY SYSTEMS

4.1. INTRODUCTION

Chronic consumption of aflatoxin-contaminated foods is a common problem in both humans and animals worldwide especially in poor developing nations of south East Asia and sub-Saharan Africa where there is poor food harvesting, processing and storage of food and food products thus allowing the growth of mold on them. Aflatoxins, their metabolites, the aflatoxin-8,9-epoxide and the generated ROS causes deleterious effects on the various body organs and body systems including the development of cancers especially the liver cancer mainly due to AFB_1 exposure.

- Aflatoxins are a group of naturally occurring carcinogens that are known to contaminate different human and animal food stuffs. Aflatoxins are poisonous by-products from soil-borne fungus *Aspergillus*, which is responsible for the decomposition of plant materials]. The occurrence of aflatoxins foods and food products vary with geographic location, agricultural and agronomic practices. The susceptibility of food product to fungal attack occurs during pre-harvest, transportation, storage, and processing of the foods. The problem of aflatoxin contamination of the food products is a common problem in tropical and subtropical regions of the world especially in the developing countries such as the sub-Saharan countries with poor practices and where the environmental conditions of warm temperatures and humidity favors the growth fungi.
- The various food products contaminated with aflatoxins include cereals like maize, sorghum, pearl millet, rice and wheat; oilseeds such as groundnut, soybean,

sunflower and cotton; spices like chillies, black pepper, coriander, turmeric and zinger; tree nuts such as almonds, pistachio, walnuts and coconut; and milk and milk products. The aflatoxins were initially isolated and identified as the causative agent in Turkey X disease that caused necrosis of the liver in 1960 and over 100,000 turkeys died in England and USA and the death was attributed to the consumption of a mould-contaminated peanut meal. Very high concentrations of aflatoxins are most often found in nutritive seeds such as maize, nuts and cereal grains in Africa and rice in China and Southeast Asia.

Table 1. Major aflatoxins produced by the *Aspergillus* species of Moulds

Difuranocoumarins	Type of aflatoxin	*Aspergillus specie(s)*
Difurocoumarocyclo-pentenone series	Aflatoxin B_1 (AFB_1)	*A. flavus, A. arachidicola, A. bombycis, A. minisclerotigenes, A. nomius, A. ochraceoroseus, A. parasiticus, A. pseudotamarii, A. rambellii, Emericella venezuelensis*
	Aflatoxin B_2 (AFB_2)	*A. arachidicola, A. flavus, A. minisclerotigenes, A. nomius, A. parasiticus*
	Aflatoxin B_{2a} (AFB_{2a})	*A. flavus*
	Aflatoxin M_1 (AFM_1)	*A. flavus, A. parasiticus*; metabolite of aflatoxin B_1 in humans and animals and comes from a mother's milk
	Aflatoxin M_2 (AFM_2)	Metabolite of aflatoxin B_2 in milk of cattle fed on contaminated foods
	Aflatoxin M_{2A} (AFM_{2A})	Metabolite of AFM_2
	Aflatoxicol (AFL)	*A. flavus*, metabolite of AFB_1
	Aflatoxicol M_1	Metabolite of AFM_1
Difurocoumarolac-tone series	Aflatoxin G_1 (AFG_1)	*A. arachidicola, A. flavus, A. minisclerotigenes, A. nomius, A. Parasiticus*
	Aflatoxin G_2 (AFG_2)	A. arachidicola, A. flavus, A. minisclerotigenes, A. nomius, A. parasiticus
	Aflatoxin G_{2A} (AFG_{2A})	Metabolite of AFG_2
	Aflatoxin GM_1 (AFG_1)	*A. flavus*
	Aflatoxin GM_2 ($AFGM_2$)	Metabolite of AFG_2
	$AFGM_{2A}$	Metabolite of $AFGM_2$
	Aflatoxin B_3 (AFB_3)	*Aspergillus* species not defined
	Parasiticol (P)	*A. flavus*
	Aflatrem	*A. flavus, A. minisclerotigenes*
	Aspertoxin	*A. flavus*
	Aflatoxin Q_1 (AFQ_1)	Major metabolite of AFB_1 in *in-vitro* liver preparations of other higher vertebrates

Aflatoxins are a group of approximately 20 related fungal metabolites produced primarily by the fungi *Aspergillus flavus* and *A. parasiticus* . Aflatoxins belongs to a group of difuranocoumarins that are classified into two broad groups according to their chemical structure and they include the difurocoumarocyclopentenone series (AFB_1, AFB_2, AFB_2A, AFM_1, AFM_2, AFM_2A and aflatoxicol) and the difurocoumarolactone series (AFG_1, AFG_2, AFG_2A, $AFGM_1$, $AFGM_2$, $AFGM_2A$ and AFB_3).

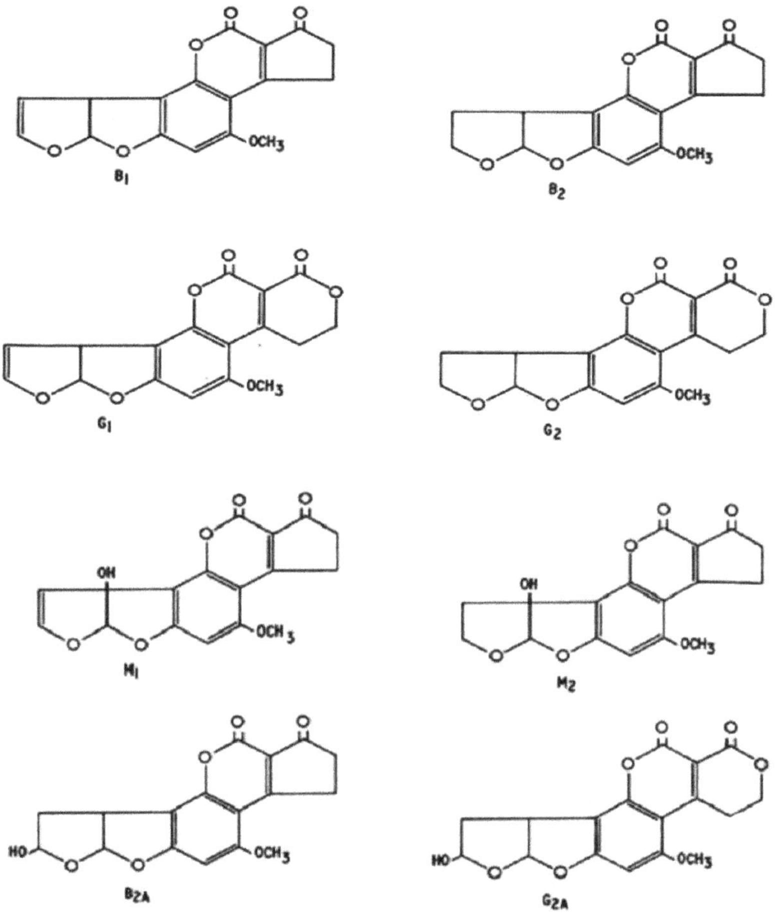

Fig. 1. Structures of the major aflatoxins B_1, B_2, G_1, G_2, M_1, M_2, B_2A and G_2A

The four major naturally known aflatoxins produced by the *Aspergillus* species of mold include AFB_1, AFB_2, AFG_1 and AFG_2 where the "B" and "G" refer to the blue and green fluorescent colors produced under UV light on thin layer chromatography plates, while the subscript numbers 1 and 2 indicate major and minor compounds, respectively. Whereas the B designation of aflatoxins B_1 and B_2 result from the exhibition of blue fluorescence under UV-light, while the G designation refers to the yellow-green fluorescence of the relevant structures under UV-light. The metabolic products of aflatoxins, M_1 and M_2 were first isolated from milk of lactating animals fed on Moldy grains contaminated with aflatoxin hence, the M designation]. These toxins have closely similar structures (Fig. 1) and form a unique group of highly oxygenated, naturally occurring heterocyclic compounds. Aflatoxins B_2 and G_2 were established as the dihydroxy derivatives of B_1 and G_1, respectively. Whereas, aflatoxin M_1 is 4-hydroxy aflatoxin B_1 and aflatoxin M_2 is 4-dihydroxy aflatoxin B_2. Of the four major aflatoxins (B_1, B_2, G_1 and G_2), G_2 occurs in high quantities though less toxic while AFB_1 is the most toxic of all the aflatoxin. The World Health Organization (WHO) classifies AFB_1 as a class 1 carcinogen. The aflatoxins display potency of toxicity, carcinogenicity, mutagenicity in the order of $AFB_1 > AFG_1 > AFB_2 > AFG_2$. The extent of toxicity depends on the organ affected especially the liver. The lethal toxicity of aflatoxin B_1 varies in different animals from extremely susceptible (Sheep, Rat, Dog) to resistant species (Monkey, Chicken, Mouse). However, there are no toxicity in humans though epidemiological data from studies in Africa, South Africa, South-East Asia and India implicate aflatoxins in the incidence of liver cancer especially the hepatobiliary carcinoma and death of children due to malnutrition, kwashiorkor and marasmus . Aflatoxins have been associated with various diseases like aflatoxicosis and other health problems in humans, livestock and domestic animals globally.

4.2 ABSORPTION, DISTRIBUTION, METABOLISM, EXCRETION AND MECHANISMS OF ACTION OF AFLATOXINS

Aflatoxins are highly liposoluble compounds and are readily absorbed from the site of exposure usually through the gastrointestinal tract and respiratory tract into blood stream. Human and animals get exposed to aflatoxins by two major routes (a) direct ingestion of aflatoxin-contaminated foods or ingestion of aflatoxins carried over from feed into milk and milk products like cheese and powdered milk as well as other animal tissues mainly as AFM_1 (b) by inhalation of dust particles of aflatoxins especially AFB_1 in contaminated foods in industries and factories. After entering the body, the aflatoxins are absorbed across the cell membranes where they reach the blood circulation. They are distributed in blood to different tissues and to the liver, the main organ of metabolism of xenobiotics. Aflatoxins are mainly metabolized by the liver to a reactive epoxide intermediate or hydroxylated to become the less harmful aflatoxin M_1 . In humans and susceptible animal species, aflatoxins especially AFB_1 are metabolized by cytochrome P450 (CYP450) microsomal enzymes to aflatoxin-8,9-epoxide, a reactive form that binds to DNA and to albumin in the blood serum, forming adducts and hence causing DNA damage. Various CYP450 enzymes isoforms occur in the liver and they metabolize aflatoxin into a reactive oxygen species (aflatoxin-8,9-epoxide), which may then bind to proteins and cause acute toxicity (aflatoxicosis) or to DNA and induce liver cancer . The predominant human CYP450 isoforms involved in human metabolism of AFB_1 are CYP3A4 and CYP1A2. Both enzymes catalyze the biotransformation of AFB_1 to the highly reactive exo-8,9-epoxide of AFB_1. CYP $_1A_2$ is also capable of catalyzing the epoxidation of AFB_1 to yield a high proportion of endo-epoxide and hydroxylation of AFB_1 to form aflatoxin M_1 (AFM_1), which is a poor substrate for epoxidation and less potent than AFB_1

. This is generally considered as the major detoxification metabolic pathway for aflatoxins. The CYP_3A_4 is the major CYP450 enzyme responsible for activation of AFB_1 into the epoxide form and also form AFQ_1, a less toxic detoxification metabolite. The CYP3A5 metabolizes AFB1 mainly to the exo-epoxide and some AFQ_1. However, polymorphism studies with CYP3A5 have indicated that, this enzyme isoform is not expressed by most people especially in Africans. Studies in Gambian children showed that aflatoxin cross the placenta and transported to the fetus and the new born where they can cause detrimental effects. The CYP_3A_7 is a major CYP450 enzyme isoform in human fetal liver and metabolizes AFB1 to the 8, 9- epoxide that may cause fetal defects to the developing fetus.

The epoxidation of AFB1 to the exo-8, 9-epoxide is a critical step in the genotoxic pathway of this carcinogen. The binding of AFB1 to DNA and DNA adduction by AFB1 exo-8,9 epoxide has been reported to cause a functional changes of DNA conformation. The epoxide is highly unstable and binds with high affinity to guanine bases in DNA to form afltoxin-N_7-guanine. The aflatoxin-N_7-guanine has been shown to be capable of forming guanine (purine) to thymine (pyrimidine) transversion mutations in DNA and hence affecting the p53 suppressor gene in the cell cycle. The p53 gene is important in preventing cell cycle progression when there are DNA mutations, or signaling apoptosis. The mutations have been reported to affect some base pair locations more than others especially in the third base of codon 249 of the p53 gene in the region corresponding to the DNA binding domain of the corresponding protein and this appears to be more susceptible to aflatoxin-mediated mutations than nearby bases. AFB_1 induces the transversion of base G to base T in the third position of codon 249 and similar mutations have been observed in hepatocellular carcinoma (HCC) in high AFB1 contaminated food in regions in East Asia and Africa .

Epoxide hydrolase and glutathione-S-transferase (GST) are both involved in hepatic detoxification of activated AFB_1, but the GST-catalyzed conjugation of glutathione to AFB1-8,9-epoxides is thought to play the most important role in preventing epoxide binding to target macromolecules like DNA and various cell proteins. Glutathione pathway is reported to play a vital role in the detoxification of AFB_1 . The $AFB_{1\ 8,9}$ exo and endoepoxides are conjugated by glutathione to form AFB-mercapturate and the reaction is catalyzed by glutathione S-transferase (GST) . The glutathione-aflatoxin conjugate is transported from the cells with an ATP-dependent multidrug-resistance protein through an accelerated process. Despite a preference for conjugating the more mutagenic AFB1 exo-epoxide isomer, the relatively low capacity for GST-catalyzed detoxification of bio-activated AFB_1 in lung may be an important factor in the susceptibility of the lung to AFB_1 toxicity .The exo and endo epoxide can also be converted non-enzymatically to AFB1-8,9-dihydrodiol which in turn can slowly undergo a base-catalysed ring opening reaction to a dialdehyde phenolate ion. AFB_1 dialdehyde can form Schiff bases with lysine residues in serum albumin forming aflatoxin-albumin complex. Also the aflatoxin dialdehyde are reduced to a dialcohol in a NADPH-dependent catalyzed reaction by aflatoxin aldehyde reductase (AFAR) . However the guanine alkylation by aflatoxin B_1 produces exo-8,9-epoxide which is the reactive form and a carcinogen to the liver and the reaction is more than 2000 times more efficient in DNA than in aqueous solution, (Fig. 2).

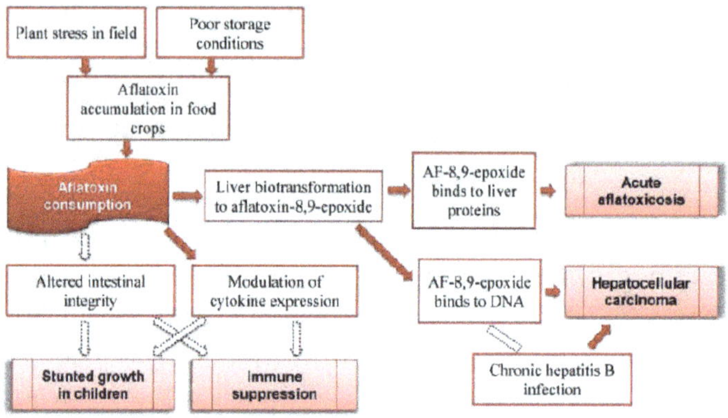

Fig. 2. Aflatoxin disease pathways in humans

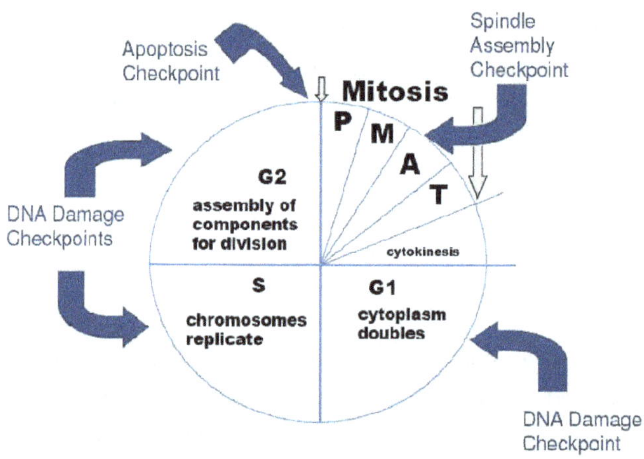

Fig. 3. Various check points that can be damaged by binding of aflatoxins and AF-8,9-epoxide causing the deregulation of the cell cycle; P –prophase, M-Metaphase, A- Anaphase, T- Telophase, S- Synthetic DNA phase, G1 and G2 – Gaps (growth phase).

4.3. EFFECT OF AFLATOXINS ON MITOCHONDRIAL STRUCTURE

AFB causes ultrastuctural changes in mitochondriaand also induces mitochondrial directed apoptosis thus reducing their function. Also the aflatoxins may affect the telomere length and the various check point in the cell cycle causing further damage to the regulatory processes of the cell cycle. Also the extent of aflatoxin binding to DNA and its damage, the level of different proteins changes from cell cycle and apoptotic pathways such as c-Myc, p53, pRb, Ras, protein kinase A (PKA), protein kinase C (PKC), Bcl-2, NF-kB, CDK, cyclins and CKI contribute to the life or death decision making process that may contribute to the deregulation of the cell proliferation leading to cancer development (Fig. 3).

4.4. GLUTATHIONE IN DETOXIFICATION OF AFLATOXINS AND THEIR METABOLITES

However like in hepatic detoxification of aflatoxins and other chemicals, GSH act as antioxidant and has many functions in membrane maintenance and stability as well as in reducing oxidative stress factors and the high reactive oxygen species (ROS) produced from the process of lipid peroxidation .The increased depletion of GSH leads to abnormally high levels of ROS found in cells affected by aflatoxin due to uncoupling of metabolic processes resulting from the lack of GSH for GSH-peroxidase catalysis of O_2 to H_2O_2 leading to lipid peroxidation and compromised cell membranes. Its reduction further enhances the damage to critical cellular components (DNA, lipids, proteins) by the 8,9 epoxides. However the most serious adverse effects of the AFB1-8,9-epoxide metabolite is that it reacts with amino acids in DNA and forms an adduct. The adduct are fairly resistant to DNA repair processes and this causes gene mutation that leads to liver cancers especially the hepatocellular carcinomas.

4.5. CYTOPLASMIC REDUCTASE IN DETOXIFICATION OF AFB_1

Also in the hepatocytes, AFB_1 are converted to other different classes of metabolites by cytoplasmic reductase such as aflatoxicol and by microsomal mixed-function oxidase system to form AFM_1, $AGFQ_1$, AFP_1 and AFB_1 -epoxide (the most toxic and carcinogenic derivative) and these metabolites may be deposited in various body tissues as well as in edible animal products. These metabolites other than the AFB_1 are less toxic and are conjugated with other molecules that enhance their rapid elimination from the body . The metabolite AFQ_1 has very little cancer-causing potential and they are usually excreted in urine with little effect on the body.

4.6. EFFECT OF AFLATOXINS ON PROTEIN SYNTHESIS

The aflatoxin binds and interferes with enzymes and substrates that are needed in the initiation, transcription and translation processes involved in protein synthesis. They interacts of with purines and purine nucleosides and impair the process of protein synthesis by forming adducts with DNA, RNA and proteins. Aflatoxin also inhibits RNA synthesis by interacting with the DNA-dependent RNA polymerase activity and thus causes degranulation of endoplasmic reticulum. Also the reduction in protein content in body tissues like in skeletal muscle, heart, liver and kidney could be due to increased liver and kidney necrosis . AFB_1 is a potent mutagenic, carcinogenic, teratogenic, and immunosuppressive and all these may interfere with normal process of protein synthesis as well as inhibition of several metabolic systems thus causing damages to various organs especially the liver, kidney and heart.

4.7. AFLATOXINS AND CANCER

Aflatoxins especially AFB_1, AFG_1 and AFM_1 are the most toxic, naturally occurring carcinogens known with AFB1 the most hepatocarcinogenic compound, causing various cancers of the liver and other body organs in humans and animals. Aflatoxin's cancer-causing potential is due to its ability to produce altered forms of DNA adducts. The primary disease associated with aflatoxin intake is hepatocellular carcinoma (HCC, or liver cancer). This disease is the third-leading cause of cancer death globally, with about 550,000–600,000 new cases each year. The incidence of liver cancer has been consistently higher in men than in women with a sex ratio ranging from 2 to 3 in most countries. Eighty-three percent of these cancer deaths occur in East Asia and sub-Saharan Africa. Hepatocellular carcinoma (HCC) is one of the most common cancers worldwide with extremely poor prognosis. The majority of cases occur in south-east Asia and sub-Saharan Africa where the major risk factors of chronic infection with hepatitis B and C viruses (HBV and HCV) as well as dietary exposure to aflatoxins are a problem. Aflatoxin B_1, the most commonly occurring and potent of the aflatoxins is associated with a specific AGG to AGT amino acid transversion mutation at codon 249 of the p53 gene in human HCC, providing mechanistic support to a causal link between exposure and disease. Liver cancer has an increasing incidence that parallels the rise in chronic hepatitis B (HBV) and hepatitis C (HCV) infection. Chronic infection with hepatitis B virus (HBV) or hepatitis C virus (HCV) can progress to advanced liver disease, including cirrhosis and hepatocellular carcinoma (HCC), a form of primary liver cancer. HCC is the third leading cause of cancer-related mortality worldwide . The data show that individuals positive for the hepatitis B virus and exposed to aflatoxin in the diet are about 60 times of risk for developing hepato-biliary carcinoma or liver cancer especially in poor developing countries worldwide. Reports have shown that a number of interactions exist between HBV and aflatoxins in development of hepatocellular carcinoma in humans. They may include the fixation of AFB_1-induced mutations in the presence of liver regeneration and hyperplasia induced by chronic HBV infection, the predisposition of HBV-infected hepatocytes to aflatoxin induced DNA damage, an increase in susceptibility to chronic HBV infection in aflatoxin exposed individuals and oxidative stress exacerbated by co-exposure to aflatoxins and chronic hepatitis infection(Fig. 4).

In humans, epidemiological studies in Africa, Southeast Asia, USA and other countries of the west where there is a high incidence of hepatocellular carcinoma, have revealed an association between cancer incidence and the aflatoxin content of the diet. Aflatoxin B_1 (AFB_1) is a major risk factor in the pathogenesis of liver cancer in Asia and sub-Saharan Africa. Aflatoxin B_1 is a potent liver carcinogen in a variety of experimental animals. It causes liver tumours in mice, rats, fish, marmosets, tree shrews and monkeys following administration by various routes. Types of cancers described in research animals include hepatocellular carcinoma (rats) colon and kidney (rats), cholangiocellular cancer (hamsters), lung adenomas (mice), and osteogenic sarcoma, adenocarcinoma of the gall bladder and carcinoma of the pancreas (monkeys).

4.8 HEALTH EFFECTS OF AFLATOXINS ON HUMAN AND ANIMALS

Aflatoxicosis is a condition caused by aflatoxins in both humans and animals. It occurs in two general forms (1) the acute primary aflatoxicosisis produced when moderate to high levels of aflatoxins are consumed. Specific acute episodes of disease may include hemorrhage, acute liver damage, edema, alteration in digestion, absorption and/or metabolism of nutrients, and possibly death. Acute dietary exposure to AFB1 has been implicated in epidemics of acute hepatic injury. Evidence of acute aflatoxicosis in humans has been reported worldwide especially in the third world countries like Taiwan, Uganda, India, Kenya and many others. The chronic primary aflatoxicosis results from ingestion of low to moderate levels of aflatoxins . The effects are usually subclinical and difficult to recognize. Some of the common symptoms are impaired food conversion and slower rates of growth with or without the production of an overt aflatoxin syndrome. The chronic forms of aflatoxicosis include teratogenic effects associated with congenital malformations mutagenic effects where aflatoxins cause changes (mutations) in the genetic code, altering DNA and these changes can be chromosomal breaks, rearrangement of chromosome pieces, gain or loss of entire chromosomes, or changes within a gene the carcinogenic effect in which the carcinogenic mechanisms have been identified such as the genotoxic effect where the electrophilic carcinogens alter genes through interaction with DNA and thus becoming a potential for DNA damage and the genotoxic carcinogens that are sometimes effective after a single exposure, can act in a cumulative manner, or act with other genotoxic carcinogens which affect the same organs. Chronic effects of aflatoxin has been reported to impair the normal body immune function by either by reducing phagocytic activity or reduce T cell number and function as observed immunological suppression in animal model. Aflatoxins have also been reported to interfere with nutrition in a dose response relationship between exposure to aflatoxin and rate of growth in infants and children. Aflatoxins also causes nutrient modification like vitamin A or D in animal models and thus making them unavailable for the normal body physiology and hence leads to nutritional deficiencies.

The contamination of foods and feeds with aflatoxin can cause serious consequences in human and animal health. It is estimated that more than 5 billion people in developing countries worldwide are at risk of chronic aflatoxin exposure due to consumption of aflatoxin-contaminated foods and of these more than 4 billion people develop aflatoxin related liver cancer especially the hepatocellular carcinoma. Aflatoxin exposure is mainly a problem in poor and developing countries with poor regulatory authorities in food processing and storage as well as with high levels of malnutrition. Aflatoxins have also been linked with kwashiorkor and marasmus in most of the sub-Saharan countries in children. Many people in these countries experience chronic aflatoxicosis associated with long-term exposure to low to moderate levels of aflatoxin in the food supply chain. AFB_1, AFB_2 and AFM have been detected in liver, gall bladder, spleen, heart, muscle and kidney. Aflatoxin B_1 exposure results in both steatosis and accumulation of fat and necrosis or cell death of liver cells. The amount of aflatoxins consumed contributes to the mutagenic, carcinogenic, teratogenic, and immunosuppressive health effects in the body. The adverse effect of aflatoxins in humans ranges from acute hepatic toxicity to chronic disease such as liver cancer, haemorrhages, oedema, and even immediate death. Prolonged consumption of aflatoxins has also been reported to cause impaired immune function and malnutrition and stunted growth in children and a number of disabilities and death. Human studies have reported that aflatoxins cause an increase in circulating alpha tumor necrosing factor, suggesting that these mycotoxins are also immunotoxic in humans. Due to the aflatoxin body immunosuppressant, it has been associated with HIV and tuberculosis . Aflatoxins also pose a threat to developing fetuses and they are transferred from mother to infant in breast milk. Aflatoxins have been reported to be associated with a Reye-like Syndrome in Thailand, New Zealand, Czechoslovakia, the United States, Malaysia, Venezuela, and Europe.

All species of animals are susceptible to aflatoxicosis and the susceptibility of individual animals to aflatoxicosis varies considerably depending on dose, duration of exposure, species, age, sex and nutrition. AFB_1, AFB_2 and AFM have been detected in liver, gall bladder, spleen, heart, muscle and kidney of growing swine when protein and protein-free portions of the diet were separately fed. Chronic exposure of aflatoxins to animals causes immunosuppression and also interferes with protein metabolism and multiple micronutrients that are critical to health due to adduct formation. These adduct are responsible for mutations, cancer, immunosuppression, lung injury and birth defects. In animals, the aflatoxins cause liver damage, decreased milk production, reduced reproductively and suppressed immunity in animals consuming low dietary concentrations. The aflatoxicosis syndrome in animals may also be characterized by vomiting, abdominal pain, pulmonary oedema, convulsions, coma, and death with cerebral edema and fatty involvement of the liver, kidneys, and heart. In dairy and beef cattle, the signs of acute toxicosis include anorexia, depression, dramatic drop in milk production, weight loss, lethargy, gastrointestinal dysfunctions such as ascitis, icterus, tenesmus, abdominal pain, bloody diarrhoea, decreased feed intake and efficiency; weight loss, jaundice, abortion, hepatoencephalopathy, blindness, walking in circles, ear twitching, frothy mouth, photosensitization, bleeding and death. In poultry, beside inappetance, weight loss, decreased egg production, leg and bone problems, poor pigmentation, fatty liver, kidney dysfunction, bruising and death, suppression to natural immunity and susceptibility to parasitic, bacterial and viral infections can occur (Fig. 4).

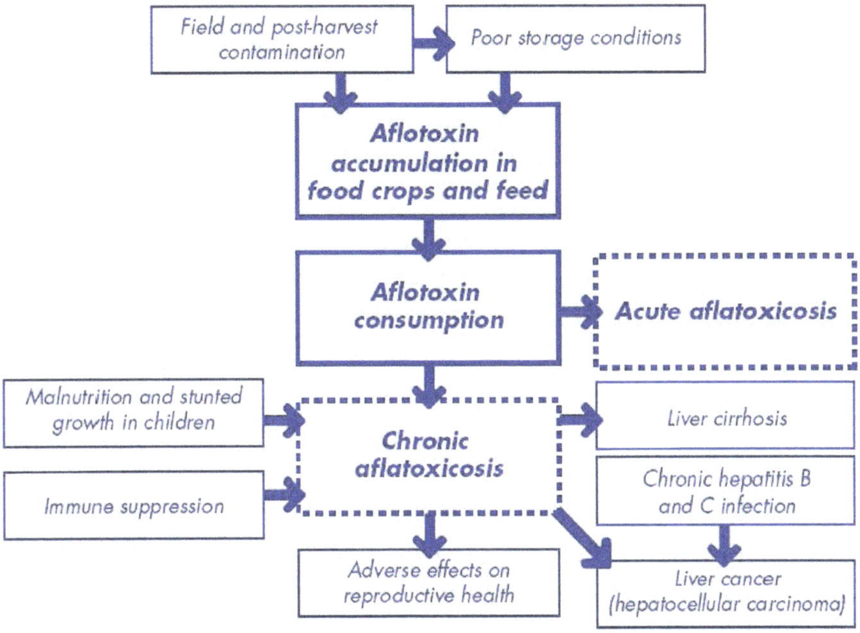

Fig. 4. Aflatoxin disease pathways in humans.

4. 9 . BIOLOGICAL EFFECT OF AFLATOXINS ON THE BODY ORGANS AND BODY SYSTEMS

Aflatoxins have been reported to affect the various body organs like the liver, kidneys, lungs, brain, testes and many endocrine and exocrine organs, the heart, skeletal muscles and the different body systems.

4.9.1. Role of Aflatoxins in Hepatic Injury and Other Body Organs and Tissues

Aflatoxins have been reported to cause liver cirrhosis as well as liver cancers. Hepatic injury can be acute or chronic form caused by a variety of toxic agents like aflatoxins, chemicals and drugs, trauma and infectious agents. The reduced level of total protein is indicative of the toxic effect of AFB1 to the liver due to the failure in synthesis of the proteins and kidney in which aflatoxins are known to impair protein biosynthesis by forming adducts with DNA, RNA and proteins, inhibits RNA synthesis, DNA-dependent RNA polymerase activity and causes degranulation of endoplasmic reticulum. Acute hepatic injury due to aflatoxin causes a rise in serum enzymes including aspartate aminotransferase, lactate dehydrogenase, glutamate dehyrogenase, gamma-glutamyltransferase and alkaline phosphatase and bilirubin that reflect liver damage as well as other biochemical changes such as proteinura, ketonuria, glycosuria and hematuria. The other frequently used liver enzymes are the alkaline phosphatase (ALP) and Gamma-glutamyltransferase and gamma-glutamyltranspeptidase (GGT and GGTP) that indicate obstruction to the biliary system, either within the liver or in the larger bile channels outside the liver. The presence of jaundice and neurological disorders due to brain damage leading to hepatic encephalopathy are associated with liver failure. Chronic liver failure leads to accumulation of metabolites in circulation such as ammonia and fatty acids that eventually lead to brain damage and hence hepatic encephalopathy. The liver failure makes it unable to detoxify ammonia, the product of protein and amino acid metabolism leading to hyperammonemia that may cross the blood brain barrier leading to increased synthesis of glutamate neurotransmitters hence leading to cytotoxicity of the brain cells and hence the hepatic encephalopathy. AFB_1 has been reported to cause pallor discoloration of liver and enlargement of liver and kidneys, congestion of liver parenchyma, cytoplasmic vaculation or fatty change of hepatocytes, necrosis of hepatocytes and newly formed bile ducts, mononuclear and heterophilic cell infiltration are reported in aflatoxin fed broiler chicks. It is also reported that there is a decrease in protein content in skeletal muscle, heart, liver and kidney in aflatoxin-fed animals due to the AFB_1's potent mutagenic, carcinogenic, teratogenic, immunosuppressive and its ability to inhibits several metabolic systems such as protein synthesis thus leading to liver, kidney and heart damage. In chicken, the activity of serum or plasma enzymes like the sorbitol dehydrogenase, glutamic dehydrogenase, lactate dehydrogenase, alkaline phosphatase, acid phosphatase, aspartate aminotransferase and alanine aminotransferase were reported to be increased in aflatoxicated chickens.

4.9.2. Effect of Aflatoxins on the Central Nervous System

In the brain or central nervous system, the neurons have a high metabolic rate but little capacity for anaerobic metabolism and subsequently, inadequate oxygen flow to the brain kills the neuronal brain cells within minutes. Some compounds damage neurons or neurotoxic and thus inhibit their function. Mycotoxins especially aflatoxins and its metabolites and other products such as the reactive oxygen species (ROS) like the AFB-8,9-epoxides may interfere with the normal functioning of the nerve cells by forming DNA adducts, protein adducts, oxidative stress factors, mitochondrial directed apoptosis of the nerve cells as well as inhibiting their synthesis of protein, RNA and DNA. Aflatoxins also cause abnormalities in mitochondrial DNA, structure and function, including defective oxidative phosphorylation in the brain cells. The oxidative stress may result in damage to critical cellular macromolecules such as DNA, lipids and proteins. Cellular fatty acids are readily oxidized by ROS to produce lipid peroxyl radicals which can subsequently propagate into MDA that may interact with cellular DNA to cause DNA-MDA adduct that may affect energy production in the brain. The role of ROS has been postulated in the development of aging and chronic degenerative diseases, inflammatory diseases and brain cancers. Aflatoxins may also deplete the myelin sheath of the nerves, an important substance that covers the nerves and hence become exposed to insults. Mycotoxins especially aflatoxins have been reported to be toxic to various aspects of brain chemistry and their function.

AFB1 also alters the levels of various biogenic amines (neurotransmitters) and their precursors in rat and mouse brains. Acute AFB_1 treatment in experimental animals has been reported to cause a decrease in regional brain acetylcholinesterase enzymes that may affect the cognitive functions as well as memory and learning of the individual while chronic exposure increases adenohypophyseal acetylcholinesterase. Aflatoxin causes a decrease in dopamine, serotonin and alterations in the levels of the precursor's tyrosine and tryptophan. Deficiencies in these neurotransmitter lead to neurological symptoms such as neurocognitive decline and alteration of sleep cycle and symptoms of brain damage like dullness, restlessness, muscle tremor, convulsions, loss of memory, epilepsy, idiocy, loss of muscle coordination, and abnormal sensations. AFB1 has also been reported to increase the central and peripheral nervous system Na^+/K^+-ATPase, β-glucuronidase and β-galactosidase while inhibiting the Mg^{2+}-ATPse in experimental animals and this also is important in the normal functioning of the glutamate neurotransmitter and their NMDA receptors. The liver failure makes it unable to detoxify ammonia, the product of protein and amino acid metabolism leading to hyperammonemia that may cross the blood brain barrier leading to increased synthesis of glutamate neurotransmitters hence leading to cytotoxicity of the brain cells and hence the hepatic encephalopathy. Toxic encephalopathy was originally described in children with Reye's syndrome associated with consumption of Aflatoxin B_1 and/or salicylates and subsequently in cases of aflatoxicosis in canines and Chinese children were reported. Aflatoxins also have been linked to Reye's syndrome that is characterized by symptoms of encephalopathy and fatty degeneration of the viscera. It is a pediatric disease characterized by cerebral edema and neuronal degeneration. Toxic encephalopathy due to aflatoxins involves multiple symptoms like loss of balance, recent memory decline, headaches, lightheadedness, spaciness/disorientation, insomnia, loss of coordination. Aflatoxins have been reported to be associated with a Reye-like Syndrome in Thailand, New Zealand, Czechoslovakia, the United States, Malaysia, Venezuela and Europe. Aflatoxins especially AFB_1 have been reported to cause tumors in both the central and peripheral nervous system and several nonepithelial neurogenic tumors like the schwannomas, gliomas, meningiomas and granular cell tumors have been reported.

4.9.3. Effect of Aflatoxins on the Gastrointestinal Tract (GIT)

The gastrointestinal tract (GIT) is the main route of entry of aflatoxins as a result of consumption of aflatoxin-contaminated foods especially AFB_1. It is also the main route of excretion aflatoxin metabolites from the bile. The aflatoxins, metabolites and AF-8,9-epoxides have been reported to cause intestinal tumors especially the human colon cancers like colon carcinomas and similar results have been reported in experimental animals . Aflatoxins have also been reported to cause serious acute effects on the GIT . Aflatoxins have been implicated as potential factors in the increased incidence of human gastrointestinal and hepatic neoplasms in Africa, Philippines and China. Aflatoxins have been reported to cause digestive system effects such as diarrhea, vomiting, intestinal hemorrhage, and liver necrosis and fibrosis. Aflatoxins have been reported also to damage the integrity of the pancreas. In domestic animals, aflatoxins cause changes in the GIT physiology especially decreased rumen motility and function in cows . In birds, aflatoxins interfere with intestinal morphology, sialic acid production and apparent digestible energy .

4.9.4. Effect of Aflatoxins on the Respiratory System

Aflatoxins have reported to have serious acute effects on the respiratory systems. The respiratory tract is the only organ system with vital functional elements in constant and direct contact with the environment. Many people working in food industries as their occupational setting get exposed to aflatoxins especially AFB1 when they inhale antitoxin-contaminated dusts like during grain shelling and processing and have been reported to have a higher incidences of upper respiratory tract and lung cancers . In experimental animals, AFB_1 was reported to induce 100% pulmonary adenomas. In the respiratory tract, aflatoxins may also be converted to active metabolites like in the nasal mucosa. It is also reported that the intranasal administration of AFB_1 lead to formation of tissue-bound metabolites in subtentacular cells, bowman's glands and in neuronal cells in the olfactory mucosa but there is no evidence that AFB1 may induce tumours in olfactory bulbs. Epoxide hydrolase and glutathione-S-transferase (GST) are both involved in hepatic detoxification of activated AFB_1 but the GST-catalyzed conjugation of glutathione to AFB1-8,9-epoxides is thought to play more important role in preventing epoxide binding to target macromolecules. However, the low capacity for GST-catalyzed detoxification of bio-activated AFB_1 in lung may be an important factor in the susceptibility of the lung to AFB_1 toxicity. Nose-only inhalation exposure of rats to AFB_1 aerosols suppressed alveolar macrophage (AM). Intratracheal administration of AFB_1 also suppressed the release of tumor necrosis factor-alpha from AMs and impaired systemic innate and acquired immune defenses as well as suppression of peritoneal macrophage phagocytosis and the primary splenic antibody response thus leading to suppression of respiratory tract defenses system.

4.9. 5. Effect of Aflatoxins on the Cardiovascular System and Blood

Aflatoxins have reported to have serious acute effects on the cardiovascular systems including vascular fragility and hemorrhaging in tissues as well as heart damage and teratogenic effects. It is reported that there is a decrease in protein content of the muscles of these tissues and organs as well as inhibition of their metabolic processes attributable by the aflatoxin consumption of contaminated foods.

Effect of Aflatoxins on the Blood and Blood Cells: The aflatoxins and its metabolites as well as the generated reactive oxygen species(ROS) has been reported to have a deleterious effects on the bone and blood cells as well as induction of cancers on the hemopoietic system in bone marrow and lymphoid organs where blood, blood cells and blood components are produced . The blood system can be damaged by agents that affect blood cell production (bone marrow), the components of blood (platelets, red blood cells, and white blood cells), or the oxygen-carrying capacity of red blood cells or impair blood clotting and their poor growth rates. Oxidative damage by the AFB_1 on human lymphocytes has been reported and significant declines in both the proportion of peripheral blood lymphocytes and in the percentages of ANAE-positive peripheral blood lymphocytes (T-lymphocytes) in a dose dependent manner has been observed. Aflatoxins have been linked to anemia in pregnancy and alterations in erythrocytes during induced chronic aflatoxicosis in rabbit also have been reported . Aflatoxin causes hematopietic suppression and anemia, decrease in total erythrocytes, packed-cell volume and hemoglobin as well as toxicity to red blood cells . Aflatoxin is known to produce hemolytic anemia by decreasing the circulating mature erythrocytes and consequently the spleen appear congested because of an unusually high concentration of inorganic iron and debris from the circulation . In birds, AFB_1 is reported to causes hematological changes. Aflatoxicosis has been reported to cause lymphocytopenia and monocytopenia and increased percentage of neutrophil counts . In cattle, aflatoxins are reported to cause blood coagulation defects that may involve impairment of prothrombin, factors VII and X and possibly factor IX and similar effects are reported in dogs. Generally aflatoxins have been reported to depress growth and alter many aspects of humoral and cellular immunity and thus affecting the hematological parameters .

4.9.6. Effect of Aflatoxins on the Urinary System

The kidney is susceptible to many toxic agents due to the high amount of blood it receives and about 20-25% of blood that flows in at rest coupled with the large amounts of circulating toxicants that reach the kidneys . The kidneys also have high oxygen and nutrient requirements because of their workload and therefore filters one-third of the blood reaching them and reabsorb 98-99% of the salt and water. Different parts of the nephrone are exposed to aflatoxins especially the AFB_1 and its metabolites leading to nephrotoxicity before it is excreted in the urine. The aflatoxin induced reduction in protein content has been reported to be due to increased necrosis of the kidney. AFB_1 has been reported to cause kidney tumors in experimental animals and a mixture of AFB and AFG was observed to cause renal and hepatic tumors in 80% of hamsters. There were also renal lesions with features of megalocytosis in the proximal renal tubules. In Africa, birds exposed to AFB_1 were reported to develop fatty and hemorrhagic kidney syndrome, thickening of the glomerular basement membrane, abnormal development of glomerular epithelial cells and degenerative changes in renal tubular cells, congestion and parenchyma hemorrhage. In other animals, there was a reduction in the glomerular filtration rate, glucose reabsorption and tubular transport of electrolytes and organic anions, reduced activities of renal glutamate-oxaloacetate and pyruvate transaminases and alkaline phosphatase in rats attributed to by the aflatoxins and their metabolites as well as the generated ROS. There was induced aggregation and loss of chromatin, mitochondrial degeneration and loss of microvilli induced by AFB1 in cultured kidney cell lines.

4.9.7 Effect of aflatoxins on the endocrine system

Aflatoxin especially AFB has been reported to interfere with the functioning of the various endocrine gland by disrupting the enzymes and their substrates that are responsible for the synthesis of the various hormones. Aflatoxins and their metabolites as well as the generated ROS have been reported to cause various cancers in different endocrine glands like pituitary gland, granulosa cell tumors of the ovary and adenomas and adenocarcinomas of the adrenal gland, kidneys, thyroid gland, ovaries, testes, thyroid gland, parathyroid glands and endocrine pancreas . The plasma testosterone and luteinizing hormone (LH) concentrations have been reported to reduce in aflatoxin-fed birds . In laboratory animals, aflatoxin causes delayed maturation of both males and females. Aflatoxicosis in white leghorn males chicken decreased feed consumption, body weight, testes weight and semen volume and decreased plasma testosterone values .

4.9.8 Effect of aflatoxins on the reproductive system

In humans exposed to chronic aflatoxin-contaminated foods, it has been reported that higher concentrations of aflatoxins occur in the semen of infertile men . It is also associated with low birth weight, a risk factor for jaundice in infants as well as presence of AFM in maternal breast milk where it can cause deleterious effect in the newborns . In Nigeria, about 37% of the infertile men had aflatoxin in their blood and semen hence contributing to the incidence of infertility in Nigerians . Experimental results indicate that certain agents like aflatoxins can interfere with the reproductive capabilities of sexes, causing sterility, infertility, and abnormal sperm, low sperm count, and/or affect hormone activity in animals. Aflatoxins have been reported to disrupt the reproductive system in both male and female animals after ingestion of aflatoxin-contaminated foods. Aflatoxins also cause pathological alterations in the form of coagulative necrosis especially in the growing and mature follicles and decrease in number and size of graffian and growing follicles with increased number of atretic follicles and small areas of degenerative changes in experimental animals . AFB_1 has been reported to have a deleterious effect on the reproductive capacity of laboratory and domestic female animals where they cause reductions in ovarian and uterine sizes, increases fetal resorption, implantation loss and intra-uterine death in the aflatoxin exposed female rats . They also cause a reduction in the primary spermatocytes and spermatids and affect the morphology of the sperm cells produced . Stillbirths were reported in the 15th to the 18th days of pregnancy in rats . The levels of plasma testosterone, plasma 5a-DHT and absolute and relative testes weights were reported in experimental animals of aflatoxin-treated males remained low in all age groups and a delay in the onset of sexual maturation during aflatoxicosis . In cows, aflatoxins affected the reproductive system by causing abortion, the birth of weak, deformed calves, reduced fertility due to reduced vitamin A levels. The teratogenic effects of AFB_1 were described as enlarged eye sockets and enlarged liver of embryos . In poultry, AFB_1 cause a reduction in semen volume, testes weight, spermatocrit and plasma testosterone as well as a reduction in egg output .

4.9.10. Effect of Aflatoxins on the Immune System

Chronic consumption of aflatoxin-contaminated foods has been reported to cause immunosuppression in both humans and animals worldwide . In human, aflatoxins affect both the cellular and humoral immune responses where they alter immunological parameters in participants with high AFB_1 levels resulting in impairments in cellular immunity hence decreasing the host resistance to infections. Aflatoxin exposure has been shown to cause immune suppression, particularly in cell-mediated responses. Chronic exposures of the individual to aflatoxins depress the phagocytic efficiency of the phagocytes and the delayed hypersensitivity reactions in birds . Aflatoxins also deplete the cell populations of the thymus; reduce the bone marrow and the red and white blood cells count, macrophage numbers and the phagocytic activity of the cells . It also depresses the T-cell-dependent functions of splenic lymphocytes in mice. The natural killer cell function of the peripheral blood lymphocytes are also affected by aflatoxins especially AFB_1 . A reduction in the leukocyte immunophenotypes in peripheral blood, CD4+ T cell proliferative response, CD4+ T and CD8+ T cell cytokine profiles and monocyte phagocytic activity were reported. Children in developing countries appear to be naturally exposed to aflatoxin through their diet at levels that compromise the immune system. In general, the proportion of childhood growth stunting is directly correlated with the proportion of the population living below the national poverty line and is inversely correlated with gross domestic product per capita As is the case with liver cancer, childhood stunting is prominent in regions such as Southeast Asia and Sub-Saharan Africa, where aflatoxin exposure through consuming contaminated food is common . It has been reported that the immunosuppression and nutritional effects of chronic aflatoxin exposure may be linked to the high prevalence of HIV in Southern Africa. The CD4 proteins that have been weakened by aflatoxin exposure have been reported to correlate positively with HIV infection . Also high aflatoxin levels have been reported to increase risk of developing tuberculosis in HIV positive individuals. Persons who are exposed to aflatoxin and are HIV positive have decreased plasma vitamin A and vitamin E in the blood, although there was no interaction detected between aflatoxin and HIV infection. HIV infection is likely to increase aflatoxin exposure by two possible routes:

1. HIV infection decreases the levels of antioxidant nutrients that promote the detoxification of aflatoxin, or
2. the high degree of co-infection of HIV-infected people with hepatitis B also increases the biological exposure to aflatoxin . Aflatoxin induce immunosuppression and increases susceptibility of toxicated birds and animals to bacterial, viral and parasitic infections .

It also affects the lymphoid follicles of caecum thus depleting the lymphocytes that may contribute to the observed immunosuppression . Aflatoxin decreases the concentrations of immunoglobulins IgM, IgG and IgA in birds as well as decrease complement activity in chickens . The low dose of AFB_1 slightly decrease both mRNA and protein levels of lymphocytic IL-2, IFNγ and it preferentially affects macrophage functions as well as IL-1α, IL-6 and TNF production by these cells . Aflatoxin suppression of the immune system therefore subjects the individual to high risk of susceptible to infectious diseases like parasitic, bacterial and viral infections.

5

AFLATOXINS IN MILK

5.1 INTRODUCTION

Aflatoxin M_1 is found in milk and milk products. Strict control of Aflatoxin B_1 level in feeds for lactating dairy cattle and buffaloes is required in order to minimize the level of Aflatoxin M_1 in milk and milk products. The United States Food and Drug Administration (USFDA, 2018) and Food Safety and Standards Authority of India (FSSAI, FSS Regulation 2011) set a maximum permissible level for Aflatoxin M_1 in milk at 0.5 µg/ kg (ppb; parts per billion) of milk, which means that one ton of milk should not contain more than 500 micrograms of Aflatoxin M_1. Aflatoxin B_1 is readily transmitted from feed to milk; approximately 1.0 to 6.0% of Aflatoxin B1 present in feed is transferred to milk as Aflatoxin M_1, depending on factors such as the genetics of animals, seasonal variation, the milking process and the environmental conditions. The occurrence of Aflatoxins in commercially available milk, and milk products is of concern, as milk is a key source of nutrients for humans.

There are more than 10,000 known species of fungi. Fortunately, most of them are beneficial to man in the production of bread, cheese, antibiotics etc. There are about 50 fungi species harmful to livestock, poultry, and man known to produce toxins, which are collectively called mycotoxins. Mycotoxins are metabolites produced by fungi during metabolism of nutrients present in feeds and feed ingredients.Fungi produce mycotoxins in the field ("field toxins": fusarium toxins) or in storage ("storage toxins": Aflatoxins, ochratoxins), or both. Mycotoxins are mainly produced by the fungi genera of Aspergillus, Fusarium, Penicillia, and Claviceps. Their formation may occur when the fungi grow on crops in the field, at harvest, in storage or during the processing of feed when conditions are favorable. No region of the world escapes these silent killers, and their negative impact on animal productivity and human health is enormous. According to the United Nation's Food and Agriculture Organization (FAO), approximately 25% of World's grain supply is contaminated with mycotoxins. Recently, a 38 member scientific task-force from the Council for Agricultural Science and Technology (CAST, 2003) in the USA released a 200 page report entitled "Mycotoxins: Risks in Plant, Animal, and Human systems". According to the report, in the United States alone the economic costs resulting from mycotoxins exceed $ 1.5 billion annually.

5.2 MYCOTOXINS: AN OVERVIEW

5.2.1 Geographical Distribution of Mycotoxins

The threat of mycotoxins has been described as early as the Second World War when the soldiers from the Russian army suffered severe dermal necrosis, hemorrhages and destruction of bone marrow after eating mouldy grains (Fusarium contaminated, Fig. 1). However, it was not until 1960, when the entire turkey population of Britain was decimated in a fatal liver disease called 'Turkey X Disease', that the scientific community recognized the negative effects associated with mycotoxins. British agriculture officials later traced the source of the outbreak to Aflatoxin in a shipment of peanut (groundnut) meal that originated from Brazil.

Fig. 1: Maize cobs contaminated with Fusarium fungus

The occurrence of mycotoxins in nature is considered a global problem. However, in certain geographical areas of the world, some mycotoxins are produced more readily than others. In colder, more temperate regions such as Canada, the Northern US and most of the European countries, Aflatoxins are not considered to be a major problem except in imported feedstuffs grown in warmer southern climates. Economically, in these regions the most important mycotoxins are deoxynivalenol (DON, Vomitoxin), Zearalenone (ZEA), T-2 toxin produced by Fusarium fungi (Fig.1). In Europe, the differences in climatic conditions among the northern, middle and southern parts favor the development of different fungal species. In the maize growing areas of southern and middle Europe (Sweden, Austria and Hungary) mainly fusariotoxins (DON, ZEA, T-2 toxin) cause illness and poor performance of dairy animals.

- ***Deoxynivalenol (DON, Vomitoxin):*** Deoxynivalenol is produced by several Fusarium species of mould primarily Fusarium graminearum and may co-occur with other mycotoxins in contaminated commodities. DON in cattle and buffaloes has been associated with reduced feed intake and lower milk production when fed with feed containing more than 5 parts per million (ppm).
- ***Zearalenone:*** Zearalenone is produced primarily by Fusarium graminearum and Fusarium roseum. Zearalenone is responsible for reproductive disorders because of its estrogenic effect. In dairy animals, the clinical manifestations such as udder enlargement, decreased milk yield, vaginal discharge, continuous estrus, infertility and abortions are observed when the level of zearalenone in the feed is more than 0.5 ppm.
- ***Fumonisins:*** Fumonisins are primarily produced by Fusarium moniliforme and Fusarium proliferatum, the most toxic being fumonisin B_1. Fumonisin mycotoxins are found in a wide range of commodities from millets to grains to banana fruits. They are a major concern to food and feed producers since they affect human and animal health. The toxin causes liver damage and decreased milk production in dairy cattle at levels greater than 50 ppm in the ration. Usually, level of fumonisins exist between 5-20 ppm in feed.

5.3 GLOBAL SCENARIO OF AFLATOXINS

Aflatoxins; the most widespread of all the mycotoxins, are common in warm and humid climatic conditions like those existing in India, Latin American, Asian and African countries, southern regions of US, and certain parts of Australia. Extensive surveys conducted in India, Pakistan, Egypt and South Africa suggested that Aflatoxins are often encountered in substantial levels in feeds and feed ingredients (Fig 2).

Fig. 2: Maize naturally contaminated with Aspergillus fungus

In Latin American countries including Brazil, Peru, Mexico, Columbia, Venezuela and Argentina, reports exist on the presence of Aflatoxins. Due to the increase in global trading of feedstuffs, mycotoxins are no longer solely found in certain geographical regions but are now more widely distributed than before . Aflatoxins B_1, B_2, G_1, and G_2 refer to toxins which fluoresce blue (B_1, B_2) or green (G_1, G_2) under ultraviolet light. Among all Aflatoxins, B_1 is the more prevalent and toxic.

5.4 LIMITS OF AFLATOXINS IN FEED AND MILK IN OTHER COUNTRIES

Strictly speaking, there is no safe level. With reference to mycotoxins, the risk directly depends on the level of the major mycotoxins and also on the presence and levels of other mycotoxins in feeds. A mycotoxin level considered safe in one farm may not be safe in another farm because of differences in management and disease prevalence. Additionally, factors such as interaction of mycotoxins with pathogens, genetic variability, environmental conditions, and nutritional status etc. influence the severity of the manifestation of mycotoxicosis. In order to reduce the toxic and economic impact of mycotoxins, several countries regulate the levels of some mycotoxins in foods and feeds. Worldwide, food and feed legislation safeguards the health of consumers and the economic interests of animal producers and traders. Virtually, all countries with fully developed market economies have regulations with the exception of some African countries.

Table 1: Limits for Aflatoxin B_1 in dairy feed

Country	Aflatoxin B_1 (µg/kg or ppb)
United States (FDA)	20
European Union (for milking animals)	5
European Union (for calves)	10
India (BIS)	20

Table 2: Limits for other mycotoxins in dairy feed Limits

Mycotoxins (EU Guidelines)	Limits (mg/kg or ppm)
Deoxynivalenol: Adult animals Calves	 5 2
Zearalenone: Adults and calves	0.5
Fumonisins B1 and B2: Adult animals Calves	50 20

Table 3: Limits for Aflatoxin M_1 in milk

Country	(µg/kg or ppb)
United States (USFDA)	0.5
European Union	0.05
European Union, for baby foods/ infants	0.025
Australia (FSANZ)	0.05
Australia, for infants	0.02
India (FSSAI)	0.5

5.5 AFLATOXINS IN FEED AND FEED RAW MATERIALS IN INDIA

Aflatoxins are primarily produced by fungi of the genus Aspergillus (*Aspergillus flavus, Aspergillus parasiticus*, and *Aspergillus nomius*; Fig.3) and are found in dairy feed and human food products. Major forms of Aflatoxins found in feeds include Aflatoxins B_1, B_2, G_1 and G_2; with Aflatoxin B_1 being the most common and toxic.

Extensive surveys conducted in India suggest that Aflatoxins are often encountered in substantial levels in feeds and feed ingredients. Aflatoxins are rarely found in forages. To study the incidence of Aflatoxin B1 in different raw materials, a survey was conducted by the Division of Animal Sciences, University of Agricultural Sciences, Bengaluru. The results of the survey showed that out of 246 samples analyzed by Thin Layer Chromatography (TLC), 206 samples were found to be positive for the presence of either Aflatoxin.

Fig. 3: Growth of *Aspergillus flavus* on rice

The cereal and cereal byproducts analyzed were maize, deoiled rice bran, rice polish and wheat bran. The samples were found 88% positive for Aflatoxins. The oilseed meals analyzed were soybean meal, full-fat soy, sunflower meal, groundnut cake/ extract/ expeller, rapeseed meal and til cake. The oilseed meals contain higher incidence of Aflatoxin (96%) contamination as compared to cereals and cereal byproducts. Aflatoxin B_1 content was analysed in some samples of cattle feed (n=254) suspected to be contaminated with Aflatoxins at NDDB, Anand (NDDB, 2018). The results found that the average level of Aflatoxin B1 was 68 ppb in these samples.

5.6 METABOLISM OF AFLATOXIN IN RUMINANTS AND CONVERSION OF AFLATOXIN B_1 TO AFLATOXIN M_1

In milk, Aflatoxin appears as Aflatoxin M_1, one of its metabolites. Aflatoxin B_1 is metabolized by enzymes found primarily in the liver (Cytochrome P450) to Aflatoxin M_1. Strict control of Aflatoxin B_1 level in feeds for dairy animals is required in order to minimize the level of Aflatoxin M_1 in milk and milk products. In the US (FDA) and in India (FSSAI), it is required by the law that Aflatoxin M_1 in milk should be less than 0.5 ppb; which means that one ton of milk should not contain more than 500 micrograms of Aflatoxin M_1. In European Union, the regulations are much more stringent, and maximum levels are set at 0.05 ppb (EC, 2002). Aflatoxin is readily transmitted from feed to milk; approximately 1.0 to 6.0% of Aflatoxin B_1 present in feed is transferred to milk as Aflatoxin M_1 (Fig. 4). After Aflatoxin M_1 is formed, it is excreted in milk and urine of the dairy animals.

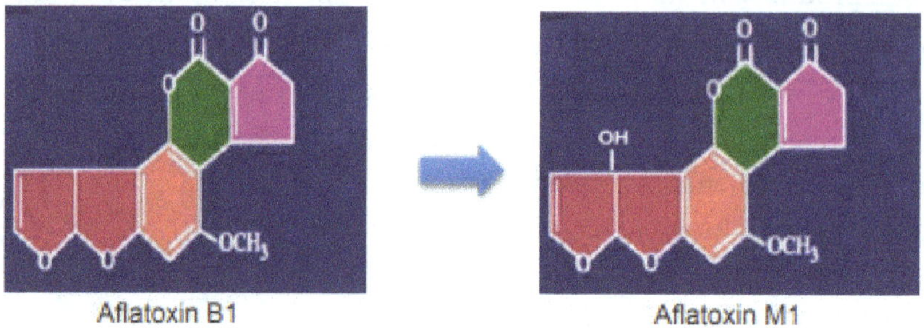

Fig. 4: Conversion of Aflatoxin B_1 to M_1

Diaz et al. (2004) have indicated that Aflatoxins appear in milk within 12 hours following oral administration of Aflatoxin B_1. The peak concentration in milk occurs after 24 hrs. Clearance is also very rapid, since Aflatoxin disappears from the milk four days after cessation of oral administration. These results confirm the rapid absorption and metabolism of Aflatoxins in ruminants. The USFDA and BIS (India) stipulate a maximum level of 20 ppb of Aflatoxin in dairy feed, which means that one ton of feed should not contain more than 20 milligrams of Aflatoxin B_1. The European Union regulations are much more stringent and the maximum level set at 5.0 ppb. Like Aflatoxin B_1, Aflatoxin M_1 is toxic, although toxicity of Aflatoxin M_1 is somewhat lower than that of Aflatoxin B_1.

5.7 IMPACT OF FEEDING AFLATOXIN CONTAMINATED FEED IN DAIRY ANIMALS

Milk producers may not be able to see visual symptoms of aflatoxicosis in the animals at low level Aflatoxin in the feed. However, high concentrations of Aflatoxins and/ or prolonged duration may cause visual symptoms in dairy animals. Feed refusal, reduced growth rate and decreased feed conversion efficiency are the predominant signs of chronic Aflatoxin poisoning. In addition, listlessness, weight loss, rough hair coat and mild diarrhea may occur. The disease may also impair reproductive efficiency, including abnormal estrous cycles (too short and too long) and abortions. Other symptoms include impaired immune system response, and increased susceptibility to diseases.

6

APPLICATION OF MYCOTOXIN BINDERS FOR CONTROL OF AFLATOXINS IN FEED

6.1 MYCOTOXIN BINDERS

Practical mycotoxin control involves following two main stages:

- Prevention of fungal growth and toxin production in feedstuffs, and
- Decontamination of existing mycotoxins in feeds.

Several strategies, to be adopted in a coordinated manner have been proposed to reduce the impact of mycotoxins, which include: plant breeding for mould resistance, efficient harvesting and storage practices to minimize contamination and decontamination process for the contaminated commodities .

6.1.1 Nutritional Modifications

Nutritional routes for protection against mycotoxins include higher levels of methionine, selenium and vitamin supplementation of affected diets. Some plant and herbal compounds including chlorophyll derivatives are also used for protection against mycotoxins. This method is partially beneficial but not cost effective.

6.1.2 Herbal Mould Inhibitors

Certain herbs and herbal extracts have been found to exert inhibitory effect on mould growth and thus toxin production. Aqueous extracts of garlic, onion, turmeric, neem etc., have been shown to exert anti-fungal activity and inhibit Aflatoxin production. This method, though beneficial to some extent, presents challenges pertaining to standardization, quality control etc.

6.1.3 Chemical Detoxification

Among the chemicals tested for their ability to detoxify/ inactivate mycotoxins, ammonia, sodium bisulfite, peroxide, acids, bases and gases are effective. However, most of the chemical methods are not practical and they do not fulfill all the requirements, specifically those concerning the safety of reaction products and the palatability of the feed.

6.1.4 Biological Methods

Advances in the field of biotechnology have opened a new avenue for tackling mycotoxicosis. A natural organic product, glucomannan-containing yeast product, a cell wall derivative of Saccharomyces cerevisiae have shown considerable binding ability with commonly occurring mycotoxins and is found beneficial in minimizing the adverse effects of mycotoxins in livestock and poultry.

6.1.5 Application of Mineral Clays

Many types of clay have been tested for counteracting mycotoxins. These include bentonites, zeolites and aluminosilicates. Several studies have demonstrated that sodium aluminosilicate, sodium calcium aluminosilicate, and sodium bentonites can adsorb Aflatoxins. A hydrated sodium calcium aluminosilicate (HSCAS) is the most widely studied mycotoxin sequestering agent among the mineral clays. Interest in its potential as a mycotoxin sequestrant was based on the ability of the silicate mineral to sequester positively charged or cationic compounds (Ramos and Hernandez, 1997).

6.2 CHARACTERISTICS FEATURES OF AN IDEAL MYCOTOXIN BINDER

The most effective method of neutralizing mycotoxins already in feed is by binding them to an inert compound before they can be absorbed from the intestines. The question that feed manufacturers, and milk producers frequently ask is: “Which mycotoxin binder is most effective?” The “ideal” features of good mycotoxin binder are:

- Ability to bind a wide range of mycotoxins;
- Low effective inclusion rate in feed;
- Rapid and uniform dispersion in the feed during mixing;
- Heat stability during pelleting, extrusion, and during storage;
- No affinity for vitamins, minerals or other nutrients.

6.3 TEST METHODS FOR AFLATOXIN ANALYSIS AT LABORATORY AND FIELD LEVELS

6.3.1 Tests to Detect Aflatoxin

There are simple, and fast tests which can be performed for testing of Aflatoxin B1 in finished feed and raw materials, and M1 in milk and milk products. Kits using ELISA

(Enzyme-Linked Immunosorbent Assay) technology are available to test at the farm as well as at the cattle feed plant level. Other analytical methods are also available such as Thin Layer Chromatography (TLC) and High Performance Liquid Chromatography (HPLC). Off-late, lateral flow based strips are available in the market for quick detection of Aflatoxins in feed and milk.

6.4 MEASURES TO BE TAKEN DURING PRE-HARVEST, HARVEST AND STORAGE UNDER FIELD CONDITIONS

The most practical approach is by good agricultural practice, which includes 'plant breeding' to develop mould resistance crops and the second approach is 'fungal bio-competition'. Application of non-toxigenic strains of Aspergillus flavus and Aspergillus parasiticus to soil in maize plots lead to reduction in colonization of toxigenic fungi in subsequent years. The non-toxigenic biocompetitive Aspergillus strains out-compete the toxigenic isolates, resulting in reducing pre-harvest contamination with Aflatoxin in groundnut and cotton.

6.4.1 Storage under Field Conditions to Prevent Fungal Growth

- Dry the raw materials immediately after harvesting to moisture content less than 13%.
- Avoid damaged and broken grains (susceptible for fungal growth).
- Avoid insect damaged grains (susceptible for fungal growth).
- Pre-clean and dry the grains before storage in silos or in bags.
- Use mould inhibitors to prevent fungal growth, if the moisture in raw materials is more than 13%.
- Store the raw materials on wooden pallets or crates and away from the walls to prevent moisture migration from the floor and walls.

6.4.2 Preventive Measures to be Taken At Cattle Feed Plant Level

- Pre-clean the maize to remove dust (contains high level of Aflatoxins) and all waste materials.
- Use mould inhibitors in finished feed in hot and humid conditions.
- Store the finished feeds on wooden pallets or crates to prevent moisture migration from the floor.
- Systematic inspection and clean-up program to keep bins, delivery trucks, and other equipment free of adhering or caked feed ingredients.
- Remove dust (contains Aflatoxins).
- Remove all waste materials.

6.5 COMMON RAW MATERIALS USED IN CATTLE FEED THAT ARE SUSCEPTIBLE TO AFLATOXINS

- ***Raw materials containing high levels of Aflatoxins***: Groundnut cake, Cottonseed cake, Copra cake, Dried Distillers Grains with Solubles (DDGS), and maize gluten meal.
- ***Raw materials containing moderate levels of Aflatoxins:*** Maize, rice polish, de-oiled rice bran and wheat bran.
- ***Raw materials containing low levels of Aflatoxins***: Rapeseed meal, soybean meal, straw and hay.

6.6 STANDARD SAMPLING PROCEDURE FOR AFLATOXIN ANALYSIS

Major problems associated with sampling for testing of mycotoxins are Mycotoxins are not evenly distributed, and Mycotoxins are present in very small amounts (ppb). To sample the feed, take at least 8 to 12 representative samples at each of 3 to 5 locations from the feed bin/ truck/ gunny bags etc. Mix the sub-samples, and take about 500 gram of composite sample. Divide the composite sample into 2 parts of 250 gram each. Store one sample in a cool, dry place, and use the other one for testing at the laboratory. The stored 250 gram sample may be used for possible confirmatory testing.

- Under practical conditions, no dairy feed is virtually free from mycotoxins. Additionally, no feed can be expected to contain only one mycotoxin. A strong recommendation is made hereby to view the problem in its natural way i.e. co-contamination of feeds and feedstuffs by mycotoxins.
- It is therefore essential to take all the necessary precautions such as:
- Prevention of fungal growth on crops in the field, at harvest of crop, during storage of feedstuffs and processing of feed.
- Time to act is not after the dairy animals are affected with mycotoxins, but before the contaminated feed is consumed by the dairy animals.
- Application of appropriate mycotoxin binder in order to achieve good productivity, safe milk and milk products for human consumption and economy.

7

POULTRY MYCOTOXICOSIS AND ITS CONTROL

7.1 INTRODUCTION

Utilization of mycotoxin contaminated raw materials presents a major problem. Detoxification as well as routine mycotoxin analysis of feed ingredients is an important step in a control programme at field level. The cost involved and reduction in nutritive value of feed are some of the constraints which limit the use of such procedures during the feed formulation. Various studies further suggest that the total elimination of moulds and their toxins is practically impossible, so there is a great need for the use of such agents that are able to bind the toxins selectively in the gut, thus limiting their bioavailability in the consumers. In addition, the possible presence of toxic residues in the poultry products (egg, meat), which enters into the food chain may have potential risk by their detrimental effects on human health.

Mycotoxins are secondary metabolites produced by various toxigenic fungi mainly belonging to three genera Aspergillus, Fusarium and Penicillium .Ubiquitous occurrence, heat stability and lipophilic nature pose serious threats in terms of toxic syndromes i.e. mycotoxicoses in animals and poultry. Toxicological spectrum of various mycotoxins is very wide encompassing different kind of toxicities *viz.* acute and chronic toxicities, carcinogenicity, genotoxicity, immunotoxicity, mutagenicity, and teratogenicity in animals and poultry. Feeds are frequently contaminated simultaneously by several moulds each of which is able to produce several mycotoxins. The synergism occurs naturally between different mycotoxins. The various in vivo and in vitro strategies have been tried to ameliorate/ detoxify/remove the toxic effects of mycotoxins/mycotoxicosis. The physical and chemical methods are impracticable in terms of safeguarding the nutritive quality of food and feed, while the biological methods like use of additives, antioxidant substances and food components; and toxin binding agents are showing promising results in amelioration/ detoxification of mycotoxins/mycotoxicosis in poultry.

Mycotoxins are biologically active, toxic metabolites produced by toxigenic fungi mainly belonging to Aspergillus, Fusarium and Penicillium species, which invade crops in the field and may grow on foods during storage under favourable conditions of temperature and humidity . The Food and Agriculture Organization estimated that about 25% of human foods and animal feeds are contaminated with mycotoxins and strong efforts have been made to decontaminate them by the use of physical and chemical adsorbents but the success made so far is limited . Like other environmental pollutants, mycotoxins also adversely affect the health and productivity in animals and poultry. In India, the economy of poultry industry is heavily affected due to wide mycotoxin exposure or contamination of various agricultural commodities. The economic losses are primarily due to the decreased growth rate, feed conversion efficacy, carcass yield, carcass quality and increased susceptibility to other diseases caused due to their immunosuppressive effects among the affected birds. Out of more than 350 mycotoxins identified in nature, aflatoxins (AF), citrinins (CIT), fumonisins (F), ochratoxins (OT) and tricothecenes are the most common and important in poultry.

7.2 COMMONLY OCCURRING MYCOTOXICOSES IN POULTRY

7.2.1 Aflatoxicosis

Aflatoxins B_1 (AFB_1), the most common mycotoxin, produced by Aspergillus flavus and A. parasiticus is primarily hepatotoxic and secondarily nephrotoxic in poultry. The LD50 single dose (mg/kg body weight) is 0.3 for ducklings, and 6.0-16.0 for chickens. According to World Health OrganisationInternational Agency for Research on Cancer (WHOIARC, 1993), aflatoxins were also considered to be a potential human carcinogen (Group 1). Aflatoxins causing immunosuppression due to damage of thymus and bursa of Fabricius make the birds susceptible to other infection like colibacillosis, chronic respiratory disease, and Ranikhet disease. Aflatoxins are also responsible for poor immuneresponse of vaccine. In layers, the aflatoxin cause drop in egg production and poor hatchability.

The significant microscopic lesions include bile duct epithelium hyperplasia, degenerative and necrotic changes in hepatocytes, nodular hyperplasia of liver parenchyma with infiltration of polymorphonuclear lymphocyte cells in portal tracts, besides considerable changes in kidneys.

7.2.2 Citrinin Toxicosis

Citrinin (CIT), a nephrotoxic mycotoxin produced by Penicillium citrinum, often co-occurs with OTA as a co-contaminant of various food commodities of poultry feed and has been implicated as a causal factor for endemic nephropathy in poultry species . It is also reported to be embryotoxic, immunotoxic mycotoxin, neurotoxic and teratogenic in various animal species. The kidney appears to be the primary site of action of CIT and produced degenerative and necrotic changes in the renal tubular epithelial cells. The enlargement in size and the increase in the relative weight of the liver during CIT toxicity in poultry might be due to hepatic degeneration and sinusoidal congestion .

7.2.3 Fumonisins Toxicosis

A number of fumonisins have been isolated and characterized, but Fumonisin B1 (FB1) remains the most toxic compound and mainly produced by fungus Fusarium verticillioides. FB1 has been reported to occur naturally in corn and caused significant toxic effects in various avian species which primary pathological changes have been reported in the liver characterized by multifocal hepatic necrosis and hepatocellular and biliary hyperplasia . Moreover, the nephrotoxic action of FB1 was also studied and the changes were predominant and consistent in the proximal convoluted tubule (PCT) lining cells. Ultrastructurally, the distortion and dilatation of cisternae of endoplasmic reticulum leading to the formation of vesicular and tubular structures was a consistent finding noticed in the Japanese quail chicks fed with 200 ppm FB1 in diet for 21 days . FB1 also reported to affect the performance and immune function in the turkey poults .

7.2.3 Ochratoxicosis

Ochratoxin A (OTA), mainly produced by Aspergillus ochraceus and Penicillium verrucosum in a wide variety of climates and geographical regions. It is primarily nephrotoxic and in terms of LD_{50}, duckling appears to be the most sensitive species to OTA followed by broiler chicks, White Leghorn chicks, Turkey poults and Japanese quail chicks, where the values are 0.5, 2.14, 3.4, 4.63, and 16.5 mg/kg body weight, respectively. OTA is a carcinogenic (Group 2B carcinogen), embryotoxic, genotoxic, hepatotoxic, immunotoxic, nephrotoxic, neurotoxic,, and teratogenic, ,mycotoxin in various animal species . In poultry, along with nephropathy, the specific effects include anaemia, decreased skeletal integrity, impaired coagulation of blood, impaired phagocytosis and reduced growth rate. It has been found to be embryotoxic and teratogenic in Japanese quail. OTA is considered as the most toxic mycotoxin for domestic fowl studied so far.

Grossly, a dry and firm gizzard sometimes with mucosal erosions, catarrhal enteritis, dehydration, emaciated carcass, and proventriculur mucosal haemorrhages have been observed. Kidneys become enlarged, pale and swollen and change in colour from the normal mahogany to tan. Liver is enlarged, friable or haemorrhagic and pale. Extensive accumulation of urates occurs on the serosal surface of several organs. Breaking strength and diameter of the tibiotarsal bones are significantly reduced. Increase intestinal fragility, with rupture is another feature. Swelling and colour changes of the kidneys have been reported as one of the most consistent lesion, while decreased serum albumin and total protein levels are the most sensitive indicator of ochratoxicosis in chicken .

Histopathologically and ultrastructurally, in kidney, degenerative/necrotic changes are most pronounced in proximal convoluted tubules (PCT) than in the distal tubules. Mitochondrial abnormality with characteristic ring forms is diagnostic in chickens and quails. Severe distension, enlargement and hypertrophy of the renal PCTs and thickening of the glomerular basement membrane are seen commonly. Liver revealed increased accumulation of cytoplasmic glycogen in the hepatocytes .

7.2.4 Trichothecenes Toxicosis

Trichothecenes, a group of over 100 fungal metabolites with the same basic structure, are produced by a number of species of the genus Fusarium. They are common contaminants of poultry feeds and feedstuffs and can produce adverse effects on poultry health and productivity. T-2 toxin has been the most extensively studied trichothecene in poultry. The primary effect of T-2 toxicosis in young broiler chicks is buccal-oral ulcerations/ necrosis. The three trichothecenes, croticin, diacetoxyscirpenol and T-2 toxin, cause oral necrosis and affect body weight gain in growing chicks. At a dietary inclusion rate of 5 μg/g of diacetoxyscirpenol, a body weight reduction of 24% resulted, whereas T-2 toxin produced an 11% reduction at the same incorporation rate. A more severe oral response was observed with diacetoxyscirpenol. No effect on body weight gain or oral inflammation or necrosis was observed with croticin fed at 10 μg/g. It is generally regarded that the presence of oral lesions in poultry is the primary means of diagnosing trichothecenes toxicosis in the field.

7.2.5 Moniliformin Toxicosis

Moniliformin is a water-soluble secondary metabolic product of Fusarium species. The worldwide occurrence of this mycotoxin has been reported as contaminant of various commodities such as corn, wheat, barley and other crops which used for the formulation of poultry feed. Moniliformin has been found to be associated with severe cardiotoxicity in chicks, poults and Japanese quails characterized by hypertrophic cardiomyopathy. Moreover, it has been found to have damaging effects on the liver and kidneys in poultry species.

7.3 MANAGEMENTAL APPROACH

7.3.1 Mycotoxin Screening

Routine analysis of feed ingredients and feed stuffs for mycotoxin contamination before the formulation of poultry ration is important. Methods of mycotoxin detection are numerous and include: 1) Microcolumns; 2) Thin Layer Chromatography (TLC); 3) Enzyme Linked Immuno-Sorbent Assay (ELISA); 4) High Pressure Liquid Chromatography (HPLC); 5) Gas Chromatography; and 6) Tandem Gas Chromatography/Mass Spectrophotometry. The TLC and ELISA methods are relatively easy and rapid technologies for mycotoxin detection in contaminated feed/food .

7.3.2 Moisture/Temperature

Monitoring and control of moisture is critical in the prevention of fungal growth and mycotoxin production. Bulk storage bins of grains must be well ventilated, and the materials must be protected from rain and wide fluctuations of temperatures. Moisture level of grains should be kept at below 13%.Aflatoxins and other mycotoxins produced by Aspergillus spp. are not likely to be produced at temperatures below 5 to 8 ^{0}C.

7.3.2 Cleaning

Periodic cleaning of all feed handling equipments with 5 to 10% bleach solution will help control mould growth as well as actually destroy, to some extent the aflatoxins present .

7.3.2 Feed Holding Time

The level and incidence of mycotoxin contamination increases after feed is manufactured. Therefore, it is important to keep the time from the manufacture of feed to when it is consumed by the birds to as short as possible. Managemental practices like withdrawal of mycotoxin contaminated feed/feed ingredients or change of feed at the farm could provide partial protection to poultry from the toxicity as well as mycotoxin residues.

7.4 CONTROL MEASURES FOR MYCOTOXINS OR MYCOTOXICOSIS

7.4.1 Pre-harvest

Pre-harvest control is practically difficult but, development of resistant crop strains by both breeding and direct genetic modifications is possible. Preharvest control measures include prevention of insect infestation, crop residues and crop rotation, irrigation and soil condition and effective drying and storage regimens.

7.4.2 Harvest

During harvesting, it is important to control factors such as timeliness, clean-up, maturity and drying of the agricultural products. Timing of harvesting greatly influences mycotoxin production, harvesting should take place as soon as the crop is fully grown and the crop cycle is completed.

7.4.3 Post-harvest

In nature, mycotoxin(s) contamination is often unavoidable. Detoxification or ameliorative methods becomes very important. The decontamination process should 1) inactivate, destroy, or remove the mycotoxin; (2) not result in the deposition of toxic substances, metabolites, or by-products in the food or feed; (3) retain nutrient value and feed acceptability of the product or commodity; (4) not result in significant alterations of the product's technological properties; (5) destroy fungal spores. In addition to these criteria, the process(es) should be readily available, easily utilized, and inexpensive. Post-harvest strategies involved various physical, chemical and biological methods to inactivate, destroy, or remove the mycotoxin .

7.5 MYCOTOXIN DECONTAMINATION METHODS

7.5.1 Physical Methods

- **Antimycotic Agents:** Antimycotic agents like sorbic acid and sorbate; propionic acid and propionate, benzoic acid, benzoates and parabens; and acetic acid and its derivatives are the chemicals that prevent mould growth and interfere with mycotoxin production. 1 % propionic acid inhibits mycotoxigenic *A. flavus, A. parasiticus, A. ochraceus* and *P. viridicatum* and mycotoxin production in stored corn . In silage, incorporation of 0.2% potassium sorbate, 0.7% methyl paraben and 0.2% sodium propionate completely inhibited fungal growth .
- Phosphine was effective in inhibiting both fungal growth and toxin production. Potassium sorbate was very effective and completely inhibited aflatoxin and ochratoxin production at the levels 0.1 to 0.2% and 0.10 to 0.15 %, respectively. Herbs and spices like cloves, cinnamon oil, mustard, allspice, garlic and oregano have been shown to have strong antimycotic properties . Low oxygen concentration and/or increased concentrations of other gases (*i.e.*, 90% CO_2) depress mould growth and mycotoxin formation.
- **Density Segregation:** Density segregation of contaminated grain and oilseeds involves sorting and delineating good versus contaminated kernels by floatation. This method can notably decrease various mycotoxin concentrations in contaminated grains.
- **Irradiation:** Gamma or electronic irradiation is highly effective for destroying the fungal spores. Fluorescent or ultraviolet (UV) rays decompose aflatoxins and ochratoxins. Applying UV light for 20 minutes at 25°C in the presence of peroxide (0.05%) decreased the concentration of aflatoxin M_1 in contaminated milk. Simple exposure of contaminated grains to sunlight (UV) substantially reduces mycotoxin levels.
- **Mechanical separation:** Toxin levels decrease as clean product is physically separated from contaminated grains. Significant decreases in aflatoxin levels from electronic- and hand-sorted peanuts have been reported and are commonly utilized.
- **Processing of food:** Most of the mycotoxins are generally stable at room temperature. Processing of food has been found to decrease the prior concentration. Wet milling, malting, brewing, cooking and dry and oil roasting are methods to eliminate the mycotoxins, effectively. Autoclaving at 1320C for 5 hours reduces the OTA concentration by 85 % . Roasting conditions decreased the aflatoxin content in the raw peanuts. OTA appears to be more readily destroyed in dry cereals than in the wet and damp cereals (unlike aflatoxin B1 and patulin). Scouring step (removal of the outer layers of the pericarp) of cereal processing

resulted in >50% reduction in OTA concentration. Simple water treatment of contaminated cereals may completely eliminate mycotoxins.

- **Solvent extraction:** Most mycotoxins can be extracted efficiently from contaminated grains using carefully selected solvent mixtures. Although effective, such treatment is considered cost prohibitive and impractical for most applications.

7.5.2 Chemical Detoxification

- **Ammoniation:** Treatment with aqueous and gaseous ammonia or ammonium hydroxide, with or without heat and pressure to destroy the mycotoxin in contaminated food and feed is currently the best and effective method. Ammoniation not only detoxified several mycotoxins (85-100% reduction), but also inhibited mould growth. Anhydrous ammonia, NH_3 (gas) or aqua-ammonia, $NH_4 0H$ (liquid), can be used. 5% ammonia for 96 hours detoxified the ochratoxins in the feed at 70^0C.
- **Ozonization:** Ozone (O_3) gas, a powerful oxidant is a non-persistent chemical, which means that it decomposes very fast. Within 20 minutes half of the ozone is decomposed into oxygen (in aqueous environments). Several studies indicate that ozone degrades aflatoxins in corn and cottonseed meals and also degrades deoxynivalenol and moniliformin.
- **Sodium hydroxide:** Warming of grain to 105^0C in the presence of 0.5% sodium hydroxide detoxified various mycotoxins in the feed.
- **Structural degradation:** Numerous chemicals including acids, bases, aldehydes, bisulfites, oxidising agents and various gases have been tested for their ability to degrade or detoxify AFB1 and other mycotoxins. Although, most of them may successfully destroy mycotoxins. They must not be used owing to the generation of toxic products and alteration of product quality.

7.5.3 Amelioration/ Biological Inactivation

- Mycotoxin-binding agents: Numerous agents like, activated carbons (charcoal), bentonites, clay, hydrated sodium calcium alumino silicate, and zeolite, have currently been used to counteract the mycotoxicosis in poultry and its carry over effect through meat and eggs. The commercial available synergistic blends of the mycotoxin binders are found to be beneficial against the commonly occurring mycotoxins in poultry species. These sorbents are nutritionally inert and reduce the bioavailability of various mycotoxins by absorption on their surface in intestinal tract.
- Activated charcoal: Generally, the absorption properties of activated charcoal (AC) are strictly dependent on the source materials and physico-chemical

parameters, such as surface area and pore size distribution. In vitro it has high ability of binding with several mycotoxins. The adsorption property of AC was found effective against aflatoxin B_1 and ochratoxin A up to 95% and 91%, respectively, during in vitro studies . The informations on the amount of activated charcoal to be added to the feed and possibly long term effects on adsorption of essential nutrients are scanty. Charcoal at 2% level had shown beneficial effects, during *in vivo* studies.

- **Bentonites**: Bentonites (hydrated aluminium silicate) are sorbents with layered (lamellar) crystalline microstructure and variable composition. Their absorption property mainly depends on the interchangeable cations (Na^+ , K^+ , Ca^{++} and Mg^{++}) present in the layers . Sodium bentonite is more effective than calcium bentonite. Bentonite could bind aflatoxin to the extent of 66% while it was of little use in adsorbing OTA. Bentonite (0.5%) could not ameliorate the effect of OTA on body weight gains, feed consumption and conversion, haematologically, biochemically and histopathologically in broiler chicks .
- **Bovine serum albumin**: Mycotoxins bind with serum albumin as observed during in vivo and in vitro studies. Bovine albumin competes with mycotoxins in the intestinal tract and helps to excrete it. Marked reduction of histological, biochemical and plasma and liver levels were recorded, during AFB_1 exposure to the day-old chick .
- **Cholestyramine**: Recent studies have indicated that cholestyramine (CHA), an ion-exchange resin with strong affinity for bile salts, was found protective against OTA induced nephrotoxicity in rats. CHA added to the diet (5%) reduced the concentration of OTA in plasma (up to 50%) and increased its clearance by way of the faeces while decreasing the amount excreted in bile and in urine . It was found to be an effective absorbent of OTA in the gastrointestinal tract of ruminants .
- **Hydrated sodium calcium alumino silicate (HSCAS):** Molecular surface of HSCAS gets saturated with water and attracts the polar structure of various mycotoxins. In young broiler chicks, HSCAS (0.5%) was effective at reducing the toxicity of aflatoxin as well as combined toxicity with OTA . Antibody titre levels against Newcastle Disease and Infectious Bursal Disease were also decreased in chicks with HSCAS supplementation to diet containing aflatoxin (500 ppb). However, its protective properties are very low towards OTA, zearelenone and nil towards trichothecenes.
- ***Polyvinylpolypyrrolidone:*** Polyvinylpolypyrrolidone, a synthetic resin, at 0.4g/kg level can bind up to 50 g/kg of AFB1 in feed. Partial amelioration was reported against AFB1, when administered with bentonite to broiler chicks .

7.5.4 Microbiological Binding Agents

Mannan oligosaccharide (MOS) extracted from the cell wall of Saccharomyces cerevisiae has shown broad-spectrum efficacy against most of the mycotoxins . Saccharomyces cerevisiae was found to have beneficial effect in poultry during mycotoxicosis and MOS was believed to be the responsible factor. MOS, esterified with glucan, bound AFB1 (up to 81.6%) and showed significant binding over zearelenone, T2 toxin and moderate binding over OTA (25.5%), during *in vitro* studies . It was found to significantly improve body weight and other parameters of economic importance . Feeding trial with the addition of MOS (0.11%) significantly increased concentration of Ig A and Ig G in turkeys .

Certain strains of Bifidobacteria, Lactic acid bacteria and Propionibacteria have cell wall structures that can bind mycotoxins and limit their bioavailability in the body. The bacterial strains such as *Bifidobacterium animalis* and *Lactobacillus acidophilus* were found to be highly effective against OTA and patulin, respectively. Certain micro-organisms (*Corynebacterium rubrum*) are also able to metabolise mycotoxins in contaminated feed or to biotransform them (*Aspergillus, Eurotium, Rhizopus*). However, these biological processes are generally slow and have a low efficiency.

7.5.6 Antioxidant Substances

The protective properties of antioxidants are probably due to their ability to act as superoxide anion scavengers, thereby protecting cell membranes from the mycotoxin induced damage.

- **Ascorbic acid (Vitamin C):** Ascorbic acid has been shown to react directly with superoxide, hydroxyl radical and singlet oxygen in addition to direct quenching of reactive free radical. The addition of vitamin C to the diet containing OTA, partially protected laying hens against the toxic effect. The OTA induced reduction in feed intake was counteracted at 25°C by vitamin supplementation. It moderated the negative effect of ochratoxicosis, in plasma electrolyte concentration and plasma aspartate transaminase activity .
- **Phenolic compounds:** The phenolic antioxidants, gallic acid, vanillic acid, protocatechuic acid, 4- hydroxybenzoic acid, catechin, caffeic acid, and chlorogenic acid were found to be effective against the fungal growth of ochratoxigenic Aspergilli and the OTA production. Moreover, several antioxidant compounds have been shown to inhibit aflatoxin biosynthesis .
- **Plant materials/Plant products:** The use of plant materials/products like leaves, bark, berries etc of certain antioxidant rich plants like Sea buckthorn which were found to confer partial protection against various mycotoxins as evident at gross, microscopic, ultrastructural and biochemical assessment . Moreover, plant materials were found to exhibit immunomodulatory activity .

- **Vitamin A:** Vitamin A possesses the antioxidant properties against the mycotoxin-induced damage. Carotenoids, mainly carotenes and xanthophylls present in carrots, palm oil and maize, not only possessed the antioxidant property but also had antimutagenic and anticarcinogenic properties and reduced the toxicities of OTA.
- **Vitamin E/Selenium:** Supplementary vitamin E administration to chickens partially counteracts the formation of lipid peroxides due to single and combined exposure to OTA and T2 toxin. Higher vitamin E intake partially ameliorated oxidative stress caused by OTA. It is known that reduced vitamin E (alpha tocopherol radicals) can be regenerated following single electron reduction by ascorbic acid. Vitamin C should, therefore, enhance the biological efficacy of vitamin E especially under stress condition. Vitamin E and selenium are involved in the formation of glutathione peroxidase, a compound vital in the cellular detoxification mechanism. Gregory and Edds (1984) reported that selenium enhanced the formation of water soluble conjugated forms of aflatoxin which promoted the clearance of the toxin and enhanced chick growth. It was concluded that the fungi may not utilize the minerals in chelated form as efficiently as in inorganic form, for aflatoxin synthesis in feed and the excess selenium either in organic or inorganic form was found to increase the aflatoxin production in maize.

7.5.7 Food Components and Additives

Numerous food components, ingredients, or additives with or without antioxidant properties have been found to have ameliorative properties against mycotoxicosis.

- **Aspartame:** Aspartame (L- aspartyl- L- phenylalanine methyl ester), a structural analogue of both OTA and phenylalanine, has been shown to have protective effect against OTA induced cytotoxicity in animals. Aspartame prevents typical cytotoxic effects of OTA including inhibition of protein synthesis, lipid peroxidation and leakage of certain enzymes, such as lactate dehydrogenase, gamma glutamyl transferase and alkaline phosphatise.
- **Crude proteins:** Raising the protein levels of diet from 14-18% to 22-26% counteracted the OTA effects. The broiler chicks fed with diet containing 5 ppm aflatoxin, increasing the crude protein level from 20 to 30% alleviated the growth and depressing effects of aflatoxin. Protein protected against the negative effects of OTA presumably because of increased concentration of phenylalanine in the diet. Increasing dietary crude protein helped alleviate but did not eliminate the adverse effects of ochratoxin A on body weight and feed conversion; mortality rates did not appear to be affected. However, increasing protein levels is a costly approach to mycotoxin control.

- **Dietary Lipids:** Inclusion of cottonseed oil at 2, 6, or 16% level into semi-purified diets containing 10 ppm aflatoxin, not only improved the body weight, but the mortality was also significantly reduced. In a further study to evaluate the effects of different types of lipids, 2 or 16% olive oil, coconut oil, safflower oil, or animal fat were included in low fat diets with 10 ppm aflatoxin. The higher levels of dietary fat reduced mortality and in some instances, improved the body weights. Lipids exerted their effects in part by interfering with absorption of the aflatoxin. Supplementation with olive oil and safflower oil, both sources of unsaturated fatty acids, also improved body weight which suggested that diets containing higher levels of linoleic acid supported better feed conversion and lower mortality in chicks fed diets with aflatoxin.
- **L-Methionine:** It has been reported to have a protective effect against many oxidant drugs. It plays an important role in preserving the structure of cell membrane and in modulating the antimicrobial activity of polymorphonuclear leucocytes in periodontal disease as well as behaving as a chemotherapeutic agent in hepatitis treatment.
- **L-Phenylalanine:** OTA has been reported to inhibit protein synthesis by competition with phenylalanine in the phenylalanine-tRNA synthetase catalysed reaction. Supplementation of 0.8% to 2.4 % phenylalanine in broiler diets containing 4 ppm OTA decreased the mortality rate from 42.5% in non-supplemented chicken to around 14%. Phenylalanine moiety of OTA competitively inhibits at least 2 enzymes, phenylalanine tRNA synthetase and phenylalanine hydroxylase, which result in altered tyrosine production from phenylalanine. Phenylalanine reduced OTA induced immunosuppression. Bailey et al. (1990) suggested that supplemental phenylalanine improved the health status of birds fed diets containing OTA.

8

APPROACHES FOR REDUCING AFLATOXIN M1 CONTENT IN MILK AND DAIRY PRODUCTS

8.1 INTRODUCTION

Using contaminated animals' milk and meat may result in human disease. Among microorganisms, fungal toxins, especially aflatoxin B^{-1} (AFB_1), are of special importance. Aflatoxin M^{-1} (AFM^{-1}) is a metabolite that is produced by conversion and hydroxylation of AFB^{-1}. Both toxins can cause acute and chronic mycotoxicosis mainly through ingestion of contaminated milk. Preventing foods contamination due to aflatoxins (AFs) is not only an expensive but also a difficult task. The best agricultural monitoring during preharvest and postharvest stages cannot eliminate the AFs, especially AFM^{-1} from milk and dairy products because of the high resistance of these toxins.

Aflatoxins (AFs) are secondary fungal metabolites that are naturally produced by some *Aspergillus* species especially *Aspergillus flavus* and *Aspergillus parasiticus*. Dairy products have favorable conditions to be contaminated by these molds at the stages of growth, processing, storage, and transportation. AFs are the most carcinogenic compounds. AF B1 (AFB1-) is an important fungal metabolite with a high potential to induce mutation and immunosuppression. In the liver, cytochrome P450 system converts AFB-1 into AF M-1 (AFM-1).Fig. 1 shows pathways involved in metabolism and elimination of AFs in the liver Fig. 1.

The presence of AFM-1 in milk and dairy products is considered to pose certain hygienic risks for mammals especially human. Level of AFM-1 in infected milk or contaminated foodstuffs is directly associated with the levels of AFB-1. To decrease or detoxify AF from infected milk, some studies have suggested the usage of an absorbent such as soil-bentonite. It is shown that such components can reduce the levels of AFs efficiently. On the other hand, a majority of studies suggested using of probiotics. Probiotics are inexpensive and safe to decrease or detoxify AF in milk and advantages of using probiotics have been demonstrated in several researches. Considering to importance of preventive approaches to decontaminate food, the point is that how many and in which condition, the mentioned methods are allowed to be used.

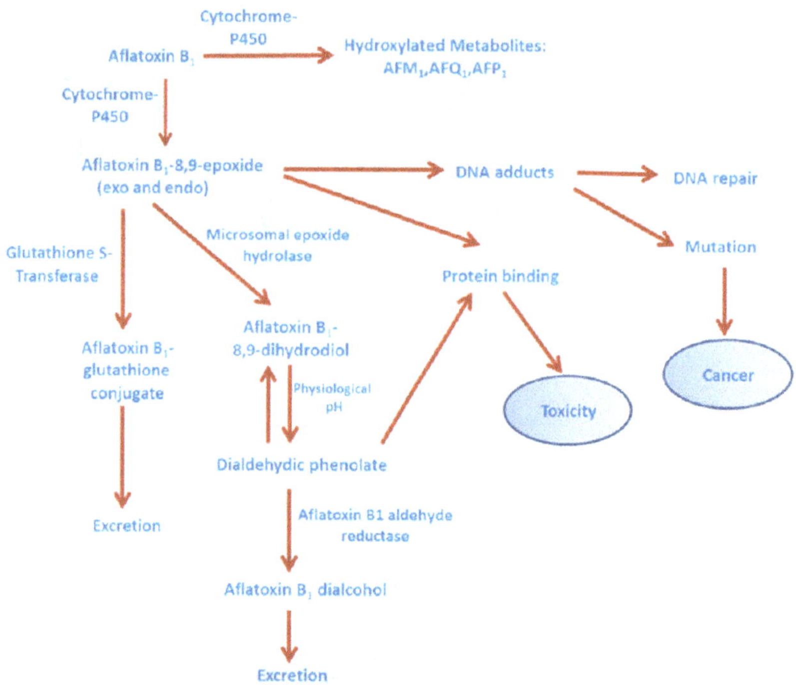

Fig. 1: Pathways that are involved in metabolism and elimination of aflatoxins in the liver

8.2 METHODS FOR REDUCING THE AFLATOXIN M-1 CONTENT IN MILK

The levels of AFM-1 in milk and dairy products of different countries are variable. It can be related to climate condition of each geographical area and differences in dairy cattle feeding system. Therefore, to decrease the level of AFM-1 in milk, AFs in both animal feed and milk should be evaluated continuously. In the last decades, by understanding the serious effects of mycotoxins on human and animals health and to protect human health, food companies in different countries attempted to adjust content of mycotoxins in food and feed. Table 1 shows the accepted level of the toxin in some countries description of Table.1.

Table 1: Maximum acceptable levels of aflatoxin M-1 in milk and milk products in some countries

Region	Maximum acceptable level (ng/1)	Type
European Union	50	Milk
Austria	50	Milk
Argentina	50	Milk
Bulgaria	500	Milk
Germany	50	Milk
Australia	20	Childrens' milk
Sweden	50	Liquid mi. products
Belgium	50	Milk
USA	50	Milk
Czech Republic	100	Childrens'
Serbia	500	Milk
Iran	50	Raw, pasteurized, and Ultra-high temperatureprocessing (UHT) milk.
France	30	Children's' milk <3 years
Turkey	50	Milk. and milk. products
Brazil	500	Milk.

Several methods have been introduced to reduce AFs in milk and dairy products. Some studies showed that preventing the crops from contamination in pre-harvest and postharvest stages is a useful strategy to eliminate fungal toxins. Other studies proposed direct methods for reducing AFM-1 in milk.

8.3 MAJOR APPROACHES TO PREVENT MILK CONTAMINATION

It is believed that prevention and control of fungal growth are an important technique to reduce AFM-1 content in milk. There is a linear correlation between AFB-1 in animals feed and quantity of AFB-1 in their milk. To prevent contamination of milk, dairy animals should be vaccinated against AFs especially AFB-1. Some studies showed that vaccination might reduce the risk of contamination of animal/dairy products by AFs. The levels of anti-AFB-1 antibodies in vaccinated heifers decrease during pregnancy and after calving and at the beginning of the milk production cycle, they return to the previous range. Such changes in levels of anti-AFB-1 antibodies after vaccination may demonstrate the effectiveness of such strategy. Another valuable method to control AFB-1 and consequently AFM-1 is to inhibit the initial contamination of feed consumed by Dairy Cattles. In a research, it was emphasized that the usage of damaged portion of crops should be prohibited and such crops must be removed during the harvesting process. Moreover, place, temperature, and moisture of storages should be controlled since they can induce the fungal growth in storehouses.

It is reported that the inhibition of contamination through livestock feed control is very important. The quality of livestock feed has an impact on the AFs content in the produced milk. During production of crop in the farm and storage of products in traditional or industrial tanks, the monitoring instruments and health considerations need to be controlled. An *in vitro* investigation proved that specific anti-AFB1- antisera could reduce the amount of AFM-1 in milk. Another research showed that using antifungal agents should be done before the formation of the toxin. Organic acids are the main inhibitors and include acetic, benzoic, formic and lactic acids. To prevent the formation of AFs, performing biological control and using absorbent clays are very useful. Overall, considering above-mentioned evidences, the preventative approaches to inhibit entrance of AFM-1 into milk are essential. It should be noted that controlling can reduce toxin but cannot remove it completely. In addition, not all of the prophylaxis methods can be used in farm or factories. Therefore, some other methods for mitigation of AFM-1 in milk should be utilized.

8.3.1 Reducing of Aflatoxin M-1 Content During Processing of Milk

Thermal treatment during processing stages is not considered as a magnificent way to reduce AFM-1 in milk. Although pasteurization and sterilization may eliminate many microorganisms from milk, they are not effective for mitigating AFs. Some researchers believe that heat does not cause significant changes in amount of AFM-1, while others have reported different levels of AFs detoxification through these techniques. For instance, in a study, it was shown that pasteurization at 62°C for 30 min could reduce the AFM-1 content in milk by 32%. Another study showed that heating might decrease AFs-content by 12%–35% (depending on the conditions). It is also reported that AFM-1 may be relatively stable during milk pasteurization process. Processing stages of milk may remove other dangerous contaminants from milk, but it cannot effectively remove AFs from infected milk.

Role of probiotics to reduce aflatoxin M-1:

The meaning of probiotic is "for life," but in biology, the term is used to call microorganisms that are cost-effective for human and animals. Probiotics have been considered to have several advantages as follows: they can potentially decrease populations of gastrointestinal pathogens, lessen the gastrointestinal pain, support the immune system, affect and improve the skin's function, regulate the intestine and reduce flatulence and bloating, induce the resistance to pollen allergens, protect DNA, proteins and lipids from oxidative damages, and maintain the normal flora of intestine while receiving antibiotics treatment. Probiotics can be used to lessen or detoxify toxins in milk, dairy products, and other foods. Such useful microorganisms can be added to cattle feed or may be utilized at the different stages of milk processing.

- ***Probiotics for biological control in the field:*** Growth of yeasts and molds can cause decay in crops and animal food (feed). Contamination of crops may lead to considerable economic damages. Probiotics as biocontrol agents may absorb toxins, compete with fungi for an ecological niche or nutrient, or destroy AF. Therefore, they can prevent fungal growth and toxin secretion in feed. The most important bacteria used as probiotics belong to bacterial genera including *Lactobacillus, Streptococcus, Pseudomonas,* Bifidobacteria and *Burkholderia.* The commercial strains can reduce AFM1- contents in phosphate-buffered saline, milk, and yogurt. Some yeast such as, *Saccharomyces* species and nontoxigenic *Aspergillus* strains, have been used as competitive biocontrol agents.
- ***Probiotics as biocontrol agents during storage of food:*** Probiotics have an important role in biocontrol of the storage houses. They have the ability to impair the growth of fungi and reduce AFs content through the above mentioned mechanisms. Below, some studies with emphasizing the effectiveness of probiotics are reviewed.In a research, *Saccharomyces Kefir* and *Lactobacillus casei* were used to reduce AFM1- in milk. The results revealed the positive significant role of these microorganisms to eliminate high level of the toxin in milk. In 2011, the role of lactic bacteria such as *Lactobacillus bulgaricus* and *Streptococcus thermophilus* for removal of AFM1- in dairy products during manufacturing was investigated. It is found that *L. bulgaricus* has more binding ability than *S. thermophilus.* Therefore, *L. bulgaricus* can reduce the amount of toxin. In another research, the beneficial effects of specific dairy strains of lactobacilli to remove AFM1- from milk were highlighted. The role of probiotics to improve liver functions as well as their positive effects on human health particularly children and the high-risk population were also emphasized. It is shown that the attachment of AFM-1 to cell wall components of probiotic microorganisms may results in the reduction of the toxin in milk. The enzymatic degradation of AFM-1 by *Lactobacillus, Pseudomonas,* and *Burkholderia* species can also inhibit fungal growth and production of Afs. The *Saccharomyces cerevisiae* is one of the yeasts that can bind to AFM1- effectively. Among bacteria, *Lactobacillus rhamnosus* L60 and *Lactobacillus fermentum* L23 have a high ability to inhibit mycelium growth of aflatoxigenic *Aspergillus* strains and reduce the AFB1- production. All of these researches show the significant roles of probiotics. The culture of lactic bacteria is an easy task. Consequently, as these microorganisms are safe, inexpensive, and have high potential ability for detoxification, they are valuable candidates for reducing AFM-1 in milk and dairy products.

Role of adsorption in reducing the aflatoxin M-1 content in milk

Toxicants enter into an organ through an adsorption process. For reducing AFs in the gastrointestinal tract and therefore diminish the rate of the toxin distribution in the tissues, adsorbing agents can be added to animal feed or use separately during their meal times. The absorbents can reduce AFs contents in milk because of their high variety and ability to bind to AFM-1 in a stable mode. Such compounds may be affected by pH and temperature. The bentonite, vermiculite, hydrated sodium calcium aluminum silicate (HSCAS), and activated carbon are the well-known absorbents.

- ***Bentonite***: Bentonite clay works such as a magnet that bonds to toxins and helps to the excretion from the body. Sodium-bentonite and calcium-bentonite are important types of bentonite clay. When activated by water, sodium-bentonite has an electromagnetic properties and can swell up six times in the size. This makes it very absorbent and helpful for drawing out toxins. Calcium-bentonite has smaller than sodium-bentonite in size. Their small size makes them to be more effective for remineralizing. These small particles can pass through the colon wall into the bloodstream where they get rid of toxins and leaves minerals behind. Bentonite is an aluminum phyllosilicate clay. The negative charges of the molecule enhance its adsorptive properties. The absorbent is able to pass through the intestinal tract and attract the positively charged toxins. The molecule is bound to the toxin by ions exchange resulting in the internalizing and absorbance of the toxic molecule to the clay molecule. Subsequently, the host body can excrete the toxin-filled clay through normal gut movements. Bentonite is capable for maximum detoxification and has a low impact on the nutritional properties of milk. While producing pelleted feeds, the clay is used to bind and flush AFM-1 and bring the level of AFM-1 below the European standard limits for the toxin. Enrichment the feed by sodium bentonite and activated carbon can reduce AFM-1 content in dairy products of early lactating goats. In a research conducted in Iran, it was shown that the bentonite can reduce AFM-1 in milk and adding the clay entrosorbent in the diet of lactating dairy cattle and goats can reduce the concentration of AFM-1 significantly. The hydrated HSCAS is a phyllosilicate clay with high effectiveness to prevent aflatoxicosis in chickens, goats, cows, and pigs.[41] It has high affinity and capacity to absorb AFB-1 in contaminated milk and oils. Using the calcium hydroxide and mono methylamine simultaneously can result in 94%–100% destruction of AFB-1. These substances can open the lactone ring. In a study, the calcium hydroxide was added to the dry diet at a concentration of 2%. Then, mono methylamine was added at a concentration of 0.5% and treatment was performed at 25°C and 10% of moisture in 4 h. As a result, the mycotoxins content was reduced by 50%.

- ***Activated carbon:*** Activated carbon is manufactured by treatment of carbonaceous source materials such as coal, coconuts, nutshells, peat, and wood and can be applied to purify liquids and gases. Activated carbons and esterified glucomannan have been shown to reduce the AFM-1 content in infected milk. It is reported that supplementation of early lactating goats by sodium bentonite and activated carbon (by 1%) can results in significant reduction of AFM-1 in milk without changing the milk composition. In a report, a mixture of activated carbon and HSCAS was added to AFB-1 contaminated feed of dairy cows (with an inclusion rate of 2%). Activated carbon and HSCAS reduced AFB-1 converting to AFM-1 in milk by 50% and 36%, respectively.

8.3.2 Chemical Approaches to Reduce Aflatoxin M-1

Chemical processes can be used to reduce, destruct, or inactivate AF in milk. They include ammoniating, acidic treatment, oxidizing, and reducing techniques. There are several issues and risks associated with these methods. First, it is difficult to detoxify the AF without reducing nutritive value and palatability. Moreover, a number of parameters such as reaction time, temperature, and moisture must be monitored. Some necessary additional cleaning treatments are expensive and time-consuming and toxic byproducts may be produced. [Fig. 2] compares effectiveness of some compounds to reduce AF content in milk [Fig. 2]

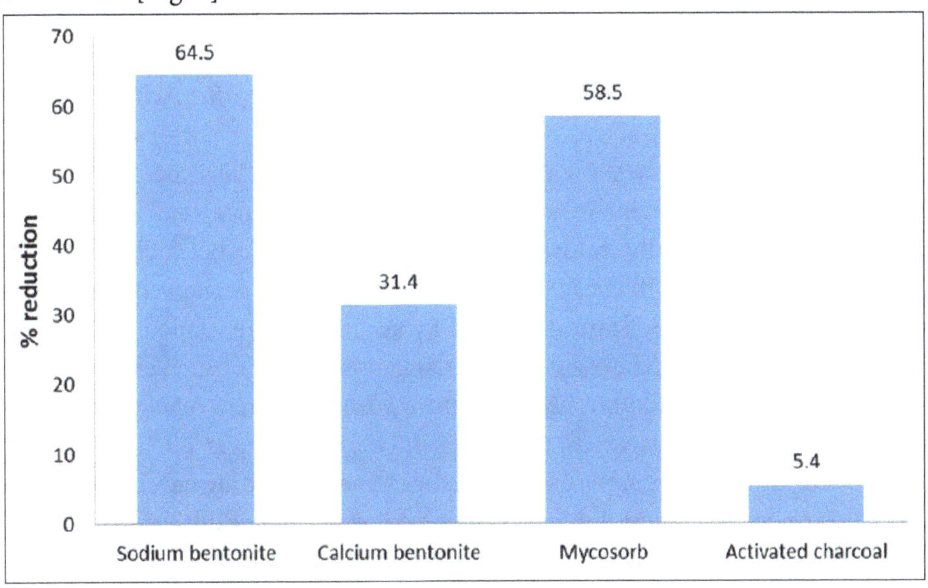

Fig. 2: Effectiveness of some compounds to reduce aflatoxin content in milk

- ***Ammonia and hydrogen peroxide***: A high number of chemicals, such as acids, bases and bisulfite oxidizing agents, have been tested for their ability to degrade or inactive AFs. Ammonia has been used to destroy AF. Ammoniation is a reaction in which ammonia is added in stoichiometric proportions of an interested compound. Ammoniation can mitigate the AFM-1 content in contaminated foods by 79%–90%. Ammonia reacts with AFs molecule by breaking oxygen bond. It can also open lactone ring of the toxin to produce compounds with lower toxicity. Hydrogen peroxide (H_2O_2) is a colorless, unstable, and oily liquid and is usually used in the aqueous solution. It is a powerful oxidizing agent used as bleach for textiles in rocket fuels. Using of H_2O_2 at different temperatures has been suggested to detoxify AFs in milk. An interesting report showed that using H_2O_2 followed by a heat treatment at 36°C for 30 min and boiling for 5 min could inactive AFs by 27.8%. Utilizing H_2O_2 followed by heat treatment at 75°C for 15 s and boiling for 5 min inactivated AFs by 28.8%. Another study showed that the addition of H2O2 or sodium hypochlorite in diet and storage at room temperature for 7 days can destruct 62% and 82% of AFB-1 content, respectively. Combined treatments such as ultra violet radiation followed by ultra-filtration can degrade AFM-1.

- ***Using plant extracts to reduce aflatoxin M-1 in milk:*** One interesting technic to diminish toxins, especially AFM-1 in foods, is using the plant extracts. Some plants such as broccoli, garlic, black cumin, and curcumin have been reported to be efficient for reduction of AFs. The mentioned extracts may also have inhibitory effect on AFs. The extracts are also known as antioxidant, antifungal, and anti-inflammatory agents because their ability to inhibit infection or inflammation inducing molecules.
    - ***Broccoli, garlic, and black cumin for reducing aflatoxin M-1:*** Broccoli extract is an attractive novel additive with indirect antioxidant properties. In a study, it was shown that the broccoli extract has the ability to reduce the AFM-1 content in milk. This is important since milk is consumed by humans especially children. Water extracts of garlic, black cumin, and carrot were attempted to use for detoxification of AFM-1. It showed that the level of AFM-1 was significantly reduced by adding the mentioned extracts to milk in ratio of 1:10. There are limited data about ability of such extracts to reduce the toxin in milk, but above mentioned finding may reflect the advantages of plant extracts to detoxify AFs in dairy products during fermentation stages. In a review, the effect of plant extracts including curcumin and Indian turmeric on AFM-1 in milk was reviewed. The results showed that such extracts can have anti-inflammatory, anticarcinogenic, antimicrobial, and gastrointestinal effects. Among the extracts, curcumin was able to enhance immunity.[49] Considering the role of herbs against aflatoxicosis, plants like coumarins, flavonoid and curcuminoids have been reported to inhibit biotransformation of AFs to their active epoxide derivatives. Using 0.05%, 1.0%, and 2.0% of turmeric concentration, the level of AF were decreased

significantly. In addition, turmeric could inhibit spore count and AFs production by A. Flavus.

- ***Antioxidants and aflatoxins reduction:*** Antioxidants can aid human body to detoxify toxins in liver and other cells, and consequently decrease the appearance of mycotoxicosis. Antioxidants, such as Vitamin C and Vitamin E, can be added in food to prevent chemical reactions in which oxygen is combined with other substances. Such reactions are harmful for human cells. Vitamin E, Vitamin C and selenium are involved in the formation of glutathione peroxidase that is vital for detoxification mechanisms in cells. using glutathione, glutathione peroxidase reduces lipid hydroperoxides to alcohols and free H_2O_2 to water. Glutathione is a very simple molecule that is produced naturally in human body. The molecule can attach to free radicals and heavy metals through its sulfur group.

9

REDUCTION OF AFLATOXIN CONTAMINATION IN GROUNDNUTS AND MAIZE

9.1 INTRODUCTION

Many agricultural commodity are susceptible to several diseases caused by fungi, viruses and insect pests. These diseases may also contaminate the crops with deadly toxins. Fungi are of particular concern. Many fungi are free-living organisms capable of surviving in the environment (soil, air and water) and can easily fit their way into crop products especially when the weather condition are suitable. Many of these fungi are free-living organisms and can survive in the environment even without the crops we grow. The toxins produced by fungi are called mycotoxins (Myco=fungus; toxin=poison).

Producing aflatoxin-free maize, groudnuts and other grains ensures food safety. The health and economic well-being of a farmer is an important investment in the agricultural sector. Since it is difficult to identify groundnuts that are contaminated, it is important to avoid consuming groundnuts that are shrievelled, broken, discolored, or show symptoms of fungal infection. As much as farmers grade groundnuts for the market, they also need to grade for consumption. Use of good crop production practices and postharvest handling techniques can effectively reduce fungal contamination and maintain the quality of the produce. Such efforts would then open up opportunities for farmers to access rewarding markets that currently have stringent quality standards.

The fungi *Aspergillus flavus* and *Aspergillus parasiticus* are the most common source of food contamination the world over. These fungi produce a special mycotoxin called aflatoxin (From *Aspergillus flavus* toxin), which, when consumed via contaminated crops/products can cause several harmful health effects on humans and livestock(Fig. 1).

Fig.1 Groundnut infected with *Aspergillus flavus*.

9.2 EFFECTS OF AFLATOXINS

Food products that are commonly contaminated by aflatoxins include cereals (maize, sorghum, pearl millet, rice and wheat), oilseeds (groundnut, soybean and sunflower), spices (chillies, black pepper, coriander, turmeric and ginger), tree nuts (almond, pistachio, walnut and coconut) and milk from cattle that have eaten contaminated feeds. Aflatoxin cannot be destroyed by cooking. Aflatoxin contamination of food products have two major effects. Firstly, eating contaminated food has several harmful health effects and secondly, aflatoxin contamination affects the quality of the grain which leads to reduced markets for contaminated grains.

9.2.1 Health effects

Eating food contaminated with aflatoxins leads to aflatoxin poisoning also called aflatoxicosis. Regular consumption of low dosages for a long time (Chronic exposure) leads to liver damage, immune suppression, malnutrition and stunted growth in children, while sudden high-level ingestion of the toxin (Acute exposure) can lead to death.

9.2.2 Income losses

Groundnut, maize and other crop products are important as a source of income when sold locally or exported, particularly to the European Union (EU). To protect humans from the harmful effects of mycotoxins, especially aflatoxin, the European Commission has proposed maximum allowable limits for aflatoxin in certain foods. The permissible limit of aflatoxin in foods for human consumption is 4-30 ppb, depending on the country involved . In the United States, 20 ppb is the maximum aflatoxin residue limit allowed in food for human consumption. Due to aflatoxin contamination the food products cannot be exported leading to financial losses. For example Malawi lost as much as 42% of its groundnut exports in 2005.

Table 1. Maximum risk levels of acceptance for total aflatoxin contamination by different countries and organizations.

Country/Organization	Maximum Risk Limit (MRL) in ppb
European Union	4
USA	20
RSA	15
Japan	0
Codex Alimentarius Commission	15

9.3. AFLATOXIN MANAGEMENT IN GROUNDNUT

9.3.1 How crops get contaminated

Aflatoxin-producing members of *Aspergillus* are common and widespread in the environment. Crops are particularly susceptible to infection by *Aspergillus* following prolonged exposure to a high-humidity environment or damage from stressful conditions such as drought, a condition that lowers the barrier to entry. The fungi can enter (colonize and contaminate) at any point of the crop value chain. The entry or infection points for these aflatoxin-producing fungi could be at any/all of these three stages:

- Preharvest entry of fungus (during plant growth).
- Entry of fungus during harvest.
- Postharvest entry of fungus (after harvest and during processing).

9.3.2 Preharvest contamination

Fungus growth and aflatoxin contamination that occurs in the field during crop growth is called preharvest contamination. Preharvest infection is usually influenced by farming practices that make crops susceptible to infection, including the following:

Repeated cultivation of host plants: Repeated cultivation of the same crop or susceptible crops species on the same piece of land supports rapid buildup of *A.flavus* populations leading eventually to preharvest contamination of crops in field.

- **Late planting:** Late planted crops will usually be affected by end-of-season drought as well as insect pest attacks, especially termites. Such insect- damaged pods create easy entry points for the fungus.
- **Drought:** Under drought stress the groundnut pods crack and facilitate entry and growth of *Aspergillus*.
- **Termite attack and poor field hygiene:** Using fields with a history of termite infestation provides an environment that supports pod damage by termites and

eventual easy entry by the fungus. Unweeded fields also support termite and pest damage of developing pods, increasing chances of *Aspergillus* infection.(Fig. 2)

Fig. 2. Termite nest in a field.

- **Plant density and poor water management:** Low plant population and limited ground cover leads to soil erosion, loss of soil moisture and destruction of soil structure that could aggravate the effects of droughts. These conditions predispose the land to increased A. flavus buildup.

9.3.3 Contamination during harvest

The manner in which a crop is handled during harvesting will determine the extent to which it is predisposed to infection. Some of the predisposing conditions that favor fungal infection during harvesting include:

- ***Poor harvesting techniques:*** Groundnuts are often harvested using hand hoes that can easily damage the nuts, creating easy entry points for the fungus. Groundnuts and bambara nuts can get infected with Aspergillus from the soil if they are harvested with soil adhering to the pods. Crops like maize, sorghum, millet and sunflower, usually harvested and dried on bare ground, may also easily get attacked by fungi present on the ground.

- ***Premature harvesting:*** Immature crops have high moisture content, a condition that favors fungal infestation. Harvesting immature nuts increases opportunity for infection by fungi (Fig. 3).

Fig. 3. Groundnut pods damaged during harvesting.

9.3.4 Postharvest Contamination

The predisposing factors to infection after harvesting the crop include:

- **Improper drying:** Drying on roofs or on the floor exposes the grains to moisture that leads to growth of fungus.
- **Improper shelling:** Practices like sprinkling water on pods to soften the shells to make shelling easy and also to improve the weight of the nuts to get more market value lead to Aspergillus flavus infection and aflatoxin contamination. Another bad practice is threshing of groundnuts in a sack.
- **Poor curing techniques:** Overdrying nuts results in the cracking of the pod and seed coat, thus exposing the nuts to infection.
- **Poor stripping:** Stripping groundnuts together with soil carries the fungus into storage and that provides a conducive environment for fungal infection and aflatoxin contamination. (Fig. 4)\
- **Sorting:** Poor grading especially the wounded nuts before storage is a source of contamination. Wounded, broken, shrivelled and cracked kernels need to be separated from healthy kernels before storage.
- **Poor storage conditions:** Storing groundnuts with high moisture content and poor storage (at household level, market/shops) that expose grains to winter rains, high humidity during the night and poor air circulation that enhances high temperatures leads to fungus growth.

- **Use of airtight containers:** Using non-porous nylon bags and other airtight storage materials decreases insect pest attack and subsequently infection by fungus.
- **Poor transport:** Transport the grains in vehicles with open roofs can expose the grain to sudden rain and moisture which leads to fungus growth.

Fig.4 A farmer sorting groundnuts. This process can reduce infection levels to acceptable standards.

9.4 REDUCTION OF AFLATOXIN CONTAMINATION

Fungal growth and aflatoxin contamination in crops occur due to several factors: some are environmental in nature while others are crop management related. Farmers have minimal control on environmental factors; however, they can improve agricultural practices used in crop production to reduce fungal infection, growth and aflatoxin production.

9.4.1 Crop management practices that reduce infection in the field

These approaches target minimization of preharvest infection by fungi. They aim at providing crops with the best possible growth environment so as to avoid infestation by insects, drought and subsequently infection by fungi (Fig. 5).

- **Early planting:** Early planting helps plants to escape end of season drought that in general predisposes pods to cracking and entry by *A. flavus*.
- **Maintaining field hygiene:** Timely weeding helps to retain soil moisture needed for proper plant growth and avoidance of dry conditions that predisposes developing pods to cracking. Termite control is also very important to prevent damage to developing pods, especially as the crop matures.

Harvesting of water in the field: Drought condition while the crop is growing in the field is a prerequisite to fungal contamination and subsequent aflatoxin contaminate Farmers should ensure that they retain moisture in their field The use of trudges (box ridges) can improve water penetration into the soils, thereby reducing exposure of the developing crop to A. Flavus infestation. Box ridges should be put in place early in the cropping season to capture enough rainwater and reduce effects of end-of-season drought. Mulching also helps to retain water in the field.

- **Soil amendments:** Application of lime to the crop supports development of strong shells (pod resistance). Strong shells provide the first line of defence against pest and fungal attack.

Fig. 5 Use of box ridges to improve rainwater harvesting.

9.4.2 Crop management practices that reduce infection during harvesting

These are management practices that reduce contamination of pods and grain by reducing and/or avoiding exposure of grain to fungi.

- Harvesting at the right stage: Premature kernels have high moisture content and support fungal infection and growth, and aflatoxin contamination. Therefore, harvesting at the right time when the crop is mature will minimize the exposure of the crop to extreme heat, sudden rain or drought, which also influence infection.
- Avoiding injuries to pods: Care should be taken when using hand hoes to avoid injury to the pods. Proper harvesting of groundnuts ensures that the hoe is focused at the plant's rooting zone to avoid damaging pods or kernels.
- Removal of soil: It is important to remove all the soil attached to the pods during harvesting to avoid carrying the fungus into stores and processing facilities.

9.4.3. Crop Management Practices that Reduce Contamination after Harvest

Postharvest handling of crops is another major point of contamination that if well managed will minimize infection of harvests by fungus and minimize aflatoxin contamination. Management practices may be implemented at two levels.

At the household level

Proper drying: Drying on roof or on soil leads to the absorption of moisture and supports fungal growth that leads to aflatoxin contamination. The use of Mandela Corks (ventilated stacking) is one of the best methods for curing the groundnuts and is specifically preferred for its ability to minimize the direct exposure of groundnuts to the sun. Mandela Corks should be stacked on a raised platform, leaving a hollow space in the middle to allow air circulation inside the structure (Fig. 6).

Fig. 6 Mandela Corks support slow curing of harvested groundnuts.

- **Proper shelling:** Sprinkling water on pods to shell the groundnut leads to fungal accumulation and aflatoxin production. It is therefore always better to avoid wetting pods during shelling. The use of mechanical shellers can also reduce labor cost.
- **Grading and sorting:** Fungi readily invade kernels with cracked or damaged pods or seed coats. Removal of such damaged, discolored, small and shrivelled pods reduces the amounts of infected produce in the lot.
- **Proper storage:** Entry of insects and moisture into storage lots allows the entry of fungus and eventually aflatoxin contamination. Grains should be stored in a dry and secure place where the entry of insects and moisture is not possible.

Tips to Minimize Storage Contamination

- Store grains in cool dry conditions.
- Stack gunny bags on wooden planks to avoid dampness.
- Ensure the storage has good roofing to prevent groundnuts from getting wet if it rains.

- Ensure the storage rooms are hygienic to avoid the contamination from the previous lot.
- Using good packaging materials for storage, gunny bags are most appropriate for groundnuts. If nylon bags are used additional perforations need to be made to facilitate air circulation or store in them only for a short time.
- Ensure groundnuts are not exposed to pest and rodents to minimize damage.
- Ensure good aeration in the storage room. (Fig.7)

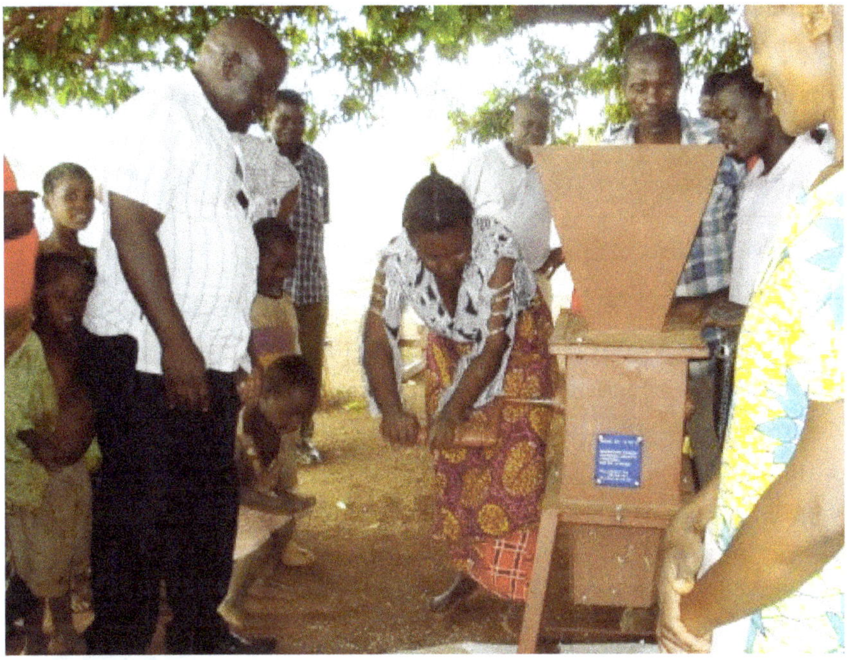

Fig.7 Demonstration of a hand-operated groundnut sheller.

At the processors' level

- Sorting before shelling: Loose shelled kernels, shrivelled, damaged or discolored pods are at higher risk for aflatoxin contamination than are large, mature and well-filled pods. Sorting alone can reduce contamination and keep it within the acceptable levels.
- Grading after shelling: After shelling, all kernels must be sorted into various grade categories based upon size, and it is better to pass through electronic color sorters to remove discolored or damaged and shrivelled kernels.
- Avoid using grade-outs: Grade-outs usually have higher amount of toxins and should not be used for consumption and animal feed. Consuming the grade-outs with high aflatoxin content will expose humans or animals to very high levels of aflatoxin that may lead to sudden death.

MANAGEMENT OF AFLATOXIN IN MAIZE

- Maize, like other cereals growing above ground, gets infected with airborne fungal spores and aflatoxin contamination during plant growth (preharvest), during harvesting and after harvest during handling (postharvest).
- Improper practices during cultivation will carry the fungus to the postharvest processing and storage stage. Improper storage also leads to fungus attack and aflatoxin contamination. Like any other crop, maize can get contaminated with not only aflatoxin but also fumonisin and other mycotoxins which are harmful for humans.
- The aflatoxin management practices for maize is similar to that for any other crop except for a few specific practices described below:

9.5. HOW CROPS GET CONTAMINATED

9.5.1 Pre-harvest Contamination

Preharvest contamination can occur in the growing plant due to:

- Damage to the cobs: Preharvest contamination in cereals like maize mostly occurs when there is a crack in growing cobs and damage due to pest attack (Fig 8).
- Airborne spores: Air dispersal of fungus is associated with infection of above ground crops like maize. Fungal infection can occur through airborne spores in the field during grain filling or during storage and handling. The fungus usually colonizes the silk and kernels when the silk is cut or the cob is damaged due to excessive drought.

Fig. 8 Insect damaged maize cob a common source of field infection by *Aspergillus*.

9.5.2 Contamination during Harvest

Fungal attack occurs during harvest mostly because of drying cobs on bare ground, allowing easy pick up of the fungus from the soil to storage facilities.

9.5.3 Postharvest Contamination

- **Improper drying:** Improper drying of cobs easily leads to retention of moisture. Cobs drying on the roof absorb more moisture while those drying on bare soil also encourage growth of the fungus which leads to aflatoxin contamination (Fig 9, 10 & 11).

Fig 9. Maize cobs drying on the roof.

Fig.10 Maize cobs drying on the ground.

Fig. 11 Maize cob contaminated with Aspergillus flavus.

- **Damaged cobs:** Drying infected and damaged cobs with the healthy cobs can lead to spread of the fungus. Hence, grading is important to avoid contamination.
- **Improper storage:** Storage of poorly dried cobs exposes them to excessive moisture and humidity and these are preconditions that support fungal infection.

9.6 REDUCTION OF AFLATOXIN CONTAMINATION

9.6.1 Pre-harvest Management

- **Pest management techniques:** Using appropriate insect management techniques will reduce formation of holes and damage to the cobs which, in turn, will reduce the entry points for the fungus.
- **Biocontrol:** Application of biocontrol agents to the crop such as Aflasafe is based on the competition between two species of Aspergillus, i.e. the non-toxigenic and toxigenic forms. This biocontrol technology makes use of carefully selected non-toxigenic strains that can safely outcompete and virtually eliminate their toxic relative effectively reducing contamination of the maize grain in fields. A single application of Aflasafe two to three weeks before maize flowering can prevent aflatoxin contamination throughout, even when grains are stored.

9.6.2 During Harvest

- **Proper harvesting:** After harvesting, cobs should not be left to dry in the field on bare soil, where they can easily pick up soilborne fungus. It is advised that cobs should be dried on polyethylene sheets laid on the ground instead of directly drying them on the ground.
- **Grading cobs:** Damaged cobs should not be mixed with healthy ones to reduce spread of spores and subsequent infection. Sorting insect-damaged cobs and cobs having poor husk covering can reduce aflatoxin contamination. Apart from this, winnowing, washing before cooking, and dehulling of maize grains are effective in achieving significant aflatoxin and other mycotoxins removal (Bankole and Adebanjo 2003).

9.6.3 Postharvest Management

- **Sanitation:** Clearing the remains of previous harvests and destroying infected crop residue reduces the spread of the fungal spores and aflatoxin contamination. Cleaning the stores before storing the new harvest also reduces aflatoxin contamination.
- **Proper storage:** Cobs should not be heaped in stores but rather packed in a clean, sealed container to avoid exposure to excessive moisture and humidity.

10

TRANSFORMATION AND CHANNELING OF AFLATOXINS INTO FOOD RAW MATERIALS IN LIVESTOCK

10.1 INTRODUCTION

Mycotoxins are harmful secondary metabolites produced by a variety of mold species that represent serious health risks to both humans and household animals and, not surprisingly, they cause both acute and chronic diseases called mycotoxicoses. The chronic pathological conditions develop over a longer period of time through the consumption of both cereals and animal products, *e.g.*, milk, meat, and eggs. They represent a risk factor to human health directly in the food chain and through biological transformations as well. Mycotoxinogenic fungi are present mainly in small grains like wheat, barley, rye, rice, triticale, and corn and also in different feedstuffs. In fact, aflatoxins were first discovered following a severe livestock poisoning incident in England involving turkeys. In addition, aflatoxins may also occur in peanuts, figs, pistachios, Brazil nuts and cottonseeds.

A number of *Aspergillus* spp. belonging to sections Flavi, Ochraceorosei and Nidulantes have the ability to produce the harmful, carcinogenic difuranocoumarin derivatives called aflatoxins. Aspergillus flavus, Aspergillus parasiticus, and Aspergillus nominus are the most often detected aflatoxigenic Aspergilli in feed. Aflatoxin producer Aspergilli are of paramount importance because the aflatoxins synthesized by them are among the strongest naturally occurring carcinogenic substances. Considering their chemical structures, aflatoxins are furanocoumarin derivatives, of which aflatoxin M1 (AFM1), a hydroxylated derivative of aflatoxin B1 (AFB1), occurs in milk and in various dairy products. AFM1 is a distinguished target in on-going mycotoxin-related research, because AFM1 consumption may be exceptionally dangerous for children especially at younger ages.

Table 1. Aflatoxin producer *Aspergillus* species detected in feed.

Country	Type of feed *Aspergillus* spp.	Isolated
Argentina	Maize silage, corn grains, cotton seed, finish. feed	*A. flavus, A. parasiticus*
Argentina	Maize silage	*A. flavus, A. parasiticus*
Brazil	Concentrated teed and maize silage	*A. parasiticus. A. nomius*
Egypt	Maize aage	*A. flat.*
France	Maize silage	*A. parasiticus*
Ghana	Com grain	*A. flaws*
Indonesia	Maize of livestock feed	*A. flavus*
Iran	Silage, concentrate, hay, TMR	*A. flavus*
Malaysia	Corn grains	*A. flame*
Malaysia	Wheat and barley	*A. flavus*
Pakistan	Feed samples	*A. flavus A. parasiticus*
Saudi Arabia	Animal feedstuff samples	*A. flavus, A. parasiticus, A. nomius.*
Serbia	Com, wheat, bade, soybean and sunflower grains	*A. flavus*
Spain	Barley grains	*A. flavus, A. parasiticus*
Tanzania	Corn grains	*A. flavus*

The risks associated with mycotoxins have an enormous economic impact, which heavily supports the need for further research in this field. The scope of future mycotoxin-linked studies should be broadened and should focus more on the prevention of mycotoxin production and the reduction of their deleterious effects. One of the major objectives of current investigations is the breeding and cultivation of novel plant varieties/hybrids more resistant to infections by mycotoxin producer fungi. Another major goal focuses on the accuracy of the storage of crops and crop products, especially silage, to control the production of mycotoxins more tightly. A further important step in mycotoxin control would be to make feeding practices more rigorous to prevent mycotoxins from entering the body of animals in the first place. Furthermore, countermeasures may also include the application of various mycotoxin binding agents mixed with the feed. Besides agricultural and technological approaches combating aflatoxins successfully, we also need to develop more sensitive and more reliable analytical methods.

To eradicate or at least to decrease mycotoxins considerably in the feed and food chain is undoubtedly a high-complexity and highly prestigious aim, which absolutely requires the effective cooperation of experts working in different fields. Such expanding co-operations will hopefully help on-going research obey the “from farm to fork” principle more. In this case, this concept means that we need to deal not only with production, storage and processing issues but also their impacts on human health as well.

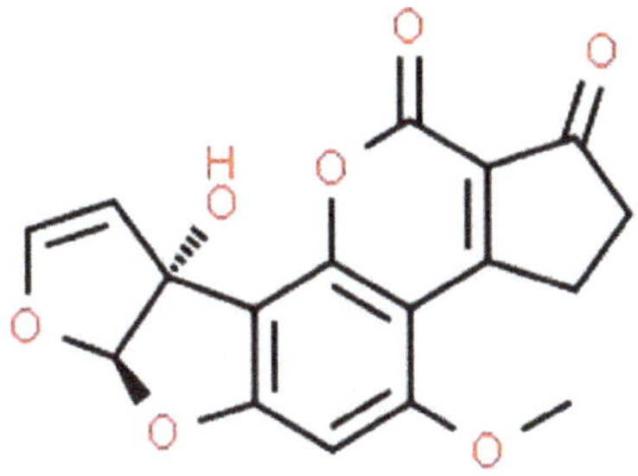

B1

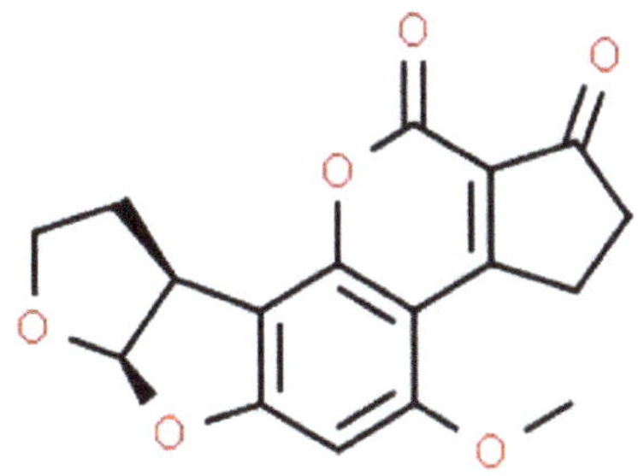

B2

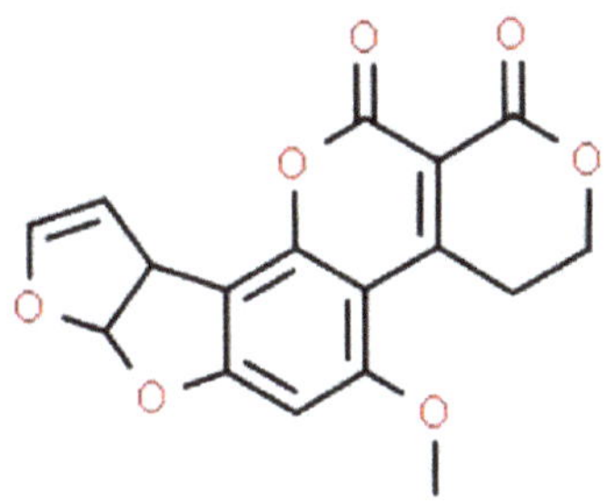

G1

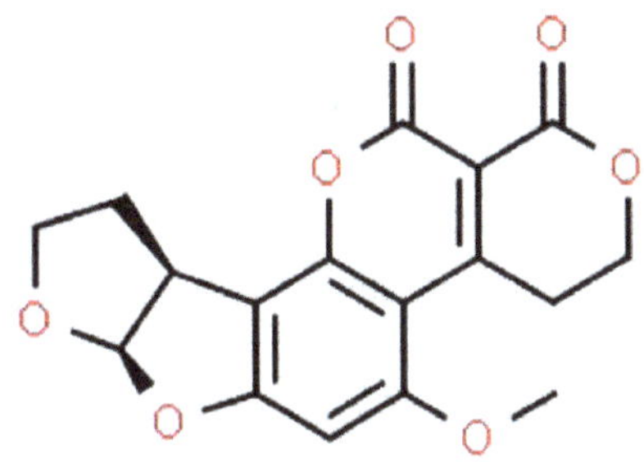

G2

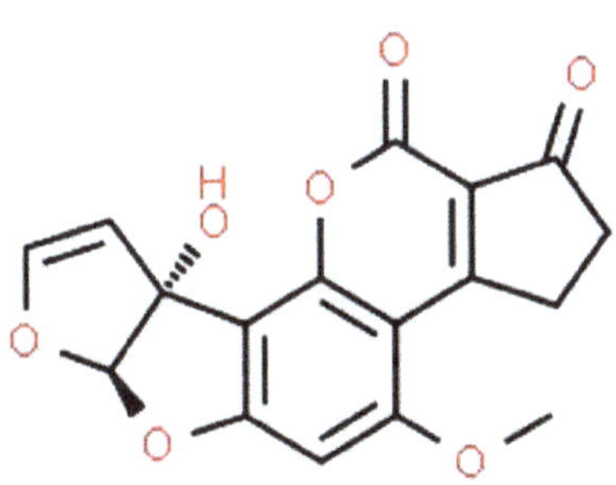

M1

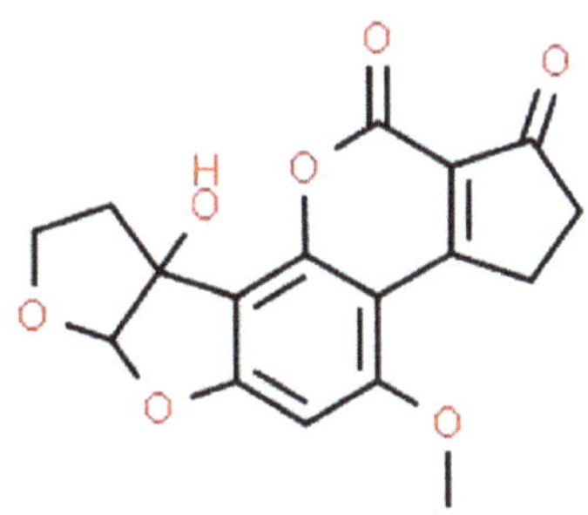

M2

Fig. 1. Chemical structures of aflatoxins most frequently found in animal husbandry.

In this review, we focus on special parts of the feed and food chain like silage production and mitigation of mycotoxins by microbial products. A special attention will be paid to novel findings, which may help the feed management in animal husbandry to prevent and alleviate aflatoxin contamination. Other major issues tackled by this review include new pieces of information on the deleterious physiological effects of aflatoxins on domestic animals, which help us further in proper risk assessment and management. Moreover, up-to-date analytical tools and methods to measure aflatoxins precisely both on farms and analytical laboratories will also be covered. We hope that shedding light on the high-complexity relations between aflatoxin producer Aspergilli, aflatoxin contaminations in feeds and feeding practices in animal husbandry will also give us new hints on the efficient control of aflatoxin contaminations in feeds and minimizing the carry-over of these harmful myctotoxins to humans through the food chain.

Aflatoxins are wide-spread harmful carcinogenic secondary metabolites produced by Aspergillus species, which cause serious feed and food contaminations and affect farm animals deleteriously with acute or chronic manifestations of mycotoxicoses. On farm, both pre-harvest and post-harvest strategies are applied to minimize the risk of aflatoxin contaminations in feeds. The great economic losses attributable to mycotoxin contaminations have initiated a plethora of research projects to develop new, effective technologies to prevent the highly toxic effects of these secondary metabolites on domestic animals and also to block the carry-over of these mycotoxins to humans through the food chain. Among other areas, this review summarizes the latest findings on the effects of silage production technologies and silage microbiota on aflatoxins, and it also discusses the current applications of probiotic organisms and microbial products in feeding technologies. After ingesting contaminated foodstuffs, aflatoxins are metabolized and biotransformed differently in various animals depending on their inherent and acquired physiological properties. These mycotoxins may cause primary aflatoxicoses with versatile, species-specific adverse effects, which are also dependent on the susceptibility of individual animals within a species, and will be a function of the dose and duration of aflatoxin exposures. The transfer of these undesired compounds from contaminated feed into food of animal origin and the aflatoxin residues present in foods become an additional risk to human health, leading to secondary aflatoxicoses.

Considering the biological transformation of aflatoxins in livestock, this summarizes :

(i) The metabolism of aflatoxins in different animal species, (ii) the deleterious effects of the mycotoxins and their derivatives on the animals, and (iii) the major risks to animal health in terms of the symptoms and consequences of acute or chronic aflatoxicoses, animal welfare and productivity. Furthermore, we traced the transformation and channeling of Aspergillus-derived mycotoxins into food raw materials, particularly in the case of aflatoxin contaminated milk, which represents the major route of human exposure among animal-derived foods. The early and reliable detection of aflatoxins in feed, forage and primary commodities is an increasingly important issue and, therefore, the newly developed, easy-to-use qualitative and quantitative aflatoxin analytical methods are also summarized in the review.

10.2 AFLATOXIN PRODUCTION IN FUNGI: BIOSYNTHESIS AND REGULATION

Considering the aflatoxin biosynthetic pathway acetate molecules are converted to norsoloinic acid at first by two fatty acid synthases, a polyketide synthase and a monooxygenase. The biosynthesis proceeds through the intermediates averantin, averufin, versiconal and branches at versicolorin B to give rise to aflatoxin B1 and G1 via the versicolorin A/sterigmatocystin and to aflatoxin B2 and G2 via the versicolorin B/dehydrosterigmatocystin pathways, respectively. The letters B and G stand for the blue and green fluorescence of these compounds observable under ultraviolet light, when separated by thin-layer chromatography. The aflatoxin biosynthetic gene cluster is sophisticatedly regulated by both local and global regulatory elements. Environmental factors like the availability of carbon and nitrogen sources, changing pH, temperature and light conditions as well as variations in the redox status of the fungal cells all have their impacts on aflatoxin production. Among environmental stresses, oxidative stress seems to play a pivotal role in the initiation of aflatoxin production. Plant–fungus interactions also affect the biosynthesis of aflatoxins e.g., through oxylipin production, which have been reviewed e.g., by Pusztahelyi et al. (2015). Undoubtedly, a deeper understanding of the elements and regulation of the aflatoxin biosynthetic gene clusters operating in aflatoxigenic fungi is an important prerequisite for the development of novel and successful mycotoxin control strategies in the future.

10.2.1 Fungal Activity and Aflatoxin Production in Stored Grains

- Aflatoxin-producing Aspergilli may originate from crop fields but post-harvest infections have also been reported. Aflatoxin production cannot be linked strictly to any specific phase of growth or processing status although poorly managed post-harvest conditions during drying and storage can result in rapid increase in mycotoxin concentrations.
- Grain drying is costly but selecting a variety or hybrid optimal for a given crop field can help farmers to harvest cereals with lower than 13–15% kernel moisture contents, which is required for safe storage (Fig. 2). Nevertheless, artificial drying is unsurmountable in most cases.

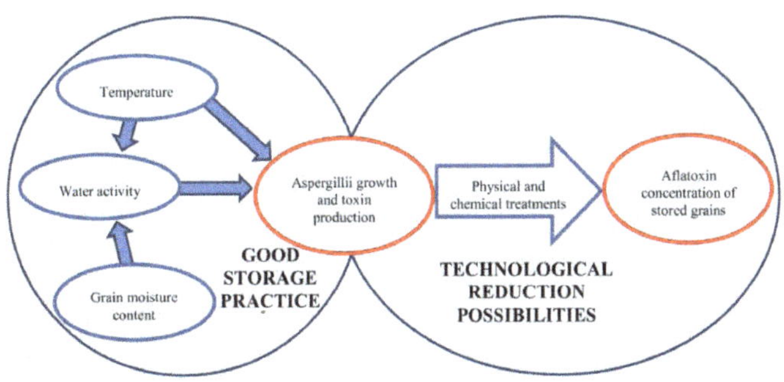

Fig. 2. Factors influencing the aflatoxin content of grains during storage.

- Obeying quality regulations, the recommended drying temperature is less than 65°C for most feed cereals and below 90°C for corn. Of course, these high drying temperatures will also have on impact on the *Aspergillus* spp., which contaminate grains. A. flavus has an outstandingly high heat tolerance in comparison to other fungi with an upper tolerance limit of 40°C. It also found that while the growth of A. flavus was suboptimal above 40°C this relatively high temperature resulted in the most intensive aflatoxin synthesis on chestnut. It also found that 60°C drying temperature still had no adverse effect on A. flavus thriving on corn kernels but raising the temperature up to 70°C significantly decreased fungal infection. Favorable effects of high drying temperature in the restriction of fungal growth were also reported for rice.
- Unfortunately, the aflatoxin molecules are highly heat-stable as their decomposing temperature is 268–269°C. As a result, simple drying technologies cannot decrease aflatoxin concentrations significantly in stored grains. On the other hand, elongated high-temperature treatments may have beneficial effects.
- The temperature, kernel moisture content and relative humidity during storage all influence the physiological processes of fungi. As demonstrated, 18 – 19°C temperature and 12 – 13% moisture content were the limiting factors for the growth and activity of the *Aspergilli*, although lower temperature (8 – 10°C) may also be permissive for growth and mycotoxin production when the grain moisture content is higher. Although these values are accepted widely in good storage practices under continental climatic conditions the relative humidity of grain silos are higher during the cold months, which results in higher water binding by the grains. Nevertheless, the lower temperature hinders increases in microbial activity, and the tolerable water activity is 0.70 for the different *Aspergillus* species. It is important to note that 'hot spots' can develop in grain heaps because of insects or increased grain physiological activity and the released heat and moisture can support fungal growth. Therefore, maintaining good hygienic practice and controlling the temperature of the grain heaps are adequate and necessary measures during storage.
- There are several procedures applicable to decrease fungal infection and mycotoxin production in kernels during storage (Table 2). Size separation by sieving and density separation by gravity table are useful measures as the lighter, smaller and broken kernels and the small components of heap may be infected or damaged by fungi and, therefore, they can be starting points for further deterioration. Not surprisingly, their removal significantly decreases aflatoxin contamination.
- Hand sorting based on visible fungal infections is a very useful tool to decrease the aflatoxin B1 (AFB1) content of corn kernels but obviously this is not a viable option in industrial scale. Another possibility is optical sorting because A. flavus

contaminated corn kernels emit bright greenish-yellowish light when illuminated by UV light enabling separation using suitable optical equipment. Unfortunately, such light emission does not occur in each case and hidden, internal fungi contaminations have no visible effects either. Nevertheless, a sorting method based on the evaluation of red and green light reflectance was also developed to separate aflatoxin containing peanuts and another one for cleaning pecans, based on fluorescence.

- It is noteworthy that a low cost multi-spectral analyzer was manufactured to screen single corn kernels at nine distinct wavelengths in the 470 – 1550 nm region for qualitative use. Although fluorescent optical techniques have higher sensitivities and specificities than near infrared spectroscopy-based and hyperspectral imaging methods near infrared spectroscopic evaluations seem to have greater capabilities to reveal both aflatoxin and fungal contaminations. Most importantly, these techniques have already been applied in automatic sorters. Color analyses can be combined easily with other visible properties. For example, the Raspberry Pi optical analytical equipment is able to sort peanuts based on their color, size, edge length and area of kernel with more than 40 kg/h sorting capacity. It is foreseeable that recent improvements in computing techniques will open new ways for visual analyses in combating both fungi and their mycotoxins.

Table 2. Summary of possibilities and examples for the reduction of the aflatoxin content of stored grains.

	Method	Effect
Removal	Cleaning and sorting by size and density	Only small Brazil nuts (smaller than 36.6 mm length and 6.3 g weight) contained AFB_1
		Corn particles passed 5.16 mm sieve contained 46 times higher more toxin than the over fraction and lower density kernels contain 50 times higher aflatoxin
	Sorting by color	UV light, fluorescent and multi spectral analysis can be used to detect contaminated kernels
	Removal of contaminated part by dehulling and polishing	Dehulling removed 92% of the initial aflatoxin content from corn kernel
		Aflatoxin residuals in corn after crushing and dehulling was almost negligible
		Dehulling decreased AF content of corn by 5.5–70%
		Dehulling and whitening of rice kernel resulted 96% decrease in AF content in polished broken grains and 79% in polished whole kernels
		Dehulling of corn kernels resulted in 88 and 92% reduction in AFB_1 and AFB_2 levels
Reduction, destruction	Thermal treatment for a long time	Heating at 100 and 150°C for 90 min decreased the AFB_1 content of soybean 41.9 and %81.2, respectively

	Method	Effect
	Oxidation by ozone	2.8 and 5.3 mg/l ozone concentration applied for 4 hours resulted 76–84% decrease in AFB_1 content of poultry feed
		66–95% AFB_1 reduction in peanut, corn and wheat kernel
	Irradiation by ionizing and non-ionizing radiation	25 kGy gamma irradiation resulted 43% decrease, microwave heating for 10 min at 1.45 kW resulted 32% decrease, direct solar irradiation for 3–30 h resulted 25–40% decrease in AFB_1 content of poultry feed
		4, 6, and 8 kGy gamma irradiation doses resulted 15–56% reduction in aflatoxin content for corn, wheat and rice kernels
		5 and 10 kGy irradiation doses resulted in 69.8 and 94.5% decreases in AFB_1 content, respectively
		Pulsed light treatment (0.52 J/cm^2/pulse in spectrum of 100–1100 nm with a xenon flash lamp) resulted 75–90% decreases in AFB_1 and AFB_2 contents of rice and rice bran
		6 and 10 kGy gamma irradiation doses resulted 90 and 95% reduction in AFB_1, respectively
		In peanuts, 5–9 kGy gamma irradiation doses result 20–43% decrease in aflatoxins, microwave radiation at 360, 480, and 600 W resulted 59–67% decrease, combined treatments have higher than 95% efficiency
	Destruction by cold plasma	Hazelnuts, peanuts, and pistachio nuts treated with air gases plasma for 20 min resulted 50% decrease in total aflatoxins, SF_6 plasma application resulted only %20reduction
		Atmospheric plasma generated with 400–1150 W power for 1–12 min resulted 46–71% decrease in AFB_1 in peanuts
		High voltage atmospheric cold plasma applied for 1 and 10 min resulted 62 and 82% reduction in AFs levels of corn.
		Atmospheric and low pressure cold plasma reduced the AFB_1 content of hazelnut by %73–72

- Dehulling, the removal of external layers of kernel surface, can be an effective tool to decontaminate grains from toxigenic fungi and significantly decreases the aflatoxin content of grains. This beneficial effect could be improved further by floating and washing before application. Polishing rice kernels is also effective to reduce aflatoxin and, hence, more than nine-fold decrease in contamination was recorded.
- Application of ozone during cereal storage is a relatively new method to improve storage conditions, based on the combined antifungal and insecticide effects of this reactive gas. Importantly, ozone treatments reduce mycotoxin contaminants

without any negative effect on the quality of the grains, and eliminate aflatoxins with high efficiency (66–95% of the original toxin concentration) in cereal grains and flours, as well as in soybean and peanut.

- Another physical method to reduce aflatoxin contaminations is irradiation. Several radiation sources have been evaluated thus far and many of them were found to be effective. For example, the advantageous effects of UV in liquid phase, gamma irradiation in corn, in other cereal kernels, in peanuts and in poultry feed have been reported in a number of publications. Direct sunlight was also effective in aflatoxin reduction in poultry feed and, in addition to exposurces to direct light, the applicability of pulsed light has also been tested and evaluated, and it has already been employed in new decontamination technologies. Meanwhile exposure to pulsed light was effective in liquid medium pulsed polychromatic light applied with a simple xenon flash lamp also resulted in significant decreases in the aflatoxin content in cereal kernels.
- Cold plasma treatment is another possible physical treatment against pathogens and fungal toxins. Cold plasma is generally a result of atmospheric dielectric discharge, and the effects of pressure (atmospheric or vacuum), air composition, humidity and flow rate, discharging power and treatment time are under continuous evaluation nowadays in different cereals and nuts. Cold plasma treatments are cost effective, ecologically neutral and have only a negligible effect on the quality of the grains when compared to classical detoxification methods.

10.2.2 Fungal Activity and Aflatoxin Production in Silage

- Climate change has a major impact on agriculture in many ways and, thereby, many studies have already been published on the effects of climate change on the growth, spread and toxin production of mycotoxigenic fungi on economically important crops.
- Aflatoxin contaminations of maize, wheat, etc. have become a major safety issue in the European agricultural industry, and aflatoxin producer Aspergillus spp. have also been detected in temperate Europe. As a consequence, mycotoxins including the Aspergillus-derived harmful aflatoxins may also contaminate European agricultural products – a foreseeable threat, which we should by no means neglect.
- Maize silage, one of the most important components in the feeding of dairy cows in Europe and worldwide, can be contaminated by several mycotoxin-producer fungi entering the feed production chain at various stages. Not surprisingly, aflatoxin contaminations can be detected occasionally both before and after ensiling. Therefore, the rigorous control of the growth of aflatoxigenic fungi is of pivotal importance, if the production of aflatoxin-free silage is to be guaranteed.

- Although microaerophilic conditions and low pH, which are typical features of silage fermentations, may prevent the growth of the majority of molds, some species of the genera *Aspergillus, Byssochlamys, Monascus, Penicillium*, and *Trichoderma* are able to survive even under ensiling conditions. To make things even worse, the aflatoxigenic capacity of the *Aspergillus* section Flavi strains derived from silage samples is remarkable. For example, del Palacio et al. (2016) demonstrated that 27.5% of these strains produced AFB1, 17.5% of them aflatoxin G2 (AFG2) and 10% synthesized aflatoxin G1 (AFG1). Interestingly, only 5% of the strains produced AFB2. In another study concomitantly performed in Pakistan, A. niger, *A. flavus, A. fumigatus*, A. ochraceous, and A. terrus were identified in both fresh fodder and corn silage. Importantly, the authors also found AFB1 in 37.5% of the fresh fodder and in 41.7% of the corn silage samples with average AFB1 concentrations of 9.5 and 8.4 μg/kg, respectively, meanwhile AFB2 was present in only two samples, and none of the analyzed samples was contaminated by AFG1 or AFG2. In Southern Brazil, aflatoxigenic A. parasiticus and A. nomius strains have been detected in the tested silage and concentrated feed samples.
- Considering the world-wide occurrence of aflatoxin contaminations (Table 3), AFB1 has been reported in corn silage in Argentina, in Brazil, and in France. Total aflatoxin contaminations have also been determined in silage samples collected in Iran and in Uruguay.

Table 3. Worldwide occurrence of aflatoxins in silage.

Country	Mycotoxin	Sample	No. of samples	No. of positive sample (Incidence%)	Mean concentration (μg/kg)	Range (μg/kg)
Argentina	AFB_1	Corn silage	35	6(17.0%)	–	1.4 – 155.8
Argentina	AFB_1	Trench silo	43	6(14.0%)	–	1.0 – 190.0
Argentina	AFB_1	Silo bag	35	19 (54.3%)	–	5.8 – 47.4
Brazil	AFB_1	Corn silage	116	15(13.0%)	33.0	2.0 – 61.0
Brazil	AFB_1	Corn silage	327	3(0.9%)	3.0	1.0 – 6.0
France	AFB_1	Corn silage	–	–	28.0	7.0 – 51.3
Iran	Total AF	Silage	42	7(16.7%)	1.24	1.1 – 27.3
Uruguay	Total AF	Wheat silage	220	–	17.0	6.1 – 23.3

10.3 MICROBIAL BIOCONTROL AND MICROBIAL DETOXIFICATION PRODUCTS FOR MYCOTOXIN MITIGATION IN ANIMAL HUSBANDRY

- In recent decades, several feasible and cost-effective strategies have entered the market aiming to mitigate the effects of feed mycotoxin contamination in animal husbandry, especially in the dairy industry. Technologies to reduce the incidence of mold and mycotoxin contaminations of silages can be employed in one of the three main phases (preharvest, harvest, ensiling) of silage production.

- During the preharvest phase, the appropriate agronomic practices may rely on (i) the use of crop varieties or hybrids, which are resistant to fungal infections, (ii) the application of pesticides and fungicides, (iii) adequate management of weeds and crop residues, (iv) the use of appropriate crop rotation, tillage, fertilization and irrigation and (v) the application of biocontrol agents, e.g., bacteria, yeasts, or atoxigenic strains of A. flavus or A. parasiticus. During the harvest phase, the most important factors that should be taken into consideration are proper harvest timing (maturity stage) and cutting height (to minimize soil contamination), as well as immediate storage of harvested feeds.
- Pre-harvest biocontrol microbes represent a promising and already widely applied method to lower mycotoxin risks in food and feed by protecting plants from pathogens and inhibiting the growth of molds during postharvest conditions. They both reduce economic loss caused by fungal infections and lower toxin levels in products. Biocontrol agents compete for nutrients and space, may secrete antifungals or even parasitize molds, and can also stimulate host plant resistance and, thereby, they mitigate the risk of plant infections and their undesirable consequences. Regarding Aspergilli infection and aflatoxin contamination, non-aflatoxigenic biocontrol Aspergillus flavus strains are most commonly applied to crops, while biocontrol yeasts species are also effective, such as the 2-phenylethanol producing Wickerhamomyces anomalus. These biocontrol agents are mostly applied to protect plants directly used in food production but may exert their effects on plant parts that are to be ensiled for feed production concomitantly.
- In the ensiling phase, attention must be payed to adequate particle size, proper silo size, immediate rapid filling, proper compaction, complete sealing (to maintain strictly anaerobic conditions), and the use of acid-based additives or microbial inoculants, e.g., lactic acid bacteria. Some specific strains in the Lactobacillus (*L. buchneri, L. fermentum, L. hilgardii, L. plantarum, L. reuteri, L. rhamnosus*), *Lactococcus* (*L. lactis*), *Leuconostoc*, and *Pediococcus* (*P. pentosaceus*) genera can inhibit or can even prevent completely the growths of various mycotoxigenic molds and their mycotoxin productions as well. It is noteworthy that there is a wide spectrum of environmental factors which influence the antifungal activity of LAB, including the type of the matrix and culture medium, the availability of nutritional compounds, the incubation time and temperature. In addition, some biological, and chemical (e.g., pH, water activity) parameters will also affect the antifungal activity in a species-specific manner. Species- and strain-specific factors are noteworthy, for example both *L. rhamnosus* and *L. plantarum*

were efficient against A. parasiticus only *L. rhamnosus* reduced the AFB1 levels produced by A. parasiticus. Quite unexpectedly, the A. parasiticus – L. plantarum interaction even stimulated aflatoxin B1 production, which makes the use of L. plantarum undesirable as a silage inoculant.

- In another study, a mixture of *P. pentosaceus* and *L. buchneri* reduced the adverse effects of rust infestation during ensiling and also decreased aerobic spoilage and aflatoxin production in maize silages with high levels of southern rust infestation. Importantly, *L. buchneri* increased the aerobic stability of the silage as well. Inoculation of corn silage with a combined inoculant of L. buchneri and Lactococcus lactis improved the aerobic stability of the silage, and the higher silage density increased the stability further. The interaction of L. buchneri, L. reuteri, L. plantarum, and *L. fermentum* strains reduced the AFB1 level, improved the stability and, furthermore, the microbiological and chemical purity of maize silage. In a most recent study by Ferrero et al. (2019), the authors examined the effect of L. buchneri, Lactobacillus hilgardii, and their combination on A. flavus contaminants and their aflatoxin production in maize silage. The results showed that the inoculation of corn silage with L. buchneri and L. hilgardii increased the aerobic stability and delayed the beginning of aerobic microbial degradation of maize silage, and indirectly reduced the risk of A. flavus emergence and aflatoxin B1 level after silage opening.
- It is reported that the AFB1 binding capacity of various silage bacteria including L. plantarum, L. buchneri, P. acidilactici, and P. pentosaceus and found that high concentration of silage bacteria could bind the AFB1 content of maize silage but population, strain, viability, and medium acidity have all affected the efficacy of binding.
- Antifungal compounds produced by LAB also reduce the mycotoxin production of molds. These LAB-produced compounds cover organic acids, carboxylic acids, phenolic compounds, including phenolic acids, fatty acids (caproic acid, decanoic acid, 3-hydroxydecanoic acid, coriolic acid, ricinoleic acid), volatile compounds (e.g., diacetyl, acetoin), cyclopeptides [e.g., cyclo(Phe-Pro), cyclo(L-Leu-L-Pro), cyclo(L-Met-L-Pro), cyclo(L-Tyr-L-Pro)], hydrogen peroxide, ethanol, reuterin, and proteinaceous compounds.
- Considering the mechanisms of actions of these antifungals, the dissociated forms of organic acids can decrease the intracellular pH within the cells, can increase the permeability of the cytoplasmic membrane, and finally can lead to the death of the fungal cells. In addition, H_2O_2 oxidizes directly the cellular proteins and the lipid components of the cellular membranes. Nevertheless, the mechanisms of the antifungal actions of hydroxy fatty acids and proteinaceous compounds have remained yet to be elucidated.

- Silage decontamination may also be applied if measures to avoid contamination were proven ineffective. Such strategies are primarily based on the adsorbents. Advantages of using adsorbent feed additives over decontamination of the final product, e.g., milk, are their safety and inexpensiveness, and that they may simply be mixed into animal feed to achieve the desired effect. These products may lower the bioavailability of mycotoxins and can help to decrease toxic effects, as well as the amount of toxin detectable in the final product (meat or milk). Such strategies may involve the use of live microbial (LAB or yeast) cultures (usually termed microbial enterosorption, biosorption), microbial or plant extracts, other organic/inorganic materials such as activated carbons or charcoals, hydrated sodium calcium aluminosilicates, and various clay-based products. LAB can not only inhibit the growth of molds but are also able to bind aflatoxins in different matrices, thereby reducing the health risks of aflatoxins. Environmental conditions have a great impact on the aflatoxin binding capabilities of LAB, which is highly species-specific. Some studies demonstrated that non-viable LAB cells had better binding capability for aflatoxin than viable LAB cells. On the contrary, reported on a higher binding efficiency by living cells. Regardless of alive or dead bacterial cells, the aflatoxin binding seems to be reversible and the bound mycotoxins are released slowly over time.

Table 4. Antifungal activity of lactic acid bacteria (LAB).

LAB	Strain	Effect
Lactobacillus buchneri	NCIMB 40 788	Decreased mold count, decreased AFB_2 and increased aerobic stability of the silage
Lb. buchneri	40788	Decreased the population of spoilage fungi, and aflatoxin production in silages
Lb. buchneri	R1102	Bound AFB_1
Lb. buchneri	LB1819	Enhanced the fermentation and aerobic stability of maize silage
Lb. buchneri	A KKP 2047 p	Reduced mold count and decreased AFB_1 amount
Lb. buchneri	NCIMB 40788	Reduced the risk of *Aspergillus flavus* outgrowth and AFB_1 production after silage opening
Lactobacillus fermentum	N KKP 2020 p	Reduced mold count and decreased AFB_1 amount
Lactobacillus hilgardii	CNCM I-4785	Reduced the risk of *Aspergillus flavus* outgrowth and AFB_1 production after silage opening
Lactobacillus plantarum	RC009	Reduce *Aspergillus parasiticus* growth rate
Lb. plantarum	PT5B	Bound AFB_1
Lb. plantarum	K KKP 593 p, S KKP 2021 p	Reduced mold count and decreased AFB_1 amount
Lactobacillus reuteri	M KKP 2048 p	Reduced mold count and decreased AFB_1 amount
Lactobacillus rhamnosus	RC007	Reduce *Aspergillus parasiticus* growth rate

LAB	Strain	Effect
Lactococcus lactis	0224	Enhanced the fermentation and aerobic stability of maize silage
Pediococcus pentosaceus	12455	Decreased the population of spoilage fungi and aflatoxin production in silages
Pediococcus acidilactici	R21	

- Based on various microbe species, sources, manufacturers, and formulations, live yeast products include several categories: yeast probiotics, Saccharomyces cerevisiae fermentation products (SCFP), dried yeast products (DY or DYP), brewery yeasts (BY), and active dry S. cerevisiae (ADSC). Compared to live bacteria-based products, these yeast products are considered and employed as direct feed additives in most cases and are not applied at the ensiling phase. Some bacterial species, e.g., Nocardia corynebacteroides (NC) are also added as direct feed additives for poultry. Microbe-derived feed additive products are also based on yeasts, and include autolyzed yeast (AZ), inactivated yeast cells (IY), distillery yeast sludge, and yeast cell wall (YCW) products.
- Live yeast or bacterial cells intended to colonize the gastrointestinal tract (GIT) of humans, or in some cases, poultry or laboratory rodents, are occasionally termed probiotics. However, especially in the case of ruminants, the use of live cells may not necessarily result in gastrointestinal colonization. The rumen's own microbiota is also to be taken into account, as it can contribute to aflatoxin detoxification and degradation (e.g., biotransformation to aflatoxicol). Aflatoxin B1 is absorbed in the rumen mainly at acidic pH, and the degradation of aflatoxins in rumen depends on both the animal species and feed type. However, it must be noted that rumen colonization by A. flavus has also been recorded, leading to toxin production in rumen liquor.
- The products SCFP, DY, BY, and ADSC consist of yeast cells, the nutrient medium on which the yeasts were grown, and the metabolites produced by the yeasts and have been shown to increase DMI, milk yield, as well as milk fat and protein yield in lactating dairy cows. However, these positive effects are attributed to adsorption of toxic substances and the modulation of the gut (prokaryote) microbiota, not to long-term gut colonization by the yeasts. Yeasts in fact are thought to play a negligible role in the microbiome of ruminants, although they may survive gastrointestinal conditions and retain their aflatoxin B1 binding ability under gastrointestinal conditions.
- Various studies have shown the effects of these live yeast products on the microbiota of the cows, however, uncovering the underlying mechanisms and a holistic understanding of dairy cow gastrointestinal health still requires further research. Interestingly, YCW has also been shown to positively modulate the gut

health in broiler chicken challenged with AFB1 or with Clostridium infection. These observations raise the possibility that yeast products, whether live or not, generally contribute to animal health both as bioadsorbents and as modulators of the gut prokaryote microbiota, as well as the immune status of the animal. Such positive effects may not only prevent toxicosis but result in increased feed intake and production.

- In poultry feedstuff, S. cerevisiae strains have been tested and made commercially available as a probiotic microbe. It must be noted though, that the intended effect of the yeasts is not necessarily gut colonization and microbiome modulation, but aflatoxin adsorption, a role, which yeasts can effectively fulfill. The applications of microbes and microbial products for mycotoxin risk mitigation are summarized in Fig. 3.

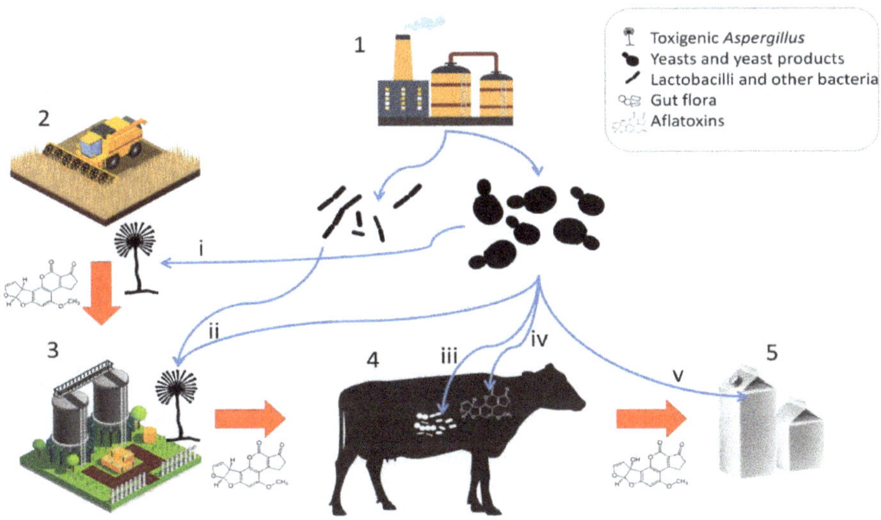

Fig. 3. Microbial products for mycotoxin mitigation in animal husbandry and their applications. Red arrows represent potential carry-over of mycotoxins or toxigenic *Aspergilli*. Blue arrows represent applications of microbes and microbe-derived products. (1) Fermentation and animal feed supplement industries; (2) crop production; (3) preparation and storage of silage and other feedstuffs; (4) livestock; (5) product; (i) pre-harvest biocontrol; (ii) antagonism in silage and feed; (iii) host gut microbiota and immune modulation, probiotic effect; (iv) enterosorption; (v) bioadsorption from product (milk). [Stock image credits: Freepik, macrovector, and vectorpocket].

- Yeast cell wall β-D-glucans, glucomannans and mannan-oligosaccharides are responsible for the mycotoxin binding abilities of these products .Some purified cell wall components have been tested in animal husbandry, such as mannan-oligosaccharides supplemented into the diet of Japanese quails affected by aflatoxicoses. However, no direct correlation between the amount of individual components and toxin binding are evident. Structural integrity and amount of the yeast cell wall is crucial in binding efficacy, while viability is not: heat-treatment can even increase adsorption capacity. Toxin binding can reach saturation rapidly and is reversible, and mycotoxins are not modified chemically during the process. It must be noted that some yeasts and bacteria are known to be able to enzymatically degrade mycotoxins if applied in viable form.

- A novel approach for the microbiological detoxification of animal feed is the screening of isolates from various environmental sources, instead of using the most widespread species, *S. cerevisiae*. Various bacteria and yeasts may exhibit toxin-binding or even toxin-degrading abilities, as well as biocontrol effects on toxigenic molds and these may be directly applied to supplement animal feed. Novel yeast species in this field include members of the genera Kluyveromyces and Pichia, both related to the widely used Saccharomyces. P. kudriavzevii has been successfully applied as a bioadsorbent feed additive to ameliorate the negative effects of AFB1 contamination on broiler chicken performance. Novel isolates of K. marxianus have also been proposed as bioadsorbents based on in vitro characterization. However, Battacone found no evidence for AFB1 detoxification in ewes fed with Kluyveromyces lactis DYP, highlighting the need for rigorous testing of novel strains in different setups and with multiple animal species.
- Apart from novel microbial strains, combined treatments of microbial and inorganic products constitute a promising strategy in ameliorating mycotoxin contamination. Recently, Jiang et al. (2018) found that both dietary clay and clay + SCFP reduced transfer of dietary AFB1 to milk as well as milk aflatoxin M1 (AFM1) concentration, while the combined treatment was the only one that also prevented the decrease in milk yield caused by AFB1. Thus, the potent adsorbing capability of inorganic products may act synergistically with the adsorbent, gut health-promoting and immunomodulatory effects of yeast products.

10.4 MICROBIAL DETOXIFICATION PRODUCTS TO COUNTERACT AFLATOXIN CONTAMINATION IN DAIRY PRODUCTS

Some studies have explored microbial aflatoxin decontamination strategies in dairy products, taking advantage of the high efficacy and relative ease of utilizing LAB and yeast, recently been reviewed. Briefly, such microbial decontamination approaches rely on heat-killed or immobilized cells, and promising results were obtained when both LAB and yeasts were applied simultaneously. Heat-treatment of bacterial cells was found to improve binding capabilities. Bacteria tested in the aforementioned studies include members of the genera traditionally considered probiotics and/or important in food production, as Bifidobacterium, Lactobacillus, or Pediococcus, and oddly, a potential pathogen, Enterococcus.

In UHT skim milk, both LAB and yeasts showed promising results, and the binding of toxins to microbial cell walls was shown to be rapid, enabling short incubation times in potential industrial applications. There are certain limitations on applying yeasts and bacteria for the decontamination of dairy products, such as the need for their subsequent removal, reversibility of binding, or even legislations on tolerated number of live or dead microbial cells in products. Nevertheless, the high toxin binding capability and the safety of heat-killed cells toward consumers compared to chemical methods makes microbial decontamination a promising strategy.

10.4.1 Aflatoxin Metabolism in Livestock

The toxicity of AFB1 is strictly related to the bioactivation and detoxification pathways operating animals in vivo (Fig. 4). Indeed, AFB1 is a "pro-carcinogen" that is activated biologically by cytochrome P450 (CYP450), a microsomal enzyme of phase I detoxification (oxidation) to the extremely reactive and electrophilic AFB1-8,9-epoxide (AFBO). This harmful AFB1 derivative is able to covalently bind to macromolecules such as DNA and proteins, thereby forming adducts, which cause acute and chronic cytotoxicity, DNA mutations and eventually expressing carcinogenic activity.

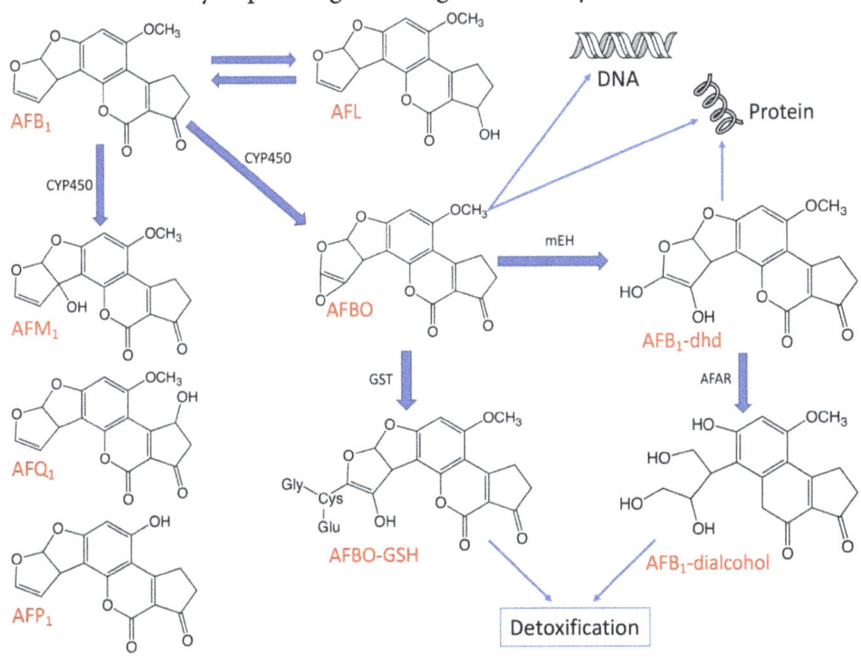

Fig. 4. The major metabolic pathways of AFB1. The aflatoxin derivatives presented here include aflatoxin B1 (AFB1), aflatoxin M1 (AFM1), aflatoxin Q1 (AFQ1), aflatoxin P1 (AFP1), AFB1-8,9-epoxide (AFBO), AFB1-8,9-dihydrodiol (AFB1-dhd) and aflatoxicol (AFL). Some enzymes taking part in the biotransformation and detoxification of aflatoxins are also indicated including cytochrome P450 (CYP450), glutathione-S-transferase (GST), microsomal epoxide hydrolase (mEH), and aflatoxin-aldehyde reductase (AFAR).

Moreover, AFBO can be hydrolyzed to AFB1-8,9-dihydrodiol (AFB1-dhd) by an epoxide hydrolase. AFB1-dhd is able to react with proteins causing cytotoxicity or, alternatively, AFBO can be metabolically detoxified via conjugation with glutathione (GSH) by glutathione S-transferase (GST), a phase II detoxification enzyme. This pathway is considered as one of the main routes of AFBO detoxification. Microsomal epoxide hydrolase (mEH) and aflatoxin-aldehyde reductase (AFAR) can also transform reactive AFB1 to AFB1-dialcohol, a real detoxified AFB1 derivative, which can be excreted in urine.

It is important to note that several isoenzymes belonging to the CYP450 supergene family metabolize AFB1 through oxidative reactions, producing various metabolites with different carcinogenic potential.

In addition to the highly reactive and toxic AFBO, the main AFB1 metabolic pathways described in animals can also give rise to relatively less toxic metabolites such as aflatoxicol (AFL) by ketoreduction or AFM1 by hydroxylation and non-toxic metabolites such as AFB2a or aflatoxin Q1 (AFQ1) by hydroxylation or aflatoxic P1 (AFP1) by demethylation

Current literature data indicate that the rates of the bioactivation and detoxification of AFB1 contribute greatly to the manifestation of its toxic effects. Furthermore, the metabolism of aflatoxins shows considerable interspatial differences and also significant variations among individuals belonging to the same species, depending for example on the age.

In conclusion, the largely different sensitivities of different animal species to the toxic effects of aflatoxins could be explained mainly with the remarkable variability of the metabolic pathways and enzymes that contribute to the bioactivation and detoxification of aflatoxins.

10.4.2 Poultry

Poultry are generally very sensitive to AFB1 and adverse health effects have been reported in turkeys, quail, chickens, and ducks but there is a great variability in species-specific sensitivities to aflatoxins. Several toxicological studies pointed at the existence of a sensitivity scale starting from the less resistant young duck and turkey, followed by quails, up to the more resistant chickens. Obviously, species-specific differences in the AFB1 biotransformation pathways, e.g., in AFB1 modifying hepatic microsomal enzymes, could explain the varying susceptibilities of the species. It has been reported in some works that the microsomal liver fractions produced only AFBO in avian species (Fig. 4), unless these animals were stimulated with CYP450 inducers. However, the ability of poultry species to metabolize AFB1 to AFM1 was reported in other works, in which the AFM1 was detected in different tissues. Lozano and Diaz (2006) reported that turkey microsomes produced 1.8–3.5 times more AFBO than quail and chicken microsomes. Furthermore, suggested that the higher resistance of chicken to AFB1 in comparison to quail could be due to a lower activation rate of AFB1 to AFBO in chicken and also to a lower affinity for AFB1 of the chicken microsomal enzymes.

The high susceptibility of poultry to AFB1 appears to be a consequence of both the high activity of phase I microsomal detoxification enzymes to form AFBO, and to a low GST efficiency as well to conjugate AFBO with GSH (phase II detoxification). Some works reported that the partial or complete lack of GST-dependent detoxification of AFBO was the major reason for the exceptionally high susceptibility of poultry including turkeys to AFB1.

Another metabolic pathway that may contribute to the extreme susceptibility of poultry to aflatoxins could be the reduction of AFB1 to AFL via cytosolic reductase because the cytosolic metabolite AFL is produced in larger quantities in turkey and duck than in quail and chicken. This fact underlines that the formation of AFL cannot be regarded as a real detoxification pathway in these birds, moreover, microsomal dehydrogenase may oxidize AFL back to AFB1, increasing the physiological half life of AFB1. Other aflatoxin metabolic pathways may also be involved in the manifestation of the high cytotoxicity of AFB1 in poultry species.

Furthermore, it has also been reported that AFB1 susceptibility correlated with age in both turkeys and broiler chickens. When livers obtained from 9, 45, and 61 day-old turkeys were compared, microsomes from younger birds were more active in AFB1 bioactivation than those from older ones (Klein et al., 2002). Moreover, Wang H. et al. (2018) underlined the efficient bioactivation of AFB1 by CYP enzymes and the deficient detoxification by GST enzymes in younger 7-day old broilers.

Aflatoxin residues were detected in various tissues mainly in liver, kidney, the organs where AFB1 is metabolized, but also in reproductive organs, in gizzard, breast and in legs. The metabolites AFB1, AFQ1, and AFL were excreted as such or as glucuronyl conjugates from bile in feces. Some of these metabolites (AFM1 and AFL) have been found in liver, kidneys and thigh muscles. The concentrations of AFB1 residues decreased in the livers and muscles of all the birds after the suspension of mycotoxin feeding, and the elimination of AFB1 from tissues was faster in older than in younger birds. The dietary exposure to aflatoxin of hens, even at low concentrations, may also cause contamination of eggs. AFB1 residues appeared in eggs after 5 days following the administration of AFB1 contaminated feedstuffs, and they accumulated in line with protracted feeding with contaminated grain. However, the amount of mycotoxin contaminants was below 0.1% of the AFB1 intake owing to the AFB1 metabolism in the birds. Few works also demonstrated the presence of hydroxylated AFB1 derivatives (AFM1 and AFQ1) in eggs.

10.4.3 Pigs

Pigs are considered relatively susceptible to AFB1. Studied AFB1 metabolism in liver of different species in relation to the susceptibility to the toxic effects. The piglet's liver showed a relatively lower cytosolic GST activity to convert AFB1-epoxide to AFB1-glutathione conjugate product, thus favoring the formation of AFB1-DNA adducts.

The metabolism and tissue distribution of AFB1 in pigs were studied by Lüthy et al. (1980), and the major excretory route was found to be the feces (51–65% of the dose administered) but also urine was also an important excretory route. Actually, both AFM1 and AFB1 were detected in pig urine samples and AFM1 was always found at higher concentrations in all studies. In fact, urinary AFB1 and AFM1 are often used as biomarkers for aflatoxin exposure in pigs.

10.4.4 Ruminants

Ruminants are generally more resistant to the toxic effects of mycotoxins than monogastric animals, which could be explained mainly by AFB1 degradation or bioconversion by rumen microorganisms. Controversely, some studies reported on that aflatoxins were generally poorly bioconverted in the rumen, with an overall decrease of only 10% Moreover, AFB1 was incubated with intact rumen fluid or fractions of rumen protozoa and bacteria from sheep and cattle in the presence or absence of milled feed and the result clearly indicated that rumen fluid had no effect on AFB1. Another study showed that AFB1 metabolism in rumen fluid was influenced by the animal species and the type of feed. In fact, rumen microbes from Korean native goats exhibited a greater degradation capacity for AFB1 in comparison to Holstein steers. These diverging observations might be the consequence of varying rumen microbe profiles.

AFM1 is the most prominent metabolite formed in bovine hepatocytes within the first hours of incubation whereas AFB1-dhd becomes determinative after a prolonged incubation. These two metabolites are mainly formed by CYP1A and CYP3A hepatic monooxygenase activities. According to Larsson et al. (1994), several extrahepatic tissues of sheep can also bioactivate AFB1 very efficiently and can conjugate the bioactivated AFB1 with GSH as well.

Following the channeling of AFB1 in ruminants, the ingested aflatoxins may be degraded, at least in part, to AFL, AFM1 and many other hydroxylated metabolites by certain rumen microbes or may be sequestered by some rumen fluid components such as chlorophyllin structures as well as bacterial and yeast cell walls The remaining fraction is rapidly adsorbed in the gastro-intestinal tract by passive diffusion and then is extensively metabolized in the liver to AFM1, which enters the systemic circulation or is conjugated to glucuronic acid, and afterward excreted via bile, urine or milk. Obviously, different levels of feed contamination may lead to different carry-over rates, which are also influenced by other physiological factors such the health status of animals including the status of the liver and its enzymatic activities. AFB1, AFM1 and AFL have been detected in liver, kidney and muscle tissue of bovine. AFM1 is excreted via urine at a greater extent than through milk but the physiological factors regulating the relative uptake by kidneys and mammary glands are still unknown. AFM1 has been detected in both the milk and urine of cattle and dairy ewes 6 h after AFB1 ingestion, and its concentration decreased rapidly after withdrawal of aflatoxin from diets. Fecal excretion of FB1 results from a lack of absorption by the GIT or a highly efficient elimination by the biliary system in the form of conjugated metabolites.

Goats were administrated with [14C]-AFB1, and urine, milk and feces were collected after 120 h. AFM1 was found in milk at the highest concentration meanwhile AFQ1 and AFL were found only in trace quantities in milk Other studies on goats also indicated that the absorption of AFB1 in the GIT of adult ruminants was very fast, as was its hydroxylation to AFM1 and release into the blood The short interval between AFB1 administration and the detection of its metabolite in milk confirmed that the absorption of the toxin took place already in the rumen in goats.

10.5 AFLATOXINS IN FOODS OF ANIMAL ORIGIN

Aflatoxins are generally considered as the most important mycotoxins due to their carcinogenic properties, their persistence in food commodities once formed, and the wide range of food commodities that may be contaminated by them. Aflatoxins contaminating feeds pose a direct threat to livestock health and, indirectly, also affect human nutrition and health by reducing livestock productivity and via transfer from feed to foods of animal origin, namely milk, meat and eggs, even if milk is the only food of animal origin with relevant aflatoxin carry-over.

Aflatoxins, particularly AFM1, are of public health concern because they are efficiently excreted into milk, even if they may also contaminate other foods of animal origin at low levels and, therefore, the associated risks are considered to be minor. Not surprisingly, many countries have set maximum levels of aflatoxins (AFB1 or total aflatoxins, AFM1) in food commodities and animal feeds, with the main aims to protect animal health and to prevent aflatoxin contamination of animal-derived foods. This review does not provide a systematic overview on aflatoxins in foods of animal origin but summarizes the discussions on the potential public health concerns specifically related to aflatoxins residues in these food commodities. In livestock animals, the best estimate transfer factors for mycotoxins in kidney, liver, muscle, fat, milk and egg were reported by MacLachlan (2011), and they clearly showed that no significant residues coming from aflatoxin contaminants of livestock feed are present in meat and eggs.

In the case of human dietary exposure from dairy products, aflatoxins are considered the most important mycotoxins and, based on data belonging to Food and Feed Safety Alert, 93% of the overall mycotoxin notifications referred to aflatoxins, whereas dioxins, dioxin-like polychlorinated biphenyls and AFM1 were the most frequently reported chemical issues in dairy products. When ruminants were fed with contaminated feed, the AFB1 consumed by the animals was partly degraded by the forestomach before reaching the circulatory system, and the remaining part was transformed by the liver into monohydroxy derivative forms, mainly to AFM1, and, in smaller quantities, also to AFM2, AFM4 and AFL. Afterward, AFM1 was secreted into the milk through the mammary glands AFM1 has only from 2 to 10% of the carcinogenic potency of AFB1 but it possesses the same liver toxicity. The ability of ruminants to convert the AFB1 ingested with feedstuff to AFM1 and to excrete this derivative in milk varies within broad limits in large and small ruminants and ranges between 0.35 and 3% in cows, 0.018 and 3.1% in goats and between 0.08 and 0.33% in sheep. This remarkable variability in AFB1 biotransformation observed in these species can be explained with differences in the activity of hepatic enzymes involved in the biotransformation and detoxification processes considering both their expression and catalytic activity The average conversion value was 2.5% in high yielding dairy cows, which produced a daily amount of about 40 L of milk, were tested. Importantly, Veldman et al. (1992) found a direct relationship between the carry-over rate and the milk yield with a maximal 6.2% carryover rate. AFM1 is the most commonly detected aflatoxin in milk and the excretion of AFM1 depends on a range of factors including diet composition, rumen degradation and liver biotransformation capacities, the duration of lactation as well as on the animal breed and udder health status. In dairy cows ingesting AFB1 contaminated feedstuffs, the excretoin of AFM1 occured in 12 – 24 h and up to 2 – 3 days in milk, whereas the AFM1 clearance in milk depended on several factors, mainly on the amount of ingested AFB1 and the duration of mycotoxin consumption with an excretion for a variable period of about 5 – 7 days from the ending of AFB1 assumption by cows.

Well-reported variations in AFM1 contamination were observed in milk worldwide, which were dependent on several factors like geographical area, environmental and climatic conditions including seasons and weather, as well as on the diversity and level of development of farming systems and the consumption of feed concentrates and green forage. In recent years and independently of the type of commodity, the occurrence of AFM1 in milk and dairy products was lower in Europe (for example in Italy, Portugal, Turkia, and Croatia) than in Asia or South America, where higher mycotoxin frequencies up to 100% were reported. In Europe, low levels of AFM1 contamination were reported in milk, and only 0.06% of the analyzed samples were above the European limit of 0.05 μg/kg milk. Nevertheless, when such incidents occur a widespread AFM1 contamination of milk may develop, which has to be taken into account and adequately considered and controlled. In addition, risk managers should also consider that aflatoxin concentrations in milk may vary within the year and may also depend on the geographical location and climatic conditions. Finally, AFM2 has also been investigated in milk with different outcomes varying from its absence to a not negligible occurrence in powdered, UHT and pasteurized milk samples.

The AFM1 contamination of dairy products is classified as an indirect contamination. For example, when the milk used in cheese-making was contaminated by aflatoxins, AFM1 unevenly distributes between whey and curd, because AFM1 prefers to bind to milk proteins, first of all to casein. For this reason, AFM1 is more concentrated in the curd and cheese than in the milk itself, which was used for cheese-making. Therefore, AFM1 levels were 3 – 8 times higher in certain dairy products than in the milk, and stable AFM1 residues were detected in the final dairy products like milk powder even after heat processing. In addition, the total amount of AFM1 does not change significantly during the cheese-making and cheese maturation processes but these steps influence the AFM1 and protein concentration ratios as a result of skimming and water loss. Although many studies on the contaminations of dairy products by AFM1 are available only few of them present any data estimating concentration factors for AFM1 in different cheeses. However, 2.5 – 3.3 and 3.9 – 5.8 times higher concentrations of AFM1 calculated on a weight basis were recorded in soft and hard cheeses, respectively, than those AFM1 concentrations found in the milk, from which these cheeses were made. In Europe, the food business operator has to justify and provide the specific concentration or dilution factors for AFM1 in the processed foodstuffs during official controls performed by the competent authority.

In this context, AFM1 contaminating milk should be unremitting to our attention and we should also take a special care of infants avoiding their exposures to AFM1 via milk and infant formulas Kerekes et al. (2016) emphasized the importance of regular control of produced milk and also the introduction of an appropriate action limit in combination with immediate corrective actions at the farm level. In fact, feed producers have to manage and control the feed ingredients intended for the production of feed for the lactating animals for risk mitigation. Feed ingredients should be selected based on their quality characteristics, whereas farmers, when the AFM1 content of milk exceeds the legal limit, have to withdraw milk consignments and also have to remove contaminated feedstuffs.

As far as the aflatoxin residues detected in edible tissues of bovine, pigs and poultry are concerned, these AFB1 entry routes do not contribute significantly to human aflatoxin exposures. Nevertheless, data on the aflatoxin contents in the edible tissues of bovine species are scarce and it is generally assumed that aflatoxins are partly degraded in the rumen and they are rapidly metabolized in the liver after absorption from the intestines. The transfer rates of aflatoxins into the edible tissues of pigs are very low owing to the rapid pre-systemic and hepatic metabolisms, and the aflatoxin residues in pork are therefore not considered as of public health concern. Similarly, poultry with low levels of aflatoxin contaminations do not seem hazardous to humans although the presence of aflatoxin-residues in poultry liver is well-documented. Importantly, a rapid decrease in AFB1 residues was observed in poultry muscles and liver after 3–7 days of uncontaminated dietary, significantly reducing the risk for human health. However, AFL is the main component of total AF residues in poultry with highest contents in liver. In the case of laying hens, aflatoxins and their metabolites, particularly AFB1 itself and AFL, can also be carried over to eggs but very discrepant transmission ratios were reported in this case. Recent studies demonstrated very low amounts of aflatoxin residues in eggs, merely between 0.01% and 0.07% of the aflatoxin intake. AFB1 residues appeared in eggs after 5 days of feeding with contaminated feedstuffs and the amount of AFB1 depended on the duration of feeding with contaminated grain. Similar to dairy products, the presence of aflatoxins in eggs may be indicative of the aflatoxins contamination of the feed.

10.6 AFLATOXICOSES AND ANIMAL SUSCEPTIBILITY

- In general, mycotoxicosis refers to syndromes appearing after ingestion, skin contact or inhalation of toxic secondary metabolites produced by toxigenic molds belonging to the genera *Aspergillus*, *Fusarium*, and *Penicillium* as well as to some other fungal taxa. Within mycotoxicoses, aflatoxicosis refers to any disease caused by the consumption of foods and feeds contaminated with aflatoxins. It is well-known that AFB1 is a potent mutagenic, carcinogenic, teratogenic, and immunosuppressive fungal secondary metabolite and all these effects may be linked to the interference of AFB1 and its derivatives with the synthesis of proteins, the inhibition of various metabolic pathways or to the onset of oxidative stress. All these disadvantageous physiological effects will lead consequently to damages in various organs, especially in the liver, kidney, and the heart.
- Aflatoxicoses may emerge in any livestock but literature reports on outbreaks mostly in poultry, pigs, equine, sheep, and cattle. The exposure of domestic animals to AFB1 mainly occurs through the ingestion of contaminated feeds, however, skin contacts or inhalation exposures might also contribute. It is well-known that ruminants are among the least susceptible animals to the negative effects of mycotoxins in comparison to monogastrics. However, the rumen has a saturable capacity of detoxifying aflatoxins by microflora, depending on (i) variations in the diet, (ii) the consequences of metabolic diseases, such as rumen acidosis, (iii) rumen barrier alterations as a result of animal diseases, and also (iv) the actual concentrations of aflatoxins present in the animal feed. Consequently, clinical manifestations of aflatoxicoses in ruminants are associated typically with aflatoxins that are not degraded at all or not completely degraded by the rumen microflora.
- Most of the data we have already had in our hands on mycotoxin toxicity are coming from experimental studies with purified compounds in otherwise healthy animals, which knowledge may help us with the early and reliable diagnosis of mycotoxicoses. However, when natural episodes of mycotoxicoses occur, versatile signs of disease could appear depending on the environmental conditions and also on several other features of the animals involved, including nutrition, sex and breed. For this reason, the diagnosis of mycotoxicoses is often difficult but it should rely on observing the clinical symptoms on the affected animals and also on analyzing the feed involved in the intoxication (Council for Agricultural Science and Technology [CAST], 2003). Given aflatoxins could act

in synergy with other mycotoxins and also with other disease-provoking agents and, therefore, additional apparently unrelated pathological symptoms and even diseases are observed and reported in the affected animals. Furthermore, most mycotoxicoses including aflatoxicoses may present non-pathognomonic features and, consequently, there are no definitive diagnostic symptoms to orient farmers and veterinarians to assign aflatoxin exposures unequivocally to the death of animals. Obviously, even other otherwise unrelated diseases may trigger similar responses in the domestic animals to those of aflatoxins (Richard, 2008).

- Aflatoxins do not affect all animals uniformly. Some animal species are inherently more resistant, such as sheep, goats and cattle, whereas other animals are more susceptible like swine, chickens, turkeys, and ducklings. In addition, considerable breed differences are documented within a given species, and the physiological responses to the adverse effects of aflatoxins are also influenced by age, sex, diet, and weight, exposure to infectious agents, and the presence of other mycotoxins or other pharmacologically active substances. In addition, when mycotoxins are present simultaneously, some interactive effects, classified as additive, antagonistic or synergistic, could also occur.

10.7 ANIMAL EXPOSURE TO AFLATOXINS

- The exposure of animals to aflatoxins may trigger biological reactions that could be classified as acute, overt diseases with high morbidity and mortality, or, as it is usually the case, chronic, insidious disorders that impairs animal productivity. When livestock ingest aflatoxins the health effects could be acute, with severe consequences and evident signs of disease or even may be lethal when these toxins are abundantly consumed, even if this event is rare under farm conditions. The timing of the proper diagnosis is a crucially important factor because the suspicious contaminated feed is likely consumed well before it can be tested (Council for Agricultural Science and Technology). The earliest clinical signs and lesions observed in turkey "X" disease, hepatitis "X" of dogs, and similar cases of acute aflatoxicoses were anorexia, lethargy, hemorrhages, hepatic necrosis, and bile duct proliferation. Furthermore, the aflatoxins' impact on animals should not be limited to the extreme effects of aflatoxicoses because it is related mainly to the chronic toxicity caused by the consumption of sublethal doses and to the fact that low levels of chronic exposures may result in cancer.
- Considering the chronic effects of aflatoxins, hidden pathological alterations with reduced ingestion, productivity and fertility were implied, including lowered milk, meat, and egg productions, decreased weight gains and/or unclear changes in animal growth, feed intake reductions or feed refusals, alterations in nutrient absorption and metabolism, various typologies of damages to vital body organs, disadvantageous effects on the reproduction and endocrine systems and also

suppression of the immune system with subsequently increased disease incidence. The economic consequences of chronic aflatoxicoses are many times larger than those of the rare acute cases with immediate morbidity and lethality (Council for Agricultural Science and Technology).

10.7.1 Hepatotoxic, Carcinogenic and Mutagenic Effects

- Among the major devastating effects of aflatoxins on animals, these harmful metabolites specifically target the liver and, hence, are proved to be primarily hepatotoxic. In acute aflatoxicosis, the emerging clinical symptoms of acute hepatic injury include coagulopathy, increased capillary fragility, hemorrhage and prolonged clotting times. Gross liver changes are caused by hemorrhage, centrilobular congestion, and fatty changes in surviving hepatocytes. Death of the poisoned animal may occur within hours or a few days after exposure. In broiler chicks, hemorrhagic anemia syndrome develops as characterized by massive hemorrhagic lesions in major organs and musculature even if the anemia could be considered as a secondary effect of severe hypoproteinemia caused by primary liver damage (Council for Agricultural Science and Technology. However, changes in extrinsic coagulation factors as determined by increased fibrinogen concentration were also reported in lambs. In addition, in broiler chicks, other reported clinical signs of aflatoxicosis were glomerular hypertrophy, hydropic degeneration of tubuler epithelium in kidneys and increases in the number of mesengial cells, as well as atrophy and lymphoid depletion in the thymus and bursa of Fabricius.
- Even in chronic aflatoxicosis, most of the effects can be attributed to hepatic injury but with milder symptomes and icterus can also be observed. The pathological alterations in the liver mostly consist of degenerative changes and circulatory disturbances and also include a yellow to brassy color, enlarged gall bladder, diluted bile, histological signs of fatty changes in the hepatocytes, bile duct proliferation and periportal necrosis. In chronic aflatoxicosis, the signs are so protean that the episode may go undiagnosed for long periods of time. Because aflatoxins metabolized in the liver, the histological changes are observed primarily within this organ. Not surprisingly, centrilobular hepatic necrosis or hepatocellular vacuolar change and bile duct proliferation are consistent lesions in cow, sheep, goat and swine. Hepatic fibrosis has been reported in all species when the animals did not die from acute aflatoxicosis. In Piedmontese calves, an outbreak of hepatic encephalopathy consequent to aflatoxin intoxication is to be mentioned: neurological signs varying from comatose or depressed mental status, spinal hyporeflexia, wasting and proprioceptive deficits, and compulsive behavior characterized by anteropulsion and right circling in large circles.

- Aflatoxins are also carcinogenic in animals and aflatoxin B1 is the most powerful liver carcinogen known for rats. AFB1 and AFG1 possess an unsaturated bond at the 8,9 position on the terminal furan ring (Fig.s 1, 4), and epoxidation at this position results in a reactive species, which induces oxidative stress of tissues, depletes antioxidants, forms DNA adducts and, hence, initiates malignant transformations. AFB2 and AFG2 are relatively less toxic unless they are metabolically oxidized first to AFB1 and AFG1 in vivo. Chronic exposure to low doses of aflatoxins is one of the major risk factors in the etiology of hepatocellular carcinoma, and all animal models exposed to AFB1 have developed this type of cancerous desease thus far.
- Aflatoxins have been reported to cause other malignancies as well, including adenomas of esophagus, trachea, kidney and lungs, carcinoma of the pancreas and osteogenic sarcomas. However, the carcinogenicity in farmed animals cannot be detected because of the relatively short period of time, in which the animals are fed prior to marketing. In addition, the chronic form of aflatoxicosis includes teratogenic effects in animals, which are associated with congenital malformations and, in the fetuses, multiple skeletal anomalies as incomplete ossification of skull bones and failure of ossification of long and flat bones, as well as delay in the intramembranous ossification process, defects in the vertebrae formation or their reduction in size. Other mutagenic effects of aflatoxins cover mutations in genes, alterations of DNA by chromosomal breaks, rearrangement of chromosome pieces or even acquisition or loss of entire chromosomes.

10.7.2 Immunotoxic Effects

- Although aflatoxins are primarily known as hepatotoxins and hepatocarcinogens, they have notable immunotoxic effects as well making animals more susceptible to many bacterial, viral, fungal and parasitic infections, as well as to the reactivation of chronic infections or reductions in vaccine and therapeutic efficacies.
- Poultry (chickens and turkeys), pigs and in particular lambs are susceptible to induced immunosuppression due to aflatoxin exposure. Aflatoxins could impair both the cellular and humoral immune systems. In vitro and in vivo studies have demonstrated that AFB1 is immunotoxic, exerting its action particularly on cell-mediated immunity through (i) reducing the number of circulating lymphocytes, (ii) the inhibition or suppression of lymphocyte blastogenesis, (iii) impairing both cutaneous delayed-type hypersensitivity and graft versus host reaction and (iv) the modification of the activities of natural killer cells and of macrophage functions through the inhibition of phagocytosis, the expression and secretion of cytokines (TNF-α, IL-1β, IL-6, IL-10, and IFN-γ), and also by reducing intracellular killing as well as the spontaneous production of oxidative radicals.

- The general mechanism of the immunosuppressive effects of AFB1 appears to be directly associated with the impairment of the synthesis of proteins. In fact, AFB1 is transformed in vivo into metabolites, which are able to bind actively DNA and RNA, to impair the activity of DNA-dependent RNA polymerase and also to inhibit the synthesis of both RNA and proteins. These inhibitory mechanisms have direct and indirect effects on the proliferation and differentiation of the lymphoid system cells and on the synthesis of cytokines involved in the regulation of the immune system. An alteration of the inflammatory responses with a reduced synthesis of pro-inflammatory cytokines and an increase of anti-inflammatory cytokines was reported in weanling piglets fed for 4 weeks with low doses of aflatoxin. The effects of aflatoxins on humoral immunity are not so clear as their effects on cell-mediated immunity, and these differences are hardly recognizable between the different animal species unless higher doses of aflatoxins were introduced (Council for Agricultural Science and Technology). Suppression of humoral immunity has also been recorded after observing decreases in lymphocyte infiltration, hemagglutination and in serum protein levels. In pig, no major effects on humoral immunity were observed after AFB1 exposure but delayed and decreased ovalbumin-specific proliferation, suggesting an impaired lymphocyte activation. However, a biphasic effect of AFB1 was shown in piglets and broiler chicks, with immunosuppressive effects observable during acute exposures and with inflammatory response with stimulatory effects depending on the doses, more precisely, low doses of AFB1 caused immunosuppression meanwhile high doses of it stimulated the immune system. In details, piglets showed decreased leukocyte counts when exposed to low AFB1, and an increase in leukocytes with a high dose. This immunotoxic effect has significantly disadvantageous consequences on the health of farmed animals, via increasing both the susceptibility and the severity of infections like coccidiosis, salmonellosis and Cryptosporidium bailey infections in chicken, Erysipelothrix rhusiopathiae, Brachyspira hyodysentariae, and Escherichia coli infections in pigs, the reactivation of chronic infection by Toxoplasma, and the impairment of vaccination efficacies for Bordetella bronchiseptica and E. rhusiopathiae or with the model antigen ovalbumin in swine, as well as for fowl cholera and Mareck disease in chickens and/or turkeys.

10.7.3 Nephrotoxic Effect

- Renal damages have also been reported after long-term administration of aflatoxins with the symptoms of inflammation, cell necrosis, and toxicosis, which may increase the weight of kidneys and may induce congestion in renal sinusoids. The kidneys are one of the target organs of aflatoxins, and their toxicity is activated by oxidative stress that alters the expression of proline dehydrogenase reducing the proline levels, which induces downstream apoptotic cell death. Moderate focal to diffuse necrosis in the renal tubules and increased renal tubular cells, which may be filled with bile pigments, hyaline, and lipid, with occlusions of their lumens with local edematous changes were reported in the kidneys of aflatoxin-exposed rats.

- In poultry, the toxic effects exerted by AFB1 on renal functions included reduced concentrations of calcium, inorganic phosphate, sodium, and potassium and increased levels of urea, creatinine and uric acid. In addition, AFB1 was reported to cause severe heart damage with tachycardia, tachypnea and even death, although the exact mechanism of cardiotoxicity has not been completely known.

10.7.4 Reproductive Effects

- Not so long time ago, the harmful effects of aflatoxins on animals did not include any direct impairment of reproduction but indirect effects through other physiological systems have been considered. Nevertheless, more recent animal studies suggested that aflatoxins should also induce direct reproductive toxicity in both male and female animals based on adverse effects to both spermatozoa and oocytes. Following aflatoxin exposure in utero, monitoring growth parameters in baby animals indicated growth retardation, reduced fetal or egg weights and reduced fetal lengths of the offspring animals. In piglets exposed to maternal aflatoxicosis, growth retardation, thymic involution and impaired peripheral immune efficiency were events frequently reported and leading to early death, whereas broiler hens exposed to aflatoxin resulted in embryonic mortality and lowered the immunity in the progeny chicks.
- In addition, aflatoxins also possess spermatotoxic effects, which have an impact on the morphology and physiology of spermatozoa: AFB1 affects the male reproduction system altering spermatogenesis as well as epididymal and Leydig cell functions, and also reducing the production of testosterone and the fertility in rats, birds and cattle. In females, AFB1 reduces the fertility of oocytes by the disruption of oocyte maturation through epigenetic modifications as well as oxidative stress, excessive autophagy and apoptosis. In addition, in poultry, worsening egg production and quality, together with the deposition of aflatoxins residues in the eggs are described in both acute and chronic aflatoxicoses. The lowered egg production was attributed to the aflatoxins' effect on liver metabolism and function as well as liver lesions in layers, to the inhibited synthesis of proteins and lipogenesis, and to decreased feed intake and digestibility. It is well-known that aflatoxin causes alterations in the carbohydrate metabolism and impairments of the lipid transport, which effects result in decreased glucose levels and reduced lipid accumulations within hepatocytes, as well as pathological alterations in serum biochemistry and of most coagulation factors have been described in poultry, pigs, cattle and rabbits.

10.7.8 Gastrointestinal Dysfunctions

- Aflatoxins modulate and affect the GIT in multiple ways, the most important of which are changes in the gut morphology, the digestive ability or activity of digestive enzymes, intestinal innate immunity and gut microbiota. Nevertheless, only few reports are available in this field and the presented data are also controversial in many cases, especially for ruminants. The absorption of aflatoxins across the intestinal barrier is maximal in the upper part of the GIT in non-ruminant animals whereas in ruminants, the harmful aflatoxins like AFB1 are transformed to less toxic compounds (e.g., AFM1) or to metabolites with similar or even higher toxicity than the parent molecules (e.g., aflatoxicol). Among the overall adverse effects of aflatoxins, the most significant ones are related to the growth of animals and result in reduced performance. Aflatoxins cause reduced feed intake or even feed refusal with a subsequent decrease in body weight gain, which is determined by direct and/or indirect effects of aflatoxins on the nutrient quality, digestibility and/or absorption. During AFB1 exposure, piglets showed reduced weight gain and Japanese quail have shown a reduction in egg weight. Reduced absorption of nutrients was reported after aflatoxin exposure and, in cattle, this decreased feed efficiency contributed to the observed compromised ruminal function by reducing cellulose digestion, volatile fatty acid production and rumen motility.
- In relation to nutrient digestibility and metabolizable energy, the presence of aflatoxin in dietary was suggested to reduce the apparent digestibility of crude proteins in ducks, to increase amino acid requirements and to reduce energy utilization in terms of net protein utilization and apparent digestible and metabolizable energy in ducks and chickens. Aflatoxins modulate the activity of digestive enzymes but contradictory effects were reported for amylase, trypsin and chymotrypsin activities in pancreas and duodenum with unchanged level of nutrient digestion in the intestine. However, aflatoxins seem to have only moderate affects on or even sometimes do not affect at all the growths of animals through the alteration or modulation of digestive functions, even if, in broiler chicks feed with experimental AFB1 diet, impaired growth, major serum biochemistry measures, gut barrier, endogenous loss, and energy and amino acid digestibility were reported. The effects of AFB1 on intestinal epithelium and microbiota were investigated in some in vivo studies in broiler chicken and rodents. The density of the whole intestine was reduced in the case of low AFB1 doses but at higher doses no such changes were recorded, instead the number of apoptotic cells in the jejunum were elevated, jejunal villi presented lower height, and intestinal lesions were observed in duodenum and ileum, with leucocytic and lymphocytic infiltration. Meanwhile, reduced microbial diversity was observed in the colon

with adverse effects on lactic acid bacteria versus unchanged proportion of Firmicutes and Bacterioidetes.

- Additional symptoms of aflatoxicosis involved malnutrition. In vitro methods and animal models, predominantly, in piglets and broiler chicks, have showed that AFB1 altered bioavailability and distributions of essential metal ions as zinc, calcium, magnesium and potassium, reduced the activities of lipogenic and amino acid metabolizing enzymes leading to reduced lipogenesis, and reduced serum concentrations of 25-hydroxy vitamin D, 1,25-dihydroxy vitamin D and calcium, consequently altering renal functions and parathyroid metabolism.
- Finally, aflatoxicosis in horses showed non-specific clinical signs, such as inappetence, depression, fever, tremor, ataxia and cough. Meanwhile, at necropsy, yellow-brown liver with centrilobular necrosis, icterus, hemorrhage, tracheal exudates and brown urine were observed.
- Qualitative and Quantitative Aflatoxin Analytical Methods - Economic Significance of Analysis Since the massive death of turkeys (Turkey-X deceases) recorded in England in 1960, a wide spectrum of research has been launched and carried out to shed light on the causes of such high mortality. Deciphering the factors leading to Turkey-X disease is a fascinating illustration of how a multidisciplinary approach may help us to solve an important animal health problem. The research covered the development of new analytical tools to measure mycotoxins more precisely, the exploration of the physiological and toxicological effects of these harmful compounds as well as the efficient removal of the toxins and setting up to prevent the manifestation of and to cure the disease itself.
- Mycotoxins are mainly produced on small grains, cereals such as wheat, barley, oats, rye and triticale or on corn but animal products such as milk, meat, liver or eggs can also be contaminated by mycotoxins at various points of the feed and food chain. Because the sampling of feeds and foods for mycotoxin analysis may follow quite different protocols in different laboratories the standardization of these procedures represents a real challenge for analytics. During mycotoxin analysis, extraction and detection are crucially important issues to gain reliable analytical data, which may help us to optimize storage conditions and setting up rules to control mycotoxin production.
- The first step in the analysis is to extract mycotoxins from the sample after correct sampling and sample preparation. The former and traditional extraction methods for aflatoxin analysis gave us a sample matrix in which the HPLC analysis was too complicated to carry out because of the presence of disturbing and interfering components (Kamimura et al., 1985). Later, the clean-up immonoaffinity columns containing gel suspension of monoclonal antibodies gained ground and became popular due its high specificity. The suspension retains the aflatoxin molecules

what can be eluted cleanly, free from any disturbing compounds. Another intention is the extraction with different solvents such as carbon tetrachloride (CCl4), chlorobenzene (C6H5Cl), chloroform (CHCl3), and dichloromethane (CH2Cl2), methanol and acetonitrile.

- Analytical methods of mycotoxin surveillance are wide-ranging and may vary within broad limits across countries. As a result of a community effort having been made to unify surveillance regulations in the European Union, the European Commission (2006) Regulation (EC) No. 401/2006 laid down the requirements for both recovery and precision in different toxin concentration ranges and gives the methodology for the validation of any analytical procedure, making possible to check if it is acceptable for official analysis (EC No. 401/2006). This covers all characteristics required for an analytical method with such a specific sample background, and the list of characteristics ranges from accuracy to measurement uncertainty through the limit of detection.
- An overview on the available analytical methods can be given based on the remarkably abundant literature having been published in this field. We have a plethora of quantitative methods ranging from the different types of Thin Layer Chromatograpy-based to different varieties of HPLC to LC-MS/MS-based methodologies. In addition, we can also find good performance procedures among semi-quantitative methods like ELISA-based or biosensor-based protocols. Emerging technologies include hyperspectral imaging and aptamer-based biosensors.
- The performance parameters of different aflatoxin analytical methods are summarized in Table 5. The different methods can be characterized by several parameters such as accuracy, applicability, reproducibility, limit of detection and so on. Trucksess and Zhang (2016) argued that all practically useful analytical methods should meet the basic guidelines of reproducibility in different laboratory settings. Based on these premises, protocols that are used in different laboratories from sampling to analysis were compiled, and systems relying on certified material samples (CRMs) are also closely related to this.

Table 5. Analytical methods for aflatoxin measurement.

Type of method	Technique	LOD
Quantitative methods	Thin Layer Chromatography combined with scanner	0.1 μg/kg B_2; G_2; M_1, 0.2 μ/kg B_1; G_1;
	High Performance Liquid Chromatography, in combination with fluorescence detector	0.002 μg/kg
	Liquid Chromatography with Mass Spectrometry	0.5 μg/kg

Type of method	Technique	LOD
	Liquid Chromatography with tandem Mass Spectrometry	1 µg/kg
	Ultra High Performance Liquid Chromatography with fluorescence detector	0.02 µg/kg
	Capillary electrophoresis	1 µg/kg 0.1 ng/g
Semi-quantitative methods	ELISA	1 ng/l
	Lateral flow tests LFT	5 µg/kg
	Direct fluorescence	5 µ/kg
	Fluorescence polarization immunoassay	30 ng/ml
	Biosensors	0.05 ml 0.005 µg/l
Indirect methods	Spectroscopy	4 µg/kg
Emerging technologies	Hyperspectral imaging	10 µg/kg
	Electronic nose	5 µ/kg
	Aptamer-based biosensors ECL	0.1 pg/ml

- Currently, a number of HPLC-MS or MS/MS equipment are used world-wide to gain a detailed overview on the mycotoxin spectra in feeds and foods depending on laboratory capabilities. At the same time, ELISA methods and equipment are used for quick mycotoxin measurements. New developments in this field have been published in the latest literature. For example, a novel and promising method has been presented to detect aflatoxin B1, B2 and ochratoxin A in rice starting with dispersive liquid–liquid microextraction followed by LC and fluorescence detection.
- The impact of aflatoxins on human health is far the most important challenge, which we should keep an eye on in the whole feed and food chain. This is the reason for why aflatoxin-related research including analytics is flourishing today. Future research should aim at a deeper understanding of the high-complexity and multi-parameter processes influencing the aflatoxin contents of feeds and foods. Novel multilateral approaches are definitely needed to control mycotoxins and their disadvantageous agricultural, health care and economic impacts more effectively.

11

AFLATOXINS IN FOOD AND FEED

11.1 INTRODUCTION

Aflatoxins' contamination of crops at pre- and post-harvest conditions can be controlled to some extent by the implementation of good agricultural practices (GAPs), good manufacturing practices (GMPs) and good storage practices (GSPs). Further, the novel processing technologies involving a microwave, UV, pulsed light, electrolyzed water, cold plasma, ozone, electron beam or gamma (γ) irradiation in combination with either biological, physical, chemical or genetic engineering methods have the potential to improve the efficiency of AFs decontamination as well as to overcome the limitations of any specific technology. . Therefore, utilization of the novel technologies along with raising public awareness for implementing GAPs, GMPs and GSPs are crucial for controlling AFs contamination in food and feed to ensure food safety and security and to safeguard human and animal health.

Aflatoxins produced by the Aspergillus species are highly toxic, carcinogenic, and cause severe contamination to food sources, leading to serious health consequences. Contaminations by aflatoxins have been reported in food and feed, such as groundnuts, millet, sesame seeds, maize, wheat, rice, fig, spices and cocoa due to fungal infection during pre- and post-harvest conditions. Besides these food products, commercial products like peanut butter, cooking oil and cosmetics have also been reported to be contaminated by aflatoxins. Even a low concentration of aflatoxins is hazardous for human and livestock. The identification and quantification of aflatoxins in food and feed is a major challenge to guarantee food safety. Therefore, developing feasible, sensitive and robust analytical methods is paramount for the identification and quantification of aflatoxins present in low concentrations in food and feed. There are various chromatographic and sensor-based methods used for the detection of aflatoxins. The current review provides insight into the sources of contamination, occurrence, detection techniques, and masked mycotoxin, in addition to management strategies of aflatoxins to ensure food safety and security.

Food safety and security are among the major problems in the current climate of increasing population. These are mainly determined by three key aspects viz., (i) enough food availability, (ii) access to safe food and (iii) utilization of the food in terms of quality, nutritional and cultural purposes for a healthy life. The failure of any of these aspects leads to food insecurity and malnutrition that further influences human health, in addition to the socio-economic aspect of society. In addition, food and feed contamination by mycotoxins are one of the key factors responsible for creating food insecurity.

As per the Food and Agriculture Organization (FAO), one-fourth of the world's crop is affected by mycotoxins. The three main genera of fungi producing mycotoxins are *Aspergillus*, *Fusarium*, and *Penicillium* . Among various type of mycotoxins, aflatoxins (AFs) are highly toxic and are known to contaminate a wide variety of foods such as maize, groundnuts, dried fruits, meat and milk-based products. AFs are produced by *Aspergillus* species, namely *A. flavus*, *A. nomius* and *A. parasiticus* , in addition to its production by other species of *Aspergillus* like *A. astellatus* . These fungi usually grow in the warm and humid conditions of tropical and subtropical regions. Food processing techniques are not sufficient to eliminate AFs from contaminated food and feed due to their heat resistant nature.

The ingestion of AFs from contaminated food and feed has led to serious health complications in humans and animals. Therefore, different countries have implemented strict regulations for AFs in food and feed to maintain the health of individuals. The safe limit of AFs lies in the range of 4–30 μg/kg for human consumption. The European Union has the strictest standard level with AFB1 and total AFs not beyond 2 μg/kg and 4 μg/kg, respectively, in any product meant for direct consumption. Similarly, the maximum acceptable limit set for AFs in the United States is 20 μg/kg. Besides this, various innovative technologies and control strategies are applied for pre- and post-harvest management of AFs to enhance sustainable agricultural productivity. Though there are numerous publications on AFs in food and feed, the novelty and strength of this review lie with the enlistment of the new methods developed for AFs detection in food and feed with special reference to masked AFs. In addition, the review also focuses on the occurrence, impact of climate change along with the control strategies of AFs in food and feed to ensure food safety and security for healthy living and socio-economic development.

11.2 OCCURRENCE OF AFLATOXINS IN FOOD AND FEED

Aflatoxins are chemically difuranocoumarin derivatives with a bifuran group attached to the coumarin nucleus and a pentanone ring (in case of AFBs) or a lactone ring (in case of AFGs). The four major AFs among the identified 20 are AFB1, AFB2, AFG1 and AFG2. The B-types are produced by *A. flavus* while G-types are produced by *A. parasiticus* . The biosynthesis of AFs consists of 18 enzymatic steps with at least 25 genes responsible for producing the enzymes and regulating the biosynthetic process.

Table 1. Occurrence of Aflatoxins in food and feed around the world.

Country	Food matrix	Aflatoxin	Range (µg/kg)	Detection technique
Turkey	Almond	AFB1	1–13	TLC
Turkey	Butter	AFM1	<0.001–0.100	ELISA
Brazil	Cashew nuts	Total AFs	0.60–31.50	ELISA
United States	Chilies	AFB1	<2	ELISA and TLC
Costa Rica	Corn	Total AFs	24	ELISA and HPLC
Zimbabwe	Corn	AFB1	0.75–26.6	HPLC
India	Corn	AFB1	48–383	HPLC
Serbia	Corn	Total AFs	1.01–86.10	ELISA
Vietnam	Corn	AFB1	1.0–34.80	ELISA
Turkey	Cream cheese	AFM1	0.1–0.70	ELISA
Pakistan	Dried Fruits	AFB1	0.04–9.80	HPLC
Turkey	Feed	AFB1	0–5	LCMS/MS
Turkey	Figs	Total AFs	0.1–28.20	HPLC
Nigeria	Ginger	Total AFs	0.11–9.52	HPLC
Ethiopia	Groundnuts	Total AFs	15–11,900	HPLC
Turkey	Hazelnut	AFB1	0.07–43.60	HPLC
Serbia	Infant formula	AFM1	<0.03–0.02	HPLC
Turkey	Lentil	AFB1	0.57–1.74	HPLC
Turkey	Maize flour	AFB1	0.041–1.12	HPLC
Egypt	Meat products	Total AFs	0.47–2.10	Fluorimeter
Greece	Milk	AFM1	<0.005–0.02	ELISA
Iran	Milk (cow)	AFM1	0.006–0.18	HPLC
Brazil	Milk (cow)	AFM1	0.05	HPLC
Italy	Milk (cow/buffalo)	AFM1	0.004	HPLC
Portugal	Milk (cow)	AFM1	0.005–0.07	ELISA
Japan	Nuts	AFB1	0.17–2.59	HPLC, HPTLC
Saudi Arabia	Nuts	Total AFs	1.0–110	HPLC
Malawi	Nut-based foods	AFB1	0.1–40.60	HPLC
Zambia	Peanuts	AB1	0.015–46.60	HPLC
Taiwan	Peanut products	Total AFs	0.2–513.40	HPLC
Turkey	Red-chili powder	AFB1	0.025–40.90	ELISA
China	Rice	AFB1	0.03–20	HPLC
India	Rice	AFB1	0.1–308	Indirect competitive (icELISA)
Pakistan	Rice	AFB1	0.04–21.30	HPLC
China	Rice	AFB1	0.1–136.80	HPLC

Country	Food matrix	Aflatoxin	Range (μg/kg)	Detection technique
Tunisia	Sorghum	AFB1	0.4–25.1	HPLC
Italy	Spices	AFB1	0.59–5.38	HPLC
Malaysia	Spices	AFB1	0.58–4.64	ELISA
Tunisia	Wheat	AFB1	0.12–18	HPLC
Malaysia	Wheat	AFB1	0.55–5.07	ELISA
China	Yogurt	AFM1	0.05	HPLC
Iran	Yogurt	AFM1	0.006–0.021	HPLC

The occurrence of AFs is common in wide varieties of food and feed. Some of the most affected food and feed include peanuts, nuts, figs, corn, rice, spices and dried fruits. It has been shown that among the tested cereals, 37.6% were at least contaminated by any of the AFs. Though rice is not the high-risk commodity for AFs contamination, but AFB1 besides other mycotoxins have been detected in rice from China, Egypt, India, Iran, Malaysia, Nepal, Pakistan, Philippines, United Kingdom and United States. Therefore, AFs pose serious health issues by their ingestion from contaminated food and feed or by carryover AFs in them.

AFB1, as a potent carcinogen to humans, is associated with serious health complications. It has been a causal factor for liver cancer and acute hepatitis as well as periodic outbreaks of acute aflatoxicosis leading to death as reported with lethal aflatoxicosis in Kenya. AFs are mostly detoxified in the liver which is the reason why liver cancer is rare. After the ingestion of AFB1, a series of metabolic processes converts it to an active intermediate, AFB1-exo-8,9-epoxide, by cytochrome P450 enzymes. The detoxification reaction occurs in conjugation with glutathione transferases (GSTs). The detoxification mechanism of AFB1-exo-8,9-epoxide might be linked to the mechanism that prevents liver cancer, however, it is not yet fully. Unfortunately, on the other hand, the food and feed contamination by AFs is a persistent problem worldwide. The outbreaks due to AFs are more prone in tropical and subtropical areas, with a few in temperate regions (like the United States Midwest). In addition, the Mediterranean zones have become prone to AFs contamination due to shifting in traditional occurrence areas of AFs because of climate change i.e., increase in average temperatures, CO_2 levels and rainfall patterns. This has led to an increase in contamination of crops with fungi and AFs worldwide.

11.3 CROPS AFFECTED BY AFLATOXINS

Cereals and cereal-based products are the major foods for human consumption worldwide. Among cereals, rice and corn are mostly contaminated by AFs in natural conditions due to changes in agricultural practices. The AFs are produced both in pre- and post-harvest conditions. As reported rice to be more prone to AFs contamination as compared to other cereals. The fungal growth occurs due to improper drying of rice grains retaining higher moisture content (>14%). As a result, these fungi cause discoloration of grain and/or husk along with deteriorating the quality of the grains. Groundnut and beans, on the other hand, are frequently used in many African diets to supplement cereal diets. However, these are highly prone to AFs contamination both in field and storage conditions. The extent of fungal growth and AFs production in cereals depends on temperature, moisture, soil type, and storage conditions. In addition, spices are susceptible to AFs contamination and are significantly affected by storage and processing conditions. As reported the AFs contamination in a wide variety of spices including black pepper, cardamom, cinnamon, clove, cumin, coriander, and ginger in the Sultanate of Oman. Furthermore, Also reported the presence of AFs in eggs collected from a poultry farm and in raw cow milk in Cameroon. Hence, the affected crops allow AFs to enter the food chain, which is very much influenced by the climatic conditions.

11.4 IMPACT OF CLIMATE CHANGE ON AFLATOXIN PRODUCTION

Climate change significantly impacts on the quality and availability of staple foods for consumption. With the increasing population worldwide, a major emphasis has been put on the safety of food and feed that can address the increasing demand with the increase in the yields by protecting the crops from adverse climatic conditions. Aflatoxins contamination has affected millions of hectares of maize and peanut crops in the United States. Maize is a staple food for people living in warm climates throughout Asia, Africa, and the Americas, which are prone to the influences of climate change. The change in climate simultaneously impacts the complex communities of AF-producing fungi by altering the number of AF-producers to change its fungal community's structure. Aflatoxins contamination occurs via an initial phase during crop development and a second phase during crop maturation. The contamination is greater in warm, humid, and even hot deserts and drought conditions.

A. flavus has highly evolved physiological mechanisms to acclimatize to adverse climatic conditions and dominates other fungal species . Climate change alters the temperature and water activity (a_w) in the environment which further influences the gene expression to produce AFs. The conditions of temperature and a_w regulate the extent of fungal growth and AFs production. The AF-producing genes are grouped on the genome and express the main regulatory genes (aflR; aflS), as well as structural genes (aflD) which are influenced by the interaction of temperature × a_w conditions. As revealed the expression proportion of aflR/aflS significantly correlates with the amount of AFB1 produced. In addition, the expression of sugar transporter genes was significantly affected by the condition of temperature and a_w . Further, studied the effect of interactions of temperature and a_w on the biosynthetic regulatory gene (aflR) expression and production of AFB1 by *A. flavus* in maize. They observed the optimum growth of *A. flavus* at 30°C/0.99 a_w with no growth at 20°C/0.90 a_w. Both temperature and a_w influenced the relative aflR gene expression and AFB1 production, however, the trends for the production of AFB1 were not in accordance with the gene expression. Further, the effect of temperature (20, 27, and 35°C) and a_w (0.82, 0.86, 0.90, 0.94, and 0.98) on the growth of *A. flavus* and *A. parasiticus* along with the production of AFB1 were investigated on ground Nyjer seeds. The maximum AFB1 production was observed at 27°C/0.90 a_w for both *A. flavus* and *A. parasiticus*. In addition to this, the fungi showed optimum growth on polished rice in the range of 28–37°C/0.92–0.96 a_w. The maximum AFB1 was produced at 33°C/0.96 a_w. Based on the investigation on the possible emergence of AFB1 in cereals in the European Union as a result of climate change, for every 2°C increase in temperature, there is an increase in AFs risk in the various regions of Spain, Italy, Greece, Portugal, Bulgaria, Albania, Cyprus and Turkey. The risk for AFs contamination in maize is likely to increase in Europe due to favorable climatic conditions for *A. flavus* in the next 30 years. Therefore, proper detection methods and control strategies are crucial to combat the burning issues of AFs in food and feed.

11.5 DETECTION METHODS

The detection of AFs has been performed by the Association of Official Analytical Chemists (AOAC) official method in food and feed samples. Among the most commonly employed methods are chromatographic methods like thin layer chromatography (TLC), high performance liquid chromatography (HPLC) and liquid chromatography mass spectroscopy (LCMS), besides the enzyme-linked immunosorbent assay (ELISA). However, the drawbacks of these standard methods are that they are unsuitable for rapid and real-time applications in food and feed samples as they are tedious, time-consuming and require skilled personnel to operate. Therefore, rapid and robust methods like polymerase chain reaction (PCR) and non-destructive methods based on fluorescence/near-infrared spectroscopy (FS/NIRS) and hyperspectral imaging (HSI) have emerged for the quick and easy detection of AFs.

Utilized the PCR technique for the molecular detection of AF producing *A. flavus* from peanuts. Similarly, the avfA, omtA, and ver-1 genes encoding the major enzymes in AF-biosynthesis were used as target genes for detecting AFs using multiplex PCR. Further, PCR was employed to detect AF-producing genes in *Aspergillus* species in Iranian pistachio nuts for their aflatoxigenic effect. In addition, utilized PCR, ELISA and HPLC for the detection of AFs from *A. oryzae* isolated from different Korean foods. HSI uses the integration of both imaging and spectroscopy to record spatial and spectral characteristics of a given sample. The visible/near-infrared (VNIR) HSI has been utilized for the identification of maize kernels of different varieties. VNIR or short-wave (SWNIR) HSI techniques are feasible for the detection of AFs as well as identification of different fungal species in maiz.e Later, used the VNIR-HSI system to detect AFB1 on surfaces of maize kernels from Georgia, Illinois, Indiana and Nebraska of United States. Also used short-wave infrared (SWIR) HIS to detect AFB1 in single maize kernels. But as the image quality could not effectively classify AFB1 level qualitatively in individual maize kernels, therefore, to improve this further combined the SWIR-HSI system with chemometric data analysis for the better detection of AFB1 on the surfaces of maize kernels. Furthermore, the color-encoded lateral flow immunoassay (LFIA) technique has been used for the simultaneous detection of AFB1 as well as fumonisins in a single test line. Further enhance the sensitivity and detection of AFs in food and feed, nanoparticles (NPs) based on Au/Ag, carbon (CBNs), magnetic (MNPs), Quantum dots (QDs), up-conversion (UCNPs), metal-organic frameworks (MOFs) as well as hybrid nanostructures have been utilized. As prepared molecular imprinted polymers (FDU-12@MIPs) using structural analog of AFs. This highly selective surface was used as an extraction sorbent in conjunction with HPLC for the detection of AFs in different food and feed samples. In addition to this, the use of biosensors compared to other spectrophotometric or chromatographic methods allow for higher selectivity, direct detection with minimal sample pretreatment, minimal cost, portability and on-field analysis of mycotoxins. Also utilized an electrochemical enzyme-linked oligonucleotide array for easy and quick multi-detection of AFB1 in maize. Furthermore, assays based on aptamer have been developed for the rapid detection of AFB1. As successfully detected the AFB1 spiked in wine, methanol and corn flour samples using the simple aptamer molecular beacon assay, which has the potential for the rapid detection of AFs in the food and feed.

11.6 MASKED MYCOTOXINS AS A MAJOR CONCERN IN DETECTION

Masked mycotoxins pose a major concern in food and feed as they are not identified and detected by the usually employed detection techniques. These are the mycotoxins produced by fungi but are modified by plant enzymes during the infection stages. They are present in vacuoles in the soluble form or bound to macromolecules, therefore, are unable to be identified by routine analysis processes and referred to as masked mycotoxins. However, the modified AFs can hydrolyze back into the toxic forms during food processing and/or digestion process. Some of these modified toxins are present in different forms as complexes with matrix compounds, hence also referred to as matrix-associated mycotoxins. The masked mycotoxins have been reported to occur in Asia, Africa, America and Europe. Therefore, a high amount of masked mycotoxins prevailing in various food and feed can pose serious health issues to both humans and animals. Therefore, the detection of masked mycotoxins is an essential part to ensure food and feed safety. Masked fumonisins were determined through hydrolysis where modified forms were converted back to their free forms and subsequently analyzed and detected through LC/MS/MS. The hydrolytic process may involve either alkaline, acidic or enzymatic treatments. However, there is less information available on the masked AFs as most of the preference is given for the detection of free AFs in agricultural food and feed. Therefore, methods like *in vitro* digestion and hydrolysis, as applied in case of masked fumonisins, can be carried out for masked AFs in food and feed followed by detection with LC/MS/MS and confirmation by other methods like ELISA to ensure the food and feed safety.

11.7 CONTROL STRATEGIES OF AFLATOXINS

Implementation of advanced agricultural technologies, good agricultural practices (GAPs), good manufacturing practices (GMPs) and good storage practices (GSPs) can mitigate the mycotoxins contamination. The novel processing techniques involving a microwave, UV, pulsed light, electrolyzed water, cold plasma, ozone, electron beam and gamma (γ) irradiation treatment have the potential for AFs management and preserving and maintaining the quality of agricultural and food products. The application of ozone degrades AFs by an electrophilic attack on the double-bonded carbons (C8-C9) of the furan ring resulting in the formation of primary ozonides followed by rearrangement into monozonide derivatives like aldehydes, ketones and organic acids. The application of ozone for the degradation AF is limited in food products due to the cost factor. Similarly, the mechanism behind the AF degradation by gamma rays lies on the effects of free radicals produced during the radiolysis of water and other components that attacks the terminal furan ring of AFB1 resulting in byproducts of reduced biological activity. The degradation efficiency of gamma irradiation is more effective when combined with other technologies.

In addition to these, several synthetic and natural food additives have been studied for AFs reduction in food and feed. For examples, the use of citric acid in combination with moisture under high temperature (200°C) and pressure (8N) was effective in degrading AFs in extruded sorghum. On the other hand, the efficacy of sodium hydrosulphite ($Na_2S_2O_4$) was enhanced with increased pressure for AFs reduction in black pepper. Furthermore, as a part of biological control measures, It also reported that inoculation of antagonistic strains of *Pseudomonas, Bacillus* and *Trichoderma* spp. had a significant reduction of *A. flavus* in pre-harvest crops. The non-aflatoxin forming strains of *A. flavus* and other non-toxigenic molds are prominent biological control agents against AFs contamination. The application of each technique has its advantages and disadvantages. Therefore, biocontrol measures in synchrony with other physical and chemical methods along with improved packaging materials should be implemented to attain food safety and security.

12

MYCOTOXINS AND THEIR CONSEQUENCES IN AQUACULTURE

12.1 INTRODUCTION

- With the world's population expected to reach 8.2 billion people by 2030, and with 842 million people estimated as having been under- nourished in the period 2011–13, food supply will present a growing challenge in the next two decades. With increases in income along with demographic changes related to family size, population aging and urbanization, and consumer trends such as concerns for healthy eating and sustainable production, there will be great shifts in demand and major changes in the composition of demand. This scenario will in turn have an impact on food supply, which will need to increase and be- come more efficient if it is to grow within the constraints presented by the availability of natural resources and existing technology.
- Currently, fish consumption by population has been increasing worldwide, mainly due to the availability, access and price in relation to meat consumption, such as beef, pork, and poultry. Consequently, some concerns begin to emerge, primarily regarding the quality of fish available in the market. Chemical residues could be present in any prod- uct of animal origin causing economic losses and putting into a risk human and animal health. According to Bostock et al. (2010), the aqua- culture contributes nearly half of all food of aquatic origin intended for human consumption, as a vital part of the global food industry.
- Global production of farmed fish, shrimp, clams, and oysters more than doubled in weight and value during the 1990s while landings of wild-caught fish remained level. Fish and other aquatic animals represent an important food source for animal and human consumption; so, this demand had led to a fast development of aquaculture. The first important point is to control the feeding of fish produced,

and obviously, control fungal and mycotoxins contamination to reduce economic losses, and to min- imize hazards to human health. As aquaculture production continues to expand and intensify, both its reliance and its impact on ocean fisheries are likely to increase. The balance between farmed and wild-caught fish, as well as the total supply of fish available for human consumption will depend on future trends in aquaculture practices. In- creased aquaculture production has the potential to influence wild fish stocks via increased demand for feed.

- Mycotoxins are toxic metabolites produced by a diverse group of fungi that contaminate agricultural crops prior to harvest or during storage post-harvest and different species including humans, poultry, swine, and fish all exhibit varying levels of mortality and morbidity upon the exposure to these harmful substances. Maintaining a safe global food and feed supply is a critical issue facing society. Natural contaminants, especially mycotoxins, pose a challenge since they are found in a wide range of crops and differ significantly in chemical structure and symptomatology in humans and signs of disease in animals following exposure and can exhibit a broad range of effects including carcinogenicity, neurotoxicity, and developmental toxicity.
- In addition to public health, the presence of undesirable and dangerous substances also limits or reduces the marketing of food products in international markets. Therefore, international or- ganizations, like Food and Drugs Administration and European Commission have established maximum residue limits (MRLs) for mycotoxins. At this moment, many countries have established legislation to reduce exposure to mycotoxins, but based on scientific risk assessment, and which allows small amounts of mycotoxins in foods or feeds, if these levels are lower than what is confirmed to not affect human's and animal's health.
- According to FAO, Food and Agriculture Organization of the United Nations (2004), the development of these regulations can be influenced by both scientific and socioeconomic factors including:
    - the availability of scientifically sound toxicological data;
    - availability of occurrence data in commodities;
    - knowledge of the distribution and concentrations of toxins in commodities;
    - availability of detection methods, including conformational and analytical; and
    - governmental legislation amid countries where trade contracts exist and
    - the need for a sufficient food supply. Because of their ubiquitous nature and the fact that current standards focus on regulating the product rather than the process, mycotoxin contamination of food and feed is unavoidable.

- Food contamination by mycotoxins is a risk to human and animal health, and it is responsible for significant economic losses. It is reported that these losses are supported by all participants along the chain of production animal, or animal producers, grain handlers and distributors, processors of crops, but also by consumers in society. An important element to reduce this type of contamination is prevention, because several products can be victims of contamination along the chain of livestock production and it is not easy to identify the contaminated product.
- Mycotoxins represent a serious problem in livestock production worldwide. Its effects — including reduction of weight gain and feed efficiency worsening compromise the overall health of the animals, causing bruises on the carcass, liver and kidney damage, which can result in serious economic implications to farmers.
- It is difficult to prove that a disease is a mycotoxicosis, and even when mycotoxins are detected, it is not easy to show that they are the etiological agents in a given veterinary or human health problem. Although it is hard to define, there is sufficient evidence from animal models and human epidemiological data to conclude that mycotoxins pose an im- portant danger to human and animal health. In summary, in the absence of appropriate diagnostic criteria and reliable laboratory tests, the mycotoxicoses will remain diagnostically daunting diseases. An approach on the occurrence of mycotoxin-contaminated animal feed, as well as toxic effects that mycotoxins may produce in fish and their residues in meat and organs are outlined in this chapter.

12.2 FEED CONSUMPTION IN AQUACULTURE AND MARKET

World fish production has experienced tremendous growth, in- creasing from 20 million tonnes in 1950 to 156.2 million tonnes in 2012, of which 97% was used for direct human consumption. Per capita fish consumption increased from 9.9 kg in 1960 to 19.1 kg in 2012. The increase in production is attributed predominantly to aquaculture, which has maintained high growth rates since the 1980s. By 2012, aquaculture production had increased to 66.5 million tonnes, or about 43% of total fish supply. Productivity growth and technological progress have been essential factors underlying production growth in aquaculture.

Fishery resources are an important source of both macronutrients and micronutrients for humans. Even though globally fish accounts about 17% of animal protein intake, there is a significant difference in consumption between countries; low-income food-deficient developing countries consume on overage 10.1 kg per capita while industrialized countries consume 28.7 kg per capita. However, when considering some economically disadvantaged countries, fish contributes more than 50% of animal protein intake. In

2010, the two species of fish most produced in the world during the period of 2002–2010 were grass carp and silver carp, with a production of 4,337,114 and 4,116,835 tonnes, respectively. Other species produced in high-scale in the world aquaculture are Catla, Japanese carpet shell, Common carp, White leg Shrimp, Bighead carp, Nile tilapia, Crucian carp, Atlantic salmon, Robo Labeo and Milk fish. The countries that produce fish or other aquatic organisms are China, India, Vietnam, Indonesia, Bangladesh, Norway and Thailand. Other important producers are Egypt, Chile, Myanmar, the Philippines, Brazil, Japan, the Republic of Korea, the United States of America, Taiwan Province of China, Ecuador, Malaysia, Spain and Iran.

Whereas the FAO estimates in relation to food production and changing habits of the people, there is a necessity to encourage the production of aquatic organisms. Feed expenses are the largest part of the total cost in any livestock production, and it is not different in intensive aquaculture. Therefore, the management and food quality are of great importance in fish nutrition. Processing technology has a direct impact on feed quality in terms of physical properties and nutrient digestibility. Extrusion technology has become more common than conventional steam pelleting since the beneficial effects of extrusion processing is been well demonstrated in fish, such as the improvement in carbohydrate utilization, water stability, feed efficiency, and protein digestibility.

Among the various factors that contribute to fish nutrition, there is that one which depends directly and indirectly from the environment where they live, like subject to environmental conditions and health management difficulty. There are some differences in requirements between the cold-water fish (optimum temperature b 18 °C) and hot water (optimum temperature N 18 °C). Studies have shown that diet influences behavior, structural integrity, health, physiological functions, reproduction and growth of fish. For that reason, the determination of qualitative and quantitative requirements of essential nutrients is of fundamental significance for an adequate formulation of diets for fish.

The fingerlings between 5.0 and 20.0 g, after the first month, can be fed with extruded feeds and subsequently crushed in proportion of 5 to 8% of live weight and ration be provided at least four times a day. The fish in the juvenile state (between 80 and 250 g), should be fed three or four times a day at a rate of 3 to 5% of live weight, with extruded feed of 4 to 5 mm in diameter and grow out phase. With fishes over 250 g, the ration must be provided at a rate of 2 to 3% of live weight with 6.0 up to 10 mm in diameter. This same ration must continue until the final phase of fattening (finish), with 1 to 3 kg live weight, when fish ingest, 1 to 2% of live weight.

The ingredients used in fish feed are made up basically of grain and its by-products, pies and bran of oilseeds, flour and bran of tubers (cassava), brewery waste, crushed, miscellaneous and hay products of animal ori- gin (fish meal, bone, flesh and bone, flesh and blood). Approximate com- position of ingredients used in fish feeds must be known, because their chemical and nutritional qualities are relevant for their efficiency.

According to European Commission (EC) (2013), the aquaculture production does not pose any problem in relation to compliance with the ban on recycling within the same species to the ex- tent that existing ducting requirements for the use of fishmeal in feed for the aquaculture animals have already proved their effectiveness.

There are multiple criteria for assessing the economic impact of mycotoxins on humans and on livestock production, but the considerations include loss of human and animal life, health care and veterinary care costs, economic loss in livestock production, forage crops and feeds, reg- ulatory costs, and research cost focusing on relieving the impact and severity of the mycotoxin problem. Formulas for worldwide economic impact have been difficult to develop and, therefore, most reports on economic impact are on a single aspect of mycotoxin exposure or contamination.

However, using a ration of quality, namely, quality raw materials and no contamination by mycotoxins or pesticides, but also adopt proper food management within each breeding may improve the adverse effects to the aquatic environment. As a result, it will improve the cost–benefit ratio, likewise results in a product of great quality for human consumption.

12.3 OCCURRENCE OF MYCOTOXINS IN FEEDS FISH

Mycotoxicoses in humans or animals are characterized as food or feed related, non-contagious, non-transferable, non-infectious, and non-traceable to microorganisms other than fungi. Clinical symptoms usually subside upon removal of contaminated food or feed.

Feed supply is central to all animal production systems and any factor that affects the security of the feed supply is a significant constraint to production. The primary objective in fish nutrition is to provide a nutritionally balanced mixture of ingredients as finished feed to support the maintenance, growth, reproductive performance, flesh quality, and health of the animals at an acceptable cost. The quality of the products used in feeds for farmed fish has become a limiting aspect for activity because these feeds are ideal substrates for the growth of fungi, which, under positive conditions, may favor the synthesis of mycotoxins. The production of these toxic metabolites can occur during the growth of the crop, during post-harvest storage, or during the storage of the compounded feed. Field contamination depends strongly on several factors, including climatic conditions (rainfall, temperature, and humidity), genetic susceptibility of cultivars to fungal infection, soil type, nutritional factors and post-harvest storage.

Experts estimate an annual loss of 140 million dollars due to reduced weight gain of poultry in the United States due to contamination of feed by mycotoxins in Australia, the loss in the poultry farming is of 4 to 20 million dollars. Latest data estimate that annual losses due to ingestion of mycotoxin-contaminated animal feed in the United States and Canada, are on the order of $5 billion.

The regulatory limits for mycotoxins are extremely varied, because it depends on the toxicity of each mycotoxin and the possible combinations; the legislations of each country; the type of cereal, grains, feed or food; and the geographical and ecological distribution of the myco- toxins produced. However, many countries follow FDA guidelines, ESFA, CODEX alimentarius and EC for granted the acceptable level of mycotoxins. These guidelines suggest that the acceptable level for aflatoxin (AFB) in animal feeds and ingredients is 20 ppb, excluding cottonseed meal (300 ppb — its acceptable if the feed is for beef cattle, swine and poultry), and the acceptable levels for fumonisin (FB1 + FB2 + FB3) in animal feed for equids and rabbits is 5 ppm (no more than 20% of diet); for swine and catfish is 20 ppm (no more than 50% of diet); for breeding ruminants, poultry and mink is 30 ppm (no more than 50% of diet); for ruminants N 3 months old and mink being raised for slaughter is 60 ppm (no more than 50% of diet); for poultry being raised for slaughter is 100 ppm (no more than 50% of diet) and for all the other species or classes of livestock and pet animals is 10 ppm (no more than 50% of diet).

Petterrson (2004) affirms that mycotoxins in cereals were found in pre-harvest (barley: deoxynivalenol [DON], nivalenol [NIV], zearalenone [ZEN], HT-2 toxin [HT-2], T-2 toxin [T-2]; maize: DON, fumonisin [FB], ZEN; oats: DON, NIV, HT-2, T-2; wheat: DON, NIV, ZEN, ergot; rye and sorghum: ergot) and post-harvest (barley: ochratoxin [OTA], AFB, citri- nin [Cit]; maize: ZEN, AFB; oats: OTA, AFB, Cit; wheat: OTA, AFB, Cit; rye: OTA and sorghum: AFB).

In Table 1, is possible to observed the incidence of mycotoxins contamination of grain (maize, oats, wheat, barley; n = 4327 samples) around the world. Considering the data from Table 1, it can be affirmed that the occurrence of AFB1 found in countries like Middle East, Asia and Africa is very high and worrying. The incidence of ZEN in Northern of Asia; DON in North America and Northern of Europe and China; FBs in South America, Africa, Europe, Asia and the Middle East, but also, OTA in the Middle East were present in more than 50% of the samples analyzed.

Ninety four samples of fish feeds were collected at different locations in Nigeria and the highest incidence of fish feed contamination was re- corded for fumonisin B1 (FB1), fumonisin B2 (FB2),, aflatoxin B1 (AFB1), and ZEN with mean of 900.9, 220.6, 103.0 and 4.5 μg/kg, respectively. In Central Europe, it is reported presence of FBs and ZEN in 11 samples of commercial fish feeds. These are evidences that fish feeds are carriers of multi-mycotoxins as they are majorly formulated from plant sources that are good substrates for mold proliferation and subsequent mycotoxin production.

Table: 1 Incidence of mycotoxin contamination of sample of grain (maize, oats, wheat, barley) in the world according to Biomin (2011, 2014), during the years 2009–2010 (*) and 2012–2013 (**), respectively.

Region/years	AFB_1 (%)		ZEN (%)		DON (%)		FBs (%)		OTA (%)	
	09/10	12/13	09/10	12/13	09/10	12/13	09/10	12/13	09/10	12/13
North America	21	26	14	9	50	33	27	55	21	2
South America	13	12	28	39	21	33	76	76	16	2
Africa	50	67	8	26	17	67	58	78	12	56
Europe										
Northern	nd	32	25	34	71	87	nd	86	nd	40
Central	19	29	41	26	64	66	51	36	20	28
Southern	33	55	14	18	36	50	56	71	41	46
Asia										
North	15	14	63	57	83	79	51	47	25	15
South	88	59	14	26	22	36	56	57	49	55
South-east	71	59	37	49	34	79	55	47	28	15
Middle East	37	30	nd	19	11	46	67	78	50	31
Oceania	6	4	26	19	49	24	12	16	11	13

AFB = aflatoxins; ZEN = zearalenone; DON = deoxynivalenol; FB = fumonisin; OTA = ochratoxin A; n = 4327 (2009/10) and n = 4200 (2012/2013).

Sixty pelleted samples of fish feed were sampled in Brazil ,of total samples proved to have AFB1 (55%) and FB1 (98%), 50% showed co-occurrence of these mycotoxins. The 3.3% of the samples presented co-occurrence of the three mycotoxins analyzed (AFB1, FB1 and OTA). The simultaneous occurrence of FB1, AFB1, and OTA emphasizes the need for further research in the area to better assess the risk to the health of fish farms and their implications for the health of consumers of this meat. It also reported that AFB1 could be detected in 96.1% of the 334 tested commercial feeds and raw materials collected from Asia.

It is analyzed that the mycotoxin occurrence in final poul- try feed samples in Brazil and showed that 100% were contaminated with AFB1 and OTA. Aflatoxin levels varied between 1.3 and 80 μg/kg (65.3 ± 18.7 μg/kg, mean ± SD). OTA levels varied between 17 and 197 μg/kg (98.2 ± 22.3 μg/kg, mean value ± SD). In a study in 9 of the 65 samples in feed, traces of OTA were found, and the results indicated that the OTA concentrations were 2.3 μg/kg in feedstuffs, 5.9-12,1 μg/kg in feed samples and 2.5–16.8 μg/kg in con- centrated feed samples from China.

It also related that the quantitative analyses were ac- complished by HPLC-fluorescence detection and ZEN values in fish- feed did not exceed the guidance (not exceed 2 mg/kg of ZEN) from European Commission (EC) (2006).

Considering the results found with the presence of mycotoxins in animal feed by the authors above, it is necessary to review the policies and laws of the various countries to standardize and improve the quality of the feed fish used in fish farming, with the goal of reducing the public health risks. Performance losses of 5–10% are typical with consumption of moldy feeds even in the absence of mycotoxins. These results also show that a greater prevalence of contamination of feed ingredients mycotoxins occurs most in developing countries. Concerning the scarciety of information about the safe amount of mycotoxins in fish for human consumption, it is crucial to worry about providing a quality feed to these animals and mainly as low levels of mycotoxins.

12. 4. TOXIC EFFECTS OF MYCOTOXINS ON FISH

The toxic effects of mycotoxins in animals vary according to the type and the amount of toxin, and period that the animal was exposed. Mycotoxins have a range of other health effects including carcinogenic effects, kidney damage, gastrointestinal disturbances, reproductive dis- orders or suppression of the immune system.

By the way, the toxins as AFB1 can be genotoxic and carcinogenic, and the affected organ is the liver in animals. The toxic effect of OTA is acute, and could lead to chronic kidney cell damage and impaired function. FB, in human, may be relationship the etiological agent of esophageal cancinoma . Trichothecenes can cause chronic effects in animals, in- cluding suppression of the immune system. Estrogenic syndrome (hyperestrogen and vulvovaginitis) is the major toxic effect of ZEN as its structure resembles that of 17-beta-estradiol, an estrogen hormone.

Mycotoxins can be responsible for the induction of many disorders in fish, such as induce to cell and organ alterations, produce functional and morphological effects and in more severe cases mortality. Thereby, losses in aquaculture caused by mycotoxins in feed can be significant, resulting in economic loss on fish production.

In studies of toxic effects on fish, there is not a standard for the amount or time of exposure to toxins produced by fungi, and that is an important factor to describe the possible toxic effects and consequences; therefore, there is a need to standardize the chronic tests, mainly in relation to the duration of the test. In Tables 2–3, it can be ob- served the various effects of mycotoxins reported in fish, as well as the related doses and time that fishes were exposed. In addition, it can be analyzed by the tests described in Table 2, that the AFB is the mycotoxin more studied in fish, possibly because its natural occurrence being most common and is a know human carcinogen, probably one the most potent know.

Table 2: Toxic effects of aflatoxins related in different species of fish.

Mycotoxin	Specie	Dose	Administration	Toxic effect
Aflatoxins	Nile tilapia (*Oreochromis niloticus*)	100 μg AFB_1/kg	Feed — oral (10 weeks)	Reduced the growth
	Nile tilapia (*Oreochromis niloticus*)	200 μg AFB_1/kg	Feed — oral (10 weeks)	Mortality (16,7%)
	Nile tilapia (*Oreochromis niloticus*)	5–38.62 μg AFB_1/kg	Feed — oral (10 weeks)	Reduction in the rate of survival by up to 67%
	Nile tilapia (*Oreochromis niloticus*)	29 μg AFB_1/kg	Feed — oral (10 weeks)	Yellowing surface
	Nile tilapia (*Oreochromis niloticus*)	200 ppb AFB/kg	Feed — oral (10 weeks)	Total erythrocyte count, total leucocyte count, hemoglobin count decreased; Weight gain lowest and reduction in the rate of survival
	Tilapia Tilapia Tilapia Tilapia Tilapia Tilapia	793 and 1641 μg AFB_1/kg 793 and 1641 μg AFB_1/kg 2.5 mg AFB_1/kg 2.5 mg AFB_1/kg 245 μg AFB_1/kg 245, 638, 793 and 1641 μg AFB_1/kg	Feed — oral (5 weeks) Feed — oral (15 weeks) Feed — oral Feed — oral (20 weeks) Feed — oral (20 weeks) Feed — oral (20 weeks)	Yellowing of the body surface Darkening of body surface Affect the hematocrit and growth performance Abnormal behavior Feed efficiency rate decreased Weight gain lowest
	Rohu (*Labeo rohita*)	2.50 and 5.00 mg/kg	Intraperitoneal (i.p.)	Reduction of production of
			(10 weeks)	oxygen radicals by neutrophils
	Rohu (*Labeo rohita*)	1.25; 2.50 and 5.00 mg/kg	Intraperitoneal (i.p.) (10 weeks)	Reduction of total protein and globulin levels

Mycotoxin	Specie	Dose	Administration	Toxic effect
	Rohu (*Labeo rohita*)	10, 20 and 40 ppm	Feed — oral (8 weeks)	Total erythrocyte count, total leucocyte count, hemoglobin count and nitroblue tetrazolium decreased
	Juvenile Rainbow trout	1190 µg/kg	Feed — oral (21 days)	Mortality
	Sea bass (*Dicentrarchus labrax* L.)	0.18 mg/kg	Feed — oral (96 h)	Loss of equilibrium, rapid opercular movement, hemorrhages of the dorsal skin surface.
	Sea bass (*Dicentrarchus labrax* L.)	0.018 mg/kg	Feed — oral (6 weeks)	ALT, AST and ALP enzymes increased; Total protein; Albumin and Globulin increasing

Reduced intake of mycotoxins-contaminated feed leading to reduced weight gain was reported in mice and fish but there are also reports of not interfering in the weight gain of fish. So, according to these results, weight gain does not appear to be a sensitive parameter to detect mycotoxin contamination. As the activity of serum ALT and LDH together with lactate concentrations seems to be a sensitive indicator of the fish responses to DON (Table 3). Little is known about the ecotoxicological impact of mycotoxins at levels found in the environment and the consequence of exposure to aquatic organisms. The majority of trials conducted with mycotoxins contamination in fish are of chronic character, contaminated feed management and several different doses as can be seen in Tables 2–3. However, it is important to keep a standard to perform the tests in relation to duration of the experiment and split the toxicology tests on acute, subchronic and chronics, but also consider route of administration and dose according to each type of mycotoxins, mainly to detect toxic effects on fish.

In fact, the toxic effect of mycotoxins may be dependent of the type and quantity of mycotoxin content in the feed, but also depends on the duration of exposure, animal species, sex and age.

12.5 MYCOTOXIN RESIDUES IN FISH AND RISKS FOR PUBLIC HEALTH

In order to protect consumer's safety, rules and safety limits for several mycotoxins in certain foodstuffs are contained in certain laws or regulations by governmental agencies of different countries. For most mycotoxins, a tolerable daily intake (TDI) has been established, which estimates the quantity of mycotoxin that someone can be exposed to daily over a lifetime without significant risk to health.

The metabolism of different mycotoxins in fish is not fully eluci- dated, but there are some papers about AFBs metabolism. So, after ingestion of AFBs by the fish via dietary exposure, it is absorbed from the gastrointestinal tract into the circulatory system owing to it high lipid solubility. In the plasma, the AFBs is carried thorough the portal blood into liver where is largely sequestered by hepato- cytes. The biliary and the urinary systems represent the major excre- tion routes in trout and catfish. AFBs are biotransformed in liver cells, to several types of metabolites by microsomal mixed-function oxidases and certain cytosolic enzymes. Aquatic species have shown dissimular susceptibility to the hepatotoxic and carcinogenic effects of AFB1 that depends on the particular species.

Studied the bioaccumulation of dietary AFB1 for 15 and 20 weeks in muscle and liver of tilapia and they did not detect AFB1 residues in muscle, whereas quite high residues of AFB1 were deter- mined in liver. These results revealed that the residual AFB1 in liver in- creased, as the AFB1 concentration increased in diets, and accumulated progressively as exposure prolonged. After 15 and 20 weeks, the resid- ual AFB1 in liver of fish fed diets containing more than 638 μg/kg was significantly more than the control group, and consequently the residu- al AFB1 after 20 weeks was twice as high as the corresponding value after 15 weeks.

Some authors (Table 4) reported the presence of residues in tissues or organs of fish poisoned with mycotoxins, but still not well regulated what is the tolerable limit of waste for different species of fish, such as different tissues or organs.

Current legislation does not include limits for ZEN occurrence in food of animal-origin, since it has been suggested that the Fusarium myco- toxins are transferred from feed into meat, milk and eggs, only to a very limited extent.

However, the residue of AFB1 in biological tissues has been frequent- ly reported fish. In the majority of these studies, the residual AFB1 level in liver is much higher than in muscle. The most accepted explanation is that the liver plays a prominent role in the metabolism of AFB1. Some authors agree that the residue of AFB1 present in tissues and organs can be due to dose and time-dependent. It is reported that the feeding for a long time of sea bass on low levels of AFB1 induced not only serious health problems in exposed-fish, but also representing a high risk to fish consumers through their residues in fish musculature. In the current study, they describe that Aflatoxin B1 accumulated to a level of 4.25 ± 0.85 ppb in the muscle of sea bass fed AFB1, suggesting a significant risk for transmission of AFB1 to the human consumer, since the US Food and Drug Administration has set a 20 ppb limit for AFB1 in human foods.

It can be assessed in Table 3, in which studies with identification of mycotoxin levels in organs and tissues of fish are presented, that it is necessary to standardize the analysis techniques. It is indispens- able to detect the different mycotoxins in fish tissues and organs, as well as to estimate the risk to public health, determining the safe level of each type of mycotoxins in the different tissues for several species of fish.

Table 3 : Toxic effects of others mycotoxins related in different species of fish.

Mycotoxin	Species	Dose	Administration	Toxic effect
Ochratoxin A	Marine Sea Bass (*D. labrax* L.)	0.29 mg/kg	Feed — oral	Reduction in the rate of survival by up to 50%; congestion of the kidneys and gills
	Rainbow trout	4.67 mg/kg	Intraperitoneal	Reduction in the rate of survival by up to 50%
	Channel catfish (*Ictalurus punctatus*)	0.5–8 mg/kg	Feed — oral (8 weeks)	Reduction in weight gain feed conversion ratio; Lower survival rate; Increased incidence and severity of melanomacrophage centers in hepatopancreatic tissue and posterior kidney
Fumonisins	Nile tilapia (*Oreochromis niloticus*)	150 mg/kg	Feed — oral (4 weeks)	Reduction in weight gain; hematocrit decreased
	Channel catfish	80 mg/kg	Feed — oral (4 weeks)	Reduction in weight gain;

Mycotoxin	Species	Dose	Administration	Toxic effect
	Channel catfish	5–15 mg/kg	Feed — oral (6 weeks)	Reduction in weight gain; hematocrit, erytrocytes, hemoglobin and protein constituents decreased
Deoxynivalenol	Carp (*Cyprinus carpio* L.)	352 μg/kg, 619 μg/kg, and 953 μg/kg, final feed	Feed — oral (6 weeks)	Cytotoxic effects on immune cells; decreased pro-inflammatory responses in carp leukocytes;
	Carp (*Cyprinus carpio* L.)	352 μg/kg, final feed	Feed — oral (6 weeks)	The length of erythrocytes was reduced; increase of catalase and SOD activity
	Carp (*Cyprinus carpio* L.); Rainbow Trout and Salmon	300 ng/kg	In vitro	Damage in cell; DNA strand breaks and alterations of intracellular glutathione level
	Carp (*Cyprinus carpio* L.)	352–0.953 μg/kg, final feed	Feed — oral (4 weeks)	Increase fat disposition and severe hyperaemia in liver tissue; Reduced ALT and AST
	Atlantic salmon	3.7 mg	Feed — oral (15 weeks)	Reduced weight gain
	Rainbow trout (*Oncorhynchus mykiss*)	0.3–2.6 mg/kg	Feed — oral (56 days)	Reduced weight gain
	Channel catfish	20 mg/kg	Feed — oral (7 weeks)	Reduced feed conversion ratio
T2-toxin	Embryos of zebrafish (*Danio rerio*)	0.20–0.80 μmol/L	Feed — oral (24 h of exposure)	Abnormalities in tail formation; cardiovascular defects; mortality upper 50%
Zeaaralenone	Embryos of zebrafish (*Danio rerio*)	100–1000 μg/L	Water — oral (5 days)	Edemas, a developmental phenotype reminiscent to the heart and soul mutant and the lack of the pigmentation gap at the base of the caudal fin;
	Zebrafish (*Danio rerio*)	3.247 μg/L	Water — oral (4 days)	Estrogenic potency
	Rainbow trout (*Oncorhynchus mykiss*)	10 mg/kg	Intraperitoneal (168 h)	Interfere with important biological processes (i.e., blood coagulation or iron-storage) and different cellular components (i.e., cytoskeleton)
Fumonisins	Nile tilapia (*Oreochromis niloticus*)	150 mg/kg	Feed — oral (4 weeks)	Reduction in weight gain; hematocrit decreased
	Channel catfish	80 mg/kg	Feed — oral (4 weeks)	Reduction in weight gain;

Mycotoxin	Species	Dose	Administration	Toxic effect
	Channel catfish	5–15 mg/kg	Feed — oral (6 weeks)	Reduction in weight gain; hematocrit, erytrocytes, hemoglobin and protein constituents decreased
Moniliformin	Nile tilapia (*Oreochromis niloticus*)	70 or 150 mg/kg	Feed — oral (4 weeks)	Reduction in weight gain; hematocrit decreased;

12.6. LEGISLATION FOR SAFETY LEVEL OF MYCOTOXINS IN HUMAN FOOD, ANIMAL FEED AND ANIMAL FEED INGREDIENTS

According to FAO, Food And Agriculture Organization of the United Nations (2014) micronutrient deficiencies affect hundreds of million people, particularly women and children in the developing world. More than 200 million children worldwide are at risk of vitamin A defi- ciency, goiter or are iron deficient. Rural diets in many countries may not be particularly diverse and, thus, it is vital to have good food sources that can provide all essential nutrients in people's diets. Individuals have never consumed so much fish or depended so greatly on the fish- eries and aquaculture sector for their nutrition as they do today. But, the demand for fish is growing and there are still huge numbers of hungry and malnourished citizens in the world. Aquaculture plays an essential role in meeting these challenges.

However, to do so sustainably, it needs to become less dependent on completely wild fish for feeds and to modify culture species and prac- tices, which, in turn, will require influencing the consumer predilec- tions. Foods from the aquatic environment are a unique source of the essential long-chain omega-3 fatty acids, which are important for opti- mal brain and neurodevelopment in children (docosahexaenoic acid — DHA) and vascular health (eicosapentaenoic acid — EPA). Sufficient in- take of omega-3 fatty acids is particularly important during pregnancy and the first two years of life. Fish consumption among adults lowers the risk of coronary heart disease mortality by up to 36% owing to a combination of EPA and DHA effects.

It must be considered that contaminated food becomes threats to global health, endangering everyone, including infants, children, pregnant women, the elderly and those with an underlying disease that are partic- ularly vulnerable. Hence, it is extremely important that the safe levels of mycotoxins present in fish are known, as well as standardized extraction technique of mycotoxins from different animal tissues, being validated by governmental responsible for determining those regulations.

Table 4 Presence of residues of mycotoxins in tissues or organs of fish.

Mycotoxin	Species	Doses/ Administration period	Amount of residue tissue			Determination technique
			Muscle	Liver	Others	
Aflatoxins	Nile tilapia (*Oreochromis niloticus*)	1641 µg AFB_1/kgFeed — Oral (20 weeks)	nd	x	x	ELISA test kits (Detection limit 0.1 ppb, Brins-live pro Biotechnology Co., China)
	Rohu (*Labeo rohita*)	3 mg AFB_1/kgFeed — oral (8 days)	0.38 µg AFB_1/kg	7.83 µg AFB_1/kg	x	
	Juvenile rainbow trout	mg AFB_1/kg feed — Oral (3 h–7 days)	2000–4100 ng AFB_1/kg	x	x	Immunoaffinity column (AFLAKING, Horiba Co., Kyoto, Japan) and subjected to HPLC
	Sea bass (*Dicentrarchus labrax* L.)	Feed — oral (14 days) — 0.18 mg	4.25 ppb	x	x	Aflatest™ method; and subjected to HPLC
Emerging *Fusarium mycotoxins*	Sea bass (*Dicentrarchus labrax* L. and *Sparus aurata*)	FB_1/kg feed Natural occurrence	nd	17.7 µg ENB/ kg (sea bass); 13.4 µg ENB/ kg(sea bream)	Head (sea bream) –15.9 µg ENB/kg; sea bass –75.4 µg ENB/kg.	LC–MS/MS/LIT
Zearalanenone	Rainbow trout	Natural occurrence	nd	nd	Ovarios –7.1 ± 3.2 µg ZEN/kg	ZearalaTest$_{WB}$; Vicam, USA — and subjectedto HPLC
	Rainbow trout	Cell cultivation (24 h — 625 ng/mL)	x	nd	x	HPLC-DAD

nd — not detectable; x — author has not achieved the analysis.

All regulations are primarily made on the basis of known toxic ef- fects in animals or humans and in addition to information about toxicity exposure assessment is another main ingredient of the risk assessment. Reliable data on the occurrence of mycotoxins in various commodities and data on food intake are needed to prepare exposure assessment. Despite the dilemmas, mycotoxin regulations have been established during the past decades in many countries, and newer regulations are still being drafted. According to Fig. 1, can be observed the countries that have some kind of legislation for foodstuffs and animal feedstuffs. In Table 5, it can be observed the maximum limits allowed in the various countries of the world for some mycotoxins and for grains, animal feed and some products. Nevertheless, there are still no values permissible limits of mycotoxins for the consumer in relation to the final product. In addition, for the most part still missing regulations for safety values in food to vulnerable people, namely, lactating women, children, elder- ly or sick.

12.7 REDUCTION OF MYCOTOXINS IN FISH FEED

Food can be contaminated at any point of production and distribu- tion, and the primary responsibility lies with food producers. Mycotoxins is a group of undesirable substances in food production and its control, it is often overlooked. Mycotoxins can be produced throughout the grow- ing cycle, but predominately after the crop has matured in the field, through the harvest and storage periods. Anyway, there is no approved or sanctioned method for detoxifying most grains, there is only expected for cotton.

Control of mycotoxin contamination must start on rural property, i.e., where the grains are produced, but it is essential to conceive a spe- cial attention to the pre and post-harvest and storage of these grains. Mycotoxins are not visible to the eye, and often the grains do not show signs of mold infection, but may be contaminated with myco- toxins.

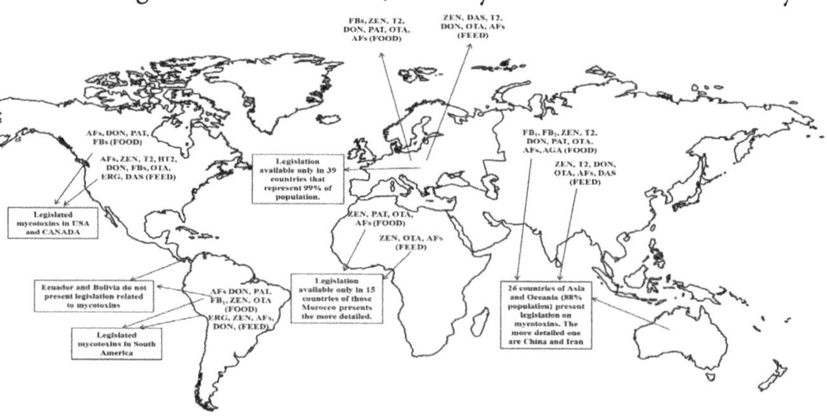

Fig. 1. Countries that have some kind of legislation for foodstuffs and animal feedstuffs. In accordance to FDA Mycotoxin Regulatory Guidance (2011a,b) and CODEX, there is the action level for AFBs and there are advisory levels for two mycotoxins that may be present in human food and animal feed ingredients: DON (vomitoxin) and FBs. The

action level for AFBs present in human food is 20 ppb; animal feed and animal feed ingredients 20–300 ppb; and advisory levels for DON present in human food is 1 ppm; animal feed and animal feed ingredients 5–10 ppm and for FBs present in human food is 2–4 ppm.

Action to take before harvest to reduce the risk of mold contamina- tion and mycotoxin production should be based on the use of certified seed or ensure that seed is free from fungal, bacterial or viral infection; avoid drought stress — irrigate if possible; sow seed as early as possible so that crop matures early; if practicing minimum or zero tillage remove crop residues; control insect, mammal, bird and virus pests; rotate crops; avoid nutrient stress.

Action to take during harvest to reduce the risk of mold contamina- tion and mycotoxin production should be based on harvest as quickly as possible; avoid field drying; transport the crop to the homestead as soon as possible; dry on platforms raised above ground; bundles of stover should also be placed on platforms to dry and not left to lie on the soil.

Action to take after harvest to reduce the risk of mold contamination and mycotoxin production should be based on drying (crop should be spread on a polythene sheet, tarpaulin or empty sacks laid on the ground or on a concrete plinth; unthreshed crop can be laid on a plat- form or in aventilated crib to dry; cobs can also be tied in pairs and suspended from a vertical frame to dry; in wet or humid conditions crop should be artificially dried in a solar dryer), threshing, shelling, winnowing (handle crop carefully to avoid broken or damaged grain; avoid beating the crop with sticks as this creates lots of damage leading to mold development unless the grain is to be used quickly and not stored) and storage (store grain in a suitable container which is raised above ground level or to do a suitable waterproof barrier; protectant to prevent insect and rodents damage; external storage structures must have a good roof with suitable overhang; sacks must be stored on a platform raised above the floor; inspected regularly and flour and feed must only be stored for short periods) .Another impor- tant point is to use quality ingredients or raw material for the prepara- tion of the fish feeds and plant ingredients and processed animal protein are suitable alternative feed- stuffs for fish feeds in aquaculture practice.

The discovery of the mycotoxin- transforming microorganisms may provide new opportunities for man- aging mycotoxin problems in aquaculture.

Mycotoxins can be considered among the most potent causes of cancer. Ingestion through the diet can pose chronic health risks that may result in liver and kidney disease and a suppression of the immune system. This review shows that in spite of inevitable and widespread, the presence of mycotoxins in feeds of fish cannot be neglected, as revealed by the sight of many researches, field outbreaks reports and pathologies related to mycotoxins, mainly because toxic effects and safety levels of mycotoxins in the different species of fish, are known on the surface.

Table 5: Permissible limits of mycotoxins in food for humans, animal feed and animal feed ingredients.

Mycotoxins (µg/kg)	Food typology			
	1	2	3	4
Aflatoxins (B_1,B_2,G_1,G_2) South America	5	20	0–3	10
EUA	20	20	nd	nd
Canadá	nd	nd	nd	20
África	11,079	42,278	10	44,105
Europa	4 0–10	4	4	10
Ásia	15–20	50	20	10–1000
Oceania	5	5	5	5
Ochratoxin A				
South America	18,537	18,537	2	nd
EUA	5	5	3	nd
Africa	nd	nd	0	300
Europa	3	3	nd	nd
Deoxynivalenol				
South America	3000	1000	200	nd
EUA	1000	nd	nd	nd
Canadá	2000	nd	nd	1000–5000
Fumonisin				
South America	nd	2000	200	nd
Zearalenone				
South America	200	400	20–200	nd

OBS: 1 = cereal and cereal products; 2 = corn; 3 = cereal-based foods for baby food andinfant formula; 4 = feed to livestock production and pet animals.

It's necessary to keep a standard to perform the tests in relation to duration of the experiment, because the toxic effect of mycotoxins may be dependent of the type and quantity of mycotoxin content in the feed, but also depends on the duration of exposure, animal species, sex and age.

An approach on the occurrence of mycotoxin-contaminated animal feed, as well as toxic effects that mycotoxins may produce in fish and their residues in meat and organs are outlined in this review. The mycotoxins of most concern from a food safety in products of animal origin in perspective include the aflatoxins (B_1, B_2, G_1, G_2 and M_1), ochratoxin A and toxins produced by Fusarium molds (FB_1, FB_2 and FB_3; trichothecenes like deoxynivalenol, T-2 and HT-2 toxin and ZEN), but in studies with different species of fish, the mycotoxin most studied is the AFBs.

The consequences of mycotoxin contamination in fish feeding, does not differ from other animal species intended for human consumption and it is directly related to losses in production, especially in reduced weight gain and feed conversion, but also, immune impairment and organ lesions are found; therefore, it is necessary to be extremely careful in concern to the production and storage of animal feed.

13

INNOVATIVE TECHNOLOGIES TO MANAGE AFLATOXINS IN FOODS AND FEEDS

13.1 INTRODUCTION

It focused on different scientific research results regarding AF control in food and feeds at pre- and post-harvest levels. It is clear that high AF levels pose human health risks and also represent a barrier to expand trade in both domestic and international contexts. Overall, it is necessary to tackle existing global food insecurity issues by adopting and implementing cutting-edge technologies. Biocontrol technologies, in conjunction with other aflatoxin-management tools such as sorting technologies, storage, irradiation, ozone fumigation, chemical and biological control along with improved packaging materials have the potential to link farmers to markets, enhance international trade, improve health conditions of people and animals, and increase food safety and security.

Food security is effectually achieved when the food pillars, including food availability, food access, food utilization, and food stability are at levels that allow all people at all times to have physical and economic access to affordable, safe, and nutritious food to meet the requirement for an active and a healthy life. When one of these four pillars weakens, then a society undermines its food security. Factors related to food insecurity and malnutrition not only influence human health and welfare, but also affect social, economic, and political aspects of society. With regards to the previous points, pre- and post-harvest losses due to mycotoxin contamination are documented as one of the driving factors of food insecurity since these substances occur along most food chains from farm to fork.

Among the different type of mycotoxins, aflatoxins (AFs) are widespread in major food crops such as maize, groundnuts, tree nuts, and dried fruits and spices as well as milk and meat products. When animal feeds are infected with AF-producing fungi, AFs are introduced into animal source food chain. AFs are toxic metabolites produced via a polyketide pathway by various species and by unnamed strains of Aspergillus section Flavi, which includes *A. flavus*, *A. parasiticus*, *A. parvisclerotegenus*, *A. minisclerotigenes*,

Strain SBG, and less commonly *A. nomius* . Normally, *A. flavus* produces only B-type aflatoxins, whereas the other *Aspergillus* species produce both B- and G-type aflatoxins. The relative proportions and level of AF contamination depends on *Aspergillus* species, growing and storage conditions, and additional factors. For instance, genotype, water or heat stress, soil conditions, moisture deficit, and insect infestations are influential in determining the frequency and severity of contamination. For M-type aflatoxins, these compounds are normally not found on crops, but their metabolites are found in both the meat and milk of animals whose feedstuffs have been contaminated by AF-B_1 and AF-B_2 .

Recently, emphasis on the health risks associated with consumption of AFs in food and feedstuffs has increased considerably. As a result of this, many experimental, clinical, and epidemiological studies have been conducted showing adverse health effects in humans and animals exposed to AFs contamination, depending on exposure. High-dose exposure of the contaminant can result in vomiting, abdominal pain, and even possible death, while small quantities of chronic exposure may lead to liver cancer. The International Agency for Research on Cancer (IARC) has classified both B- and G-type aflatoxins as Group 1 mutagens, whereas AF-M_1 is classified in Group ٢B. Furthermore, AFs may contribute to alter and impair child growth. Together with other mycotoxins, AFs are commonly suspected to play a role in development of edema in malnourished people as well as in the pathogenesis of kwashiorkor in malnourished children. Moreover, AF contamination negatively impacts crop and animal production leading not only to natural resource waste, but also decreased market value that causes significant economic losses.

Due to these effects, different countries and some international organizations have established strict regulations in order to control AF contamination in food and feeds and also to prohibit trade of contaminated products. The regulations on "acceptable health risk" usually depend on a country's level of economic development, extent of consumption of high-risk crops, and the susceptibility to contamination of crops to be regulated. Indeed, the established safe limit of AFs for human consumption ranges 4–30 μg/kg. The EU has set the strictest standards, which establishes that any product for direct human consumption cannot be marketed with a concentration of AF-B_1 and total AFs greater than 2 μg/kg and 4 μg/kg, respectively. Likewise, US regulations have specified the maximum acceptable limit for AFs at 20 μg/kg. However, if the EU aflatoxin standard is adopted worldwide, lower-income countries such as those in Asia and Sub-Saharan Africa will face both economic losses and additional costs related to meeting those standards. This situation requires alternative technologies at pre- and post-harvest levels aimed to minimize contamination of commercial foods and feeds, at least to ensure that AF levels remain below safe limits.

Implementation of innovative technologies is invaluable to address the challenges related to AFs and their effects. Reduction of AF contamination through knowledge of pre- and post-harvest managements is one of the first steps towards an appropriate strategy to improve of agricultural productivity in a sustainable way. This has direct positive effects on enhancing the quality and nutritional value of foods, conserving natural resources, as well as advancing local and international trade by increasing competitiveness. It is important to identify and document available technologies that can effectively control and minimize aflatoxin contamination to sustain healthy living and socioeconomic development. There exists ample literature on tools for AF control and their benefits. Therefore, this review compiles data on innovative pre- and post-harvest technologies developed that can manage AF contamination in foods. The benefits of these technologies are also discussed in terms of food security, human health, and economic value. Finally, implications for research and management policies addressing AF issues are highlighted.

13.2. INNOVATIVE MANAGEMENT STRATEGIES OF AF REDUCTION

A wide range of AF management options exist in literature. Depending on the "type" or mode of application, management has been classified in this review as pre-harvest stage, specifically biological control, while sorting technology, treatments with electromagnetic radiation, ozone fumigation, chemical control agents, biological control agents, and packaging material are grouped as post-harvest stage. Each of these groups of control/management options are discussed in this section.

13.2.1 Biological Control

Non-aflatoxin forming strains of *A. flavus* have been used as a biological control for long-term crop protection against AF contamination under field conditions. stated that when the spore number of nontoxigenic strains in the soil is high, they will compete with other strains, both toxigenic and other atoxigenic, for the infection sites and essential nutrients needed for growth. Moreover, soil inoculation with nontoxigenic strains has a carryover effect, which protects crops from contamination during storage. The ability of fungus to compete with closely related strains depends on several factors such as pH and soil type as well as the availability of nitrogen, carbon, water, and minerals.

The International Institute of Tropical Agriculture (IITA) and the United States Department of Agriculture - Agriculture Research Service together with other partners have been researching in Africa on non-toxigenic biocontrol fungi that act through competitive exclusion strategy. They have successfully developed several country-specific indigenous aflatoxin biocontrol products generically named as Aflasafe™ , which can be used on maize and groundnut. This product is an eco-friendly innovative biocontrol technology that utilizes native non-toxigenic strains of *A. flavus* to naturally out-compete their aflatoxin-producing cousins. Aflasafe™ has been shown to consistently reduce aflatoxin contamination in maize and groundnut by 80–99% during crop development, post-harvest storage, and throughout the value chain in several countries across Africa.

Aflasafe products have been registered for commercial use in Kenya, Nigeria, Senegal and Gambia, while products are under development in seven other African nations. Each Aflasafe™ product contains four unique atoxigenic strains of *A. flavus* widely distributed naturally in the country where it is to be applied.

Another study on biological control has been reported by who found that inoculation of antagonistic strains of fluorescent *Pseudomonas*, *Bacillus* and *Trichoderma* spp. on peanuts resulted in significant reduction of pre-harvest seed infection by *A. flavus*. also demonstrated that the extract of *Equisetum arvense* and a mixture 1:1 of *Equisetum arvense* and *Stevia rebaudiana* is effective against growth of *A. flavus* and subsequent production of aflatoxin under pre-harvest conditions. also observed 71% reduction in AF contamination in soils and in groundnuts when an AF competitive exclusion strain of *A. flavus* AFCHG2 was applied to Argentinian groundnuts. Similarly, showed that non-toxigenic strains of *A. flavus* mitigated AF contaminations in maize through pre-harvest field application. Furthermore, evaluated the efficacy of a bioplastic-based formulation for controlling AFs in maize. The results showed that bio-control granules inoculated with *A. flavus* NRRL 30797 or NRRL 21882 reduced AF contaminations up to %90 in both non-Bt and Bt hybrids.

13.2.2. Sorting Technology

Sorting processes seek to eliminate agricultural products with substandard quality. Normally sorting, especially for grains, can be achieved based on differentiation of physical properties such as colour, size, shape, and density as well as visible identification of fungal growth in affected crops. By rejecting damaged and discoloured samples, sorting operations reduce the presence of AFs as well as contaminating materials in food and feed mentioned that floating and density separation could reduce AFs in stored groundnut kernels by up to 95%. In another report, and reported that AF-contaminated groundnuts were eliminated by colour sorting processes, while fluorescence sorting was effective to reduce levels of AF contamination in pecan and pistachio nuts. These observations were validated through a recent study, which showed that AFs contamination in pistachio nuts is more than %95 reduced by colour sorting.

Nonetheless, such physical methods are often laborious, inefficient, and impractical for in-line measurements. The application of computer-based image processing techniques is one of the most promising methods for large-scale screening of fungal and toxin contaminations in food and feed. Grains and other agricultural products contain various nutritional substances that are degraded by fungal growth, which in turn influence absorbance spectra of the material. For instance, reported that scattering and absorbance characteristics are influenced by the presence of *A. flavus* in the kernel since fungal development causes the endosperm to become powdery. also showed that it was possible to quantify fungal infection and metabolites such as mycotoxins produced in maize grain by *Fusarium verticillioides* using Near Infrared Spectroscopy (NIRS). found that NIRS successfully identified kernels contaminated with AFs. Moreover, highlighted NIRS technique as a fast and non-destructive tool for detecting mycotoxins such as AF-B_1 in maize and barley at a level of 20 ppb. Nevertheless, NIRS only produces an average spectrum, which lacks in spatial information from the sample with respect to distribution of the chemical composition.

Hyperspectral imaging (HSI) is another method that can be employed to monitor both the distribution and composition of mycotoxins in contaminated food samples, especially grains. This method can produce both localized information and a complete NIR spectrum in each pixel used hyperspectral imaging (HSI) techniques to estimate AF contamination in maize kernels inoculated with *A. flavus* spores. also demonstrated the potential HSI based in the Vis/NIR range for quantitative identification and distinction of AFs in inoculated maize kernels. Mentioned that the spectral reflectance ratio 735/1005 nm, which is located in the transition between Vis and NIR, can be analysed to identify highly contaminated AF corn kernels (>100 ppb) from those contaminated lower than 10 ppb. The observation was in agreement with other studies by and they reported that *Aspergillus* fungi in maize and wheat were detectable by analysing the HSI in 1000–400 nm or the fusion of HSI and digital images.

Another image based sorting technology has been proposed by, who reported that approximately 98% of the AFs in contaminated figs were successfully detected and separated by a UV light coupled with colour detection system. This method used the viability of bright greenish-yellow fluorescence (BGYF), which is produced by *A. flavus* via the oxidative action of peroxidases in living plant tissue as an image screening technique for the classification of AF contaminated crops.

13.2.3. TREATMENTS WITH ELECTROMAGNETIC RADIATION

Gamma (γ) radiation has been considered as an effective tool for preserving and maintaining quality of agricultural and food products. Very high-energy photons generated by a gamma source such as cobalt-60 (^{60}Co) are used to destroy pathogenic and spoilage microorganisms by causing direct damage to DNA in microbial cells. An additional effect of γ-irradiation is the interaction of energy with water molecules present in substrates or foods, producing free radicals and ions that attack the DNA of microorganisms. However, the efficiency of γ-irradiation depends on many factors, namely the number and type of fungal strain, radiation dose, composition of food, and air humidity. Several studies have reported that γ-irradiation can be performed to decrease AF contamination as exhibited in.

Table 1. Reported applications of gamma irradiation against aflatoxin production in food products.

Product	Aflatoxin (s)	Moisture content %	Dose kGy	Temperature °C	Reduction %
Black pepper	AF-B_1	18	30	26–30	47
	AF-B_2				39
	AF-G_1				47
	AF-G_2				40
White pepper	AF-B_1	18	30	26–30	51
	AF-B_2				35
	AF-G_1				48
	AF-G_2				43
Ground red chillies	Total AFs	12–17	6	25–28	81–91
	AF-B_1				92–98
Raisins	AF-B_1	–	10	25	65
Maize	AF-B_1	–	10	–	95

Some authors reported that AF content could be reduced even with a low-dose γ-irradiation. For example, observed that using 5 kGy of γ-irradiation is sufficient to inhibit the growth of A. flavus and production of AF-B_1 in soybean seeds over 60 days of storage without any noticeable changes in chemical composition. Similarly, mentioned that a dose of 6 kGy reduced total AFs and AF-B1 content by more than 80% in red chilies. However, some claimed that such reductions can be achieved only using high-dose γ-irradiation. for instance showed that the γ-irradiation at dose of 10 kGy led to an approximately 65% decrease of the initial AF-B1 accumulation in raisins samples inoculated by A. parasiticus, compared to the non-irradiated sample on the same day. The experiments done on naturally contaminated maize samples by also indicated that the irradiation with a 10 kGy dose can be used to reduce the amount of AF-B1 to an acceptable level without compromising animal and human health. Nevertheless, some authors argued that even more than 20 kGy of γ-irradiation is not effective in reducing AFs. The efficacy of γ-irradiation at high doses to decontaminate black and white peppers from AF-B_1, AF-B2, AF-G_1, and AF-G_2 was reported by . They mentioned that a gamma irradiation of 30 kGy (the maximum allowable dosage as permitted by FDA) in samples at 18% moisture content was not sufficient to completely eradicate AFs.

Some reports can be found in literature about the application of ultraviolet (UV) irradiation as a non-thermal, economical technology for AF destruction in different food products. showed that AF-B1 and AF-G_1 in wheat grain were completely eliminated after UV short wave (254 nm) and long wave (362 nm) was applied for 30 min, while AF-B_2 was decreased by 50 and 74% when exposed to UV short wave and long wave for 120 min, respectively. A study of UV-C irradiation on groundnut, almond, and pistachio was performed by after treatment with UV-C at 265 nm for 15 min, all nut samples showed 100% degradation of AF-G2, while the complete elimination of AF-G1 was observed only in almond and pistachio. The level of AF-B_1 was reduced by approximately 97% after UV-C irradiation for 45 min. Showed the potential of UV-A and UV-B irradiation, which can be used to reduce mycotoxin production from A. carbonarius and A. parasiticus in grape and pistachio media.

Another non-thermal technology called pulsed light (PL) has also been used in AF reduction. Normally, PL generates short, high-intensity flashes of broad-spectrum white light. The synergy between full spectra of ultraviolet, visible, and infrared light destroys both the cell wall and nucleic acid structure of microorganisms present on the surface of either food or packaging materials in a few seconds investigated the effect of PL treatment of 0.52 J cm−2 pulse−1 on the production of AF-B_1 and AF-B_2 in rough rice inoculated with A. flavus. Application of PL treatment for 80 s reduced AF-B_1 and AF-B_2 in rough rice by 75 and 39%, respectively. Additionally, the mutagenic activity of AF-B1 and AF-B_2 was completely eliminated by PL treatment, while the toxicity of these two aflatoxins decreased significantly.

Dielectric processes of radio frequency (RF) and microwave (MW) are additional alternative methods for controlling AFs contamination in agricultural products. Used the RF to reduce AF-B1 in Perilla frutescens L. highland oil seed. They revealed that A. niger, A. flavus, and AF-B1 in seeds with an initial moisture content of 18% w.b. were highly inhibited by RF heat treatment at 90 °C for 7 min. For microwave application, 2.45 GHz MW was applied directly to hazelnuts contaminated with A. parasiticus by who then documented MW effects on post-harvest safety and quality of the product. The results showed that MW treatment for 120 s was able to reduce fungal count of A. parasiticus on in-shell hazelnut without any noticeable change in the nutritional and organoleptic properties. Unlike microbial inhibition, MW treatment was not effective to decrease AFs in hazelnuts. Tested the effect of MW application during alkaline-cooking of AF contaminated maize. A 36% reduction of AF-B1 and 58% reduction of AF-B2 were observed after the maize was treated at 1650 W power output and 2450 MHz operating frequency for 5.5 min. In addition, the effectiveness of MW heating on the reduction of AF contamination in groundnuts and respective products was evaluated by samples heated with MW up to 92 °C for 5 min resulted in a maximum AF-B1 reduction of 51.1–100%.

13.2.4. Ozone Fumigation

Ozone, the triatomic form of oxygen (O_3), is one of the most powerful disinfectants and sanitizing agents. It has been approved as Generally Recognized as Safe (GRAS) meaning it can be directly applied as an antimicrobial agent in the food industry. Normally, ozone can be produced by several methods such as electrical discharge in oxygen, electrolysis of water, photochemical, and radiochemical. A primary attractive aspect of ozone is that, after reaching its half-life (20–50 min), decomposition products do not represent any hazard for the treated materials. In post-harvest treatment, gaseous and aqueous ozone phases are applied to inactivate bacterial growth, prevent fungal decay, destroy pesticides and chemical residues, control storage pests, and degrade AFs.

The mechanisms of ozone to inhibit microbial populations in food occur via the progressive oxidation of vital cellular components. Ozone oxidizes polyunsaturated fatty acids or sulfhydryl group and amino acids of enzymes, peptides, and proteins to shorter molecular fragments. In addition, ozone degrades the cell wall envelope of unsaturated lipids resulting in cell disruption and subsequent leakage of cellular contents. The mechanism of ozone on the degradation of AF-B_1 and AF-G_1 involves an electrophilic reaction on the C8-C9 double bond of the furan ring causing the formation of ozonide. These compounds are then rearranged into monozonide derivatives such as aldehydes, ketones, acids, and carbon dioxide. Since there is no C8-C9 double bond in the structure, AF-B_2 and AF-G_2 are more resistant to ozonisation than AF-B_1 and AF-G_1. Even though the efficiency of ozone as a chemical detoxifier is high, a greater concentration is required to kill fungi or contaminated surfaces, while low concentration of ozone and short fumigation time is generally considered necessary in order to preserve product properties like colour, flavour, aroma, and vitamins.

Ozone detoxification has been found by some studies to be useful to reduce AFs in food commodities as summarized in observed that ozone treatment degraded AF-B_1 in red peppers, while no significant variation in colour quality was found. Investigated the effectiveness of gaseous ozone against microbial flora and AF-B_1 content in dried figs. The results exhibited that *Escherichia coli*, mould, and AF-B_1 were inactivated after ozone application. Using groundnut samples, demonstrated the efficacy of the fungicidal and detoxifying effects of ozone against total AFs and AF-B_1. In their study, ozone could control potential aflatoxin producing species, *A. flavus* and *A. parasiticus,* in groundnuts. The concentration of total AFs and AF-B_1 was also reduced. A study conducted showed that AF-B_1 levels in groundnuts tend to decrease with ozone application, however the ozonolysis efficiency on AF-B_1 was not further improved after 60 h. Moreover, in the sub-chronic toxicity experiment, they also found that ozone did not show any toxic effects in male and female rats. Treated groundnut samples with ozone and observed that the detoxification rate of AFs increased. In addition, the results demonstrated that ozone application did not influence the contents of polyphenols, resveratrol, acids, and peroxide in treated samples. examined the effect of ozone treatment on the degradation of AF-B_1 in maize and found that the toxicity of AF-B_1 contaminated maize was diminished by ozone treatment.

Table 2. Reported applications of ozone against aflatoxin production in food products.

Product	Aflatoxin (s)	Concentration mg L^{-1}	Time min	Reduction %
Red pepper	AF-B_1	33	60	80
		66	60	93
Dried figs	AF-B_1	13.8	30	48.8
			60	72.4
			180	95.2
Peanuts	Total AFs	21	5760	30
	AF-B_1			25
	AF-B_1	50	3600	89.4
	Total AFs	6	30	65.8
	AF-B_1			65.9
Corn	AF-B_1	90	40	88

13.2.5. Chemical Control Agents

A number of studies have determined the effect of synthetic and natural food additives on AF reduction in food products. A prime example of this effect is citric acid on AF-B_1 and AF-B_2 degradation in extruded sorghum investigated the effect of sodium hydrosulphite ($Na_2S_2O_4$) and pressure on the reduction of AFs in black pepper. The study reported that the application of 2% $Na_2S_2O_4$ under high pressure resulted in a greater percentage reduction of AF-B_1, AF-B_2, AF-G_1, and AF-G_2, without damage to the outer layer of black pepper. Nevertheless, AF-B_2 was found to be the most resistant against the applied treatment. Apart from that, it is evident that respiration from insects increases the temperature and moisture content of grains providing favourable conditions for fungal growth. For this reason, evaluated the efficacy of 2, 6-di (*t*-butyl)-p-cresol (BHT) and the entomopathogenic fungus *Purpureocillium lilacinum* on the accumulation of AF-B_1 in stored maize. The results clearly showed that the highest reduction of AF-B_1 in stored maize occurred with the combination of BHT and *Purpureocillium lilacinum*. In addition, the effects of organic acids during soaking process on the reduction of AFs in soybean media were studied. The highest reduction rate of AF-B_1 was obtained from tartaric acid followed by citric acid, lactic acid, and succinic acid, respectively. These acid treatments convert AF-B_1 to β-keto acid that subsequently transforms to AF-D_1, which has less toxicity than that of AF-B_1 reported another novel technology that has been applied to inhibit AF contamination called acidic electrolyzed oxidizing water, which is an electrolyte solution prepared using an electrolysis apparatus with an ion-exchange membrane, used to decontaminate AF-B_1 from naturally contaminated groundnut samples. The content of AF-B_1 in groundnuts decreased about %85 after soaking in the solution. Remarkably, the nutritional content and colour of the groundnuts did not significantly change after treatment.

Table 3: Reported applications of chemical agents against aflatoxin production in food products.

Product	Aflatoxin(s)	Chemical agent	Reduction %
Black pepper	AF-B_1	2% $Na_2S_2O_4$	96
	AF-B_2	Pressure = 1.5 bar	77
	AF-G_1	Temp = 121 °C	100
	AF-G_2	Time = 15 s	100
Maize	AF-B_1	BHT + P. *Lilacinum*	90
Soybean	AF-B_1	1 mol L^{-1} tartaric acid	95
		Soaking time = 18 h	
Sorghum	AF-B_1 + AF-B_2	8 mol L^{-1} citric acid	59–89
		MC = 200–300 g kg^{-1}	
Peanuts	AF-B_1	Acidic electrolyzed oxidizing water	85
		Ratio of liquid to solid = 5:1 (v m^{-1})	
		Temp = RT	
		Time = 15 min	
Peanuts	AF-B_1	1 g oriental mustard flour/50 g sample	65
	AF-B_2		86
	AF-G_1		97
	AF-G_2		100
Piadina (Italian flatbread)	AF-B_1	1 g oriental mustard flour/10 g sample	89
	AF-B_2		83
	AF-G_1		87
	AF-G_2		85
Milk	AF-M_1	1.21% calcium montmorillonite clay	68
Urine (human)	AF-M_1	3 g/day calcium montmorillonite clay, ACCS 100	44–54

To overcome the development of fungal resistance as well as residual toxicity posed by synthetic additives, the actions of some plant-based preservatives toward AF reduction have been studied in various food products. evaluated the effect of isothiocyanates, generated by enzymatic hydrolysis of glucosinolates, contained in oriental mustard flour. The findings showed that isothiocyanates reduced *A. parasiticus* growth in groundnut samples, whereas the AF-B_1, AF-B_2, AF-G_1, and AF-G_2 reduction ranged between 65 and %100. Similar results were obtained, who reported the inhibition of AFs by isothiocyanates derived from oriental and yellow mustard flours in piadina (a typical Italian flatbread) contaminated with *A. parasiticus*. These results can be explained by the electrophilic property of isothiocyanates, which can bind to thiol and amino groups of amino acids, peptides, and proteins, forming conjugates, dithiocarbamate, and thiourea structures leading to enzyme inhibition and subsequently to cell death. However, it is worth noting that ρ-hydroxybenzyl isothiocyanate (ρ-HBITC), which is formed in yellow mustard flour, is less stable than allyl isothiocyanate (AITC) from oriental mustard. In substitution of common commercial preservatives, also applied active packaging devices containing allyl isothiocyanate to avoid the growth of *A. parasiticus* and AF production in fresh pizza crust after 30 days. Another study used neem leaves (*Azadirachta indica*) to inhibit the growth of AFs in wheat, maize, and rice during storage for 9 months. Due to fungicidal and anti-aflatoxigenic properties of neem leaves, the application of 20% neem powder fully inhibited all types of aflatoxins synthesis for 4 months in wheat and for 2 months in maize, whereas the inhibition of AF-B_2, AF-G_1, and AF-G_2 was observed for 3 months in rice. Essential oils of different aromatic plants have been also used as food preservatives due to their antimicrobial properties. However, the antibiotic functions of essential oils are not yet clearly understood. Stated that the anti-aflatoxigenic activity of essential oils may be related to inhibition of ternary steps in AF biosynthesis involving lipid peroxidation and oxygenation. Determined the antifungal potential use of eugenol, a compound derived from essential oils, against AF-B_1 production in stored sorghum grain. Presented the efficacy of *Piper betle* L. essential oil against the AF-B_1 production in some dried fruits, spices, and areca nut. showed the inhibiting effect of thyme essential oil (*Thymus vulgaris*) against fungal development and AF production on *A. flavus* cultures. Likewise, reported the possible utilization of flavanones (naringin, hesperidin, and neohesperidin) obtained as by-products from the citrus industry to inhibit the production of AFs from *A. flavus*.

Overall, few studies exist about chemical control of AFs in milk and dairy products. Investigated the effect of a modified yeast cell wall extract on the excretion of AF-B_1 and AF-M_1 in faeces, urine, and milk. They observed that feed supplementation with modified extract cell walls of yeasts reduced the absorption of AF-B_1, and decreased the concentration of AF-B_1 and AF-M_1 in ewe faeces. The results indicated that this organic material could be used to protect ruminants from chronic exposure to AFs present in feeds. Another study by examined the effect of calcium montmorillonite clay (Novasil Plus, NSP) in dairy feed on dry matter intake, milk yield, milk composition, vitamin A, riboflavin, and AF-M_1. The calcium montmorillonite clay was found to reduce AF-M_1 content in milk samples without affecting milk production and nutrition qualities. Similarly, suggested that inclusion in the human diet of calcium silicate 100 (ACCS100), a calcium montmorillonite clay, may reduce aflatoxin bioavailability and potentially decrease the risk of aflatoxicosis in aflatoxin-prone areas such as in Kenya. These results can be explained by the fact that calcium montmorillonite clay binds tightly to AFs in the gastrointestinal tract, therefore reducing AFs bioavailability and distribution to the blood, liver, and other affected organs.

13.2.6. Biological Control Agents at Post-Harvest Processing Stages

Physical and chemical detoxification methods have some disadvantages, such as loss of nutritional value, altered organoleptic properties, and undesirable effects in the product as well as high cost of equipment and practical difficulties making them infeasible, particularly for lower-income countries. However, biological methods based on competitive exclusion by non-toxigenic fungal strains have been reported as a promising approach for mitigating formation of mycotoxins and preventing their absorption into the human body. Among various microorganisms, lactic acid bacteria (LAB) namely *Lactobacillus*, *Bifidobacterium*, *Propionibacterium*, and *Lactococcus* are reported to be active in terms of binding AF-B_1 and AF-M_1. The binding is most likely a surface phenomenon with a significant involvement of lactic acid and other metabolites such as phenolic compounds, hydroxyl fatty acids, hydrogen peroxide, reuterin, and proteinaceous compounds produced by LAB reported that AF binding seems to be strongly related to several factors such as LAB strain, matrix, temperature, pH, and incubation time. Found that *Lactobacillus rhamnosus* was the best strain with the ability to bind to AF-B_1 in contaminated wheat flour during bread-making process. Similar results were observed in yogurt cultured with %50 *Staphylococcus thermophiles* and *Lactobacillus bulgaricus* and %50 *Lactobacillus plantrium* with the greatest AF-M_1 reduction observed at the end of storage mentioned that LAB could inhibit AF-B_1 production in brewer's grains used as raw material for pig feed. More recently, investigated the effect of LAB against AF development in bread with the results showing that AF content was reduced %100–84 allowing up to 4 days of additional shelf life.

Other microorganisms have also been reported to bind or degrade aflatoxins in foods and feeds. Tested the AF-B_1 binding abilities of *Saccharomyces cerevisiae* strains *in vitro* in indigenous fermented foods from Ghana. The results indicated that some strains of *Saccharomyces cerevisiae* have high AF-B_1 binding capacity. These binding properties could be useful for the selection of starter cultures to prevent high AF contamination levels in relevant fermented foods. Showed that 20–38% of AF-B_1 was eliminated using probiotic culture of *Enterococcus faecium*. A study also reported the protective effect of *Bacillus subtilis* ANSB060 on meat quality due to its ability to prevent AF residue absorption in the livers of broilers fed with naturally mouldy groundnut meal. Moreover, some bacteria such as *Rhodococcus erythropolis*, *Bacillus* sp., *Stenotrophomonas maltophilia*, *Mycobacterium fluoranthenivorans*, and *Nocardia corynebacterioides* have been found to degrade AF-B_1 . Even though, many *Bacillus* species are still avoided due to their nature of producing toxic compounds recently showed that the non-toxic enzymes produced by *Bacillus subtilis* strain UTBSP1 can be used to reduce AF-B_1 from contaminated substrates.

13.2.7 Packaging Materials

In post-harvest management, packaging materials are frequently considered as the final step of product development in order to extend the preservation of food and feed products. During storage and distribution, food commodities can be affected by a range of environmental conditions, such as temperature and humidity as well as light and oxygen exposure. Overall, these factors have been reported to facilitate various physicochemical changes such as nutritional degradation and browning reactions with the latter causing undesirable colour changes. The interaction of these factors can also elevate the risks of fungal development and subsequent AF contamination. Many smallholder farmers in lower-income countries traditionally store agricultural products such as grains in containers typically made from wood, bamboo, thatch, or mud placed and covered with thatch or metal roofing sheets. Recently, metal or cement bins have been introduced as alternatives to traditional storage methods, but their high costs and difficulties with accessibility make adoption by small-scale farms limited stated that even though polypropylene (PP) bags are currently used for grains storage, they are still contaminated by fungal and AFs especially when those reused bags contain *A. flavus* spores.

Several studies have reported the application of Purdue Improved Crop Storage (PICS) bags to mitigate fungal growth and resulting AF contamination. indicated that the PICS bags successfully suppressed the development of *A. flavus* and resulting AF contamination in maize across the wide range of moisture contents in comparison to non-hermetic containers. These results correspond with who mentioned that grains stored in PICS bags absorbed less moisture than grains stored in woven polypropylene bags. This could be a result of PICS bag construction consisting of triple bagging hermetic technology with two inner liners made of high-density polyethylene (HDPE) and an outer layer woven PP. In addition, PICS bags reduced the oxygen influx and limited the escape of carbon dioxide, which can prevent the development of insects in stored grain. In Benin, Ghana, Burkina Faso, and Nigeria, used PICS bags to store locally infested maize. Although 53% of maize had AF levels above 20 ppm, samples from PICS bags tended to have less accumulation than those from woven bags. Also evaluated the efficacy of PICS bags for protecting groundnuts from quality deterioration and aflatoxin contamination caused by *A. flavus* and found that there was less toxin production in PICS bags compared to cloth bags under similar conditions.

13.3 BENEFITS OF INNOVATIVE MANAGEMENT

Many innovative management strategies that can potentially reduce AF contamination in food and feed chains have been identified by this review. These strategies have the potential to mitigate adverse effects of AF contamination on food security, public health, and economic development. An understanding of these benefits can motivate policy makers and value chain actors to explore effective ways of managing AFs during pre- and post-production processes.

13.3.1. Food Security Benefits

The quantity and quality of agricultural products are degraded by the presence of AFs, while the opposite is true when AF contamination is effectively prevented. The use of biocontrol methods for instance has been shown to reduce contamination up to 90% which potentially reduces complete loss of harvested or stored crops As mentioned earlier the use of the PICS technology for grain storage can reduce AF contamination due to the controlled environment in the hermitic bags. For subsistent households, such measures can potentially increase availability of harvested food crop for family consumption Farmers can even afford to sell their excess produce and use the proceeds to purchase other food ingredients they do not produce themselves. Moreover, applications of innovative control technologies can ensure that products are safer to consume, thereby improving utilization efficiency. By reducing significant losses during storage, the control of AF can certify that the foodstuffs are available over extended periods of time, thereby ensuring consistent food availability. Effective control of AF contamination therefore has the potential to enhance food availability, food access, food utilization, and food stability.

13.3.2. Health Benefits

AFs are a serious risk to public health, especially in low-income countries where most people consume relatively large quantities of susceptible crops such as maize or groundnuts. According to the estimation of the US Center for Disease Control and Prevention, about 4.5 billion people are chronically exposed to mycotoxins Prolonged exposure to even low levels of AF contamination in crops could lead to liver damage or cancer as well as to immune disorders. In children, stunted growth and Kwashiorkor pathogenesis are caused by breast milk consumption or direct ingestion of AF-contaminated foods. Controlling AF contamination through the application of effective technologies could potentially avoid such health risks and have significant benefits in a number of ways. First chronic diseases can be prevented to minimize pressure on the health facilities of an economy due to savings on cost of medication and treatment. People will have access to good quality food ingredients for health living and making work efficient labour force available for the economy.

13.3.3. Economic Benefits

The economic benefits of AF reduction are observed through both domestic and high-value international trade markets. At domestic and regional levels, markets might not reward reduced AF in crops, but avoiding contamination could allow, in ideal cases, to increase the volume of sales, which would lead to elevated incomes as well as greater returns to investments for producers. Farmers who successfully inhibit AF contamination can also benefit from increased income due to greater product acceptance, higher market value, or access to high-value markets. In reality, there are numerous factors that have to be enhanced in order to create premium class products such as aflatoxin control, consumer awareness, marketing channels, aflatoxin testing, and stricter enforcement of production and market regulations. When such enabling conditions are met, it has been shown that aflatoxin-conscious market can pay a premium for aflatoxin safe products even in the domestic market in Africa. Moreover, the control of AF contamination could reduce costs the associated with consequent effects on humans, such as medical treatments, primarily of individuals suffering from liver cancer, as well as indirect costs such as pain and suffering, anxiety, and reduction in quality of life associated with exposure to AFs.

At the international level, many developed countries have established regulations to limit exposure to AFs. Some countries have different limits depending on the intended use, the strictest on human consumption, exports, and industrial products Despite that stringent measures that makes phytosanitary standards seemingly more expensive, once suppliers internalize the economic costs of compliance in reality, greater economic benefits for society can be achieved. This is due to access to larger and more stable markets, and less incidence of disease. Controlling AF contamination in exportable agricultural commodities could maintain or even increase trade volumes and foreign earnings for exporting economies. Furthermore, the savings from such control measures could be channelled or invested in other economic sectors in order to generate additional income and propel growth and development.

13.4 IMPLICATIONS FOR RESEARCH AND POLICY

AFs are a critical problem for food safety in many lower-income countries where AF formation in key staple crops causes significant post-harvest losses and negative impacts on human life. Currently, several innovative AF control technologies have shown potential to improve health and economic factors for farmers and other actors in commodity value chains. However, the efficacy, safety, and quality of these technologies must be verified prior to adoption. The feasibility of using biocontrol products depends not only on safety regulations in each individual country, but also on the accessibility of such biocontrol tools like Aflasafe™ to smallholder farmers. The ability to develop and maintain biocontrol strains from local resources, particularly in the production of Aflasafe™, are highly cost-effective and facilitate availability. Meanwhile, non-profit governmental or non-governmental organizations can also promote such products, which are particularly suitable for sustainable development. Mentioned that application of biocontrol technologies in conjunction with other AF management tools can profitably link farmers to markets, improve human and animal health, and increase food safety. However, biocontrol adoption still requires a flexible system that allows the use of biopesticides together with a favorable policy and institutional supports.

Furthermore, other techniques have been developed such as sorting technologies that offer numerous advantages including (1) rapid, real-time product information via non-destructive measurement, (2) reduction of laborious and destructive analytical methods, (3) continuous monitoring, and (4) integrating into existing processing lines for control and automation. However, investment costs are usually the main factor determining whether such technologies are adopted or not. For simplicity, development of cheap and portable diagnostics techniques that are adaptable to different field networks is imperative. In addition, future research should still be conducted in cooperation with final users to achieve full adoption potential. Despite technological advances, hand sorting may still be more suitable in lower-income countries where access to equipment is limited. The culls from sorting must be disposed in a manner that they do not enter the food chain, particularly of economically vulnerable populations. Still, with regulatory approval, irradiation and ozone fumigation could effectively reduce aflatoxin levels in crops, but these interventions are less applicable due to higher costs and safety concerns. Moreover, naturally infected grains have both internal and external colonization. While external contamination can be decontaminated, ozone cannot penetrate the internal sites of colonization and AF formation. Therefore, large ozone doses for a long time might be required for effective ozone decontamination. The application of chemical and biological control agents has been shown to reduce AF contamination both in animals and humans. Nonetheless, little information is available regarding the effective doses and frequencies as well as costs and efficacies. Generally, individual countries with their own specific cultural context, especially those with higher risks of AF, face public aversion to these technologies. Regarding storage methods, there is evidence that suggests hermetic technologies like PICS triple layer bags could be cost effective against key grain storage pests. They may also provide an improved alternative for insecticide-free, long-term storage of grains with minimal grain damage. However, these PICS bags may not be suitable and affordable for small-scale farmers over very large areas. The technology is also limited to cereals and grain crops.

Although there are many initiatives that aim to reduce AF contamination in lower-income countries, the lack of regulation enforcement, or even definition of acceptable limits, does not allow for their full development and implementation. In order to reduce AF contamination, it is necessary to have policies focused on: (1) raising awareness of public health impacts associated with AF contamination to all actors along the entire value chain, including families, farmers, consumers, processors, and traders; (2) estimating the lifespan of each technology and calculating their respective social and economic costs of diminishing the contamination risk at different intervention points; (3) reducing the harmful effects of AFs by implementing the appropriate pre- and post-harvest technologies; (4) investing in infrastructure with such capacity that allows to support further activities both in order to reduce AFs and to monitor contamination levels in different agricultural products; (5) establishing of reliable and effective low-cost testing methods to monitor AF contamination levels in rural areas; and (6) providing the required data and risk management tools for driven policy reforms, which create an effective regulatory environment to ensure domestic food safety in rural and urban areas and also facilitates trade opportunities in the region. Finally, governments need to solve the issue of how agricultural businesses can be enabled to operate profitably while complying with existing standards and limits of AF contamination.

14

INTERVENTIONS TO PREVENT AFLATOXIN-INDUCED CARCINOGENESIS IN DEVELOPING COUNTRIES

14.1 INTRODUCTION

The aflatoxins are ubiquitous contaminates of the human food supply throughout the economically developing world. The adverse toxicological consequences of these compounds in populations are quite varied owing to a wide range of exposures that lead to acute effects, including rapid death, and chronic outcomes, such as hepatocellular carcinoma . Furthermore, several studies describe a variety of general adverse health effects associated with aflatoxin, such as impaired growth in children. Aflatoxin exposures multiplicatively increase the risk of liver cancer in people chronically infected with hepatitis B virus (HBV), which illustrates the deleterious impact that even low toxin levels in the diet can have on human health (Collectively, the public health impact of aflatoxin exposure is pervasive; consequently, we need to design intervention strategies that are practicable for the high-risk populations in economically developing countries.

Mycotoxins, of which aflatoxins are one class, are fungal secondary metabolites of enormous chemical diversity that occur naturally on a wide variety of cereal grains, oilseeds, and tree nuts consumed by humans. The three major genera of mycotoxin-producing fungi of concern for human health are *Aspergillus*, *Fusarium*, and *Penicillium*, and this field has been extensively and more concisely reviewed by Bennett & Klich and Etzel, respectively. The potential production of mycotoxins is invidious because fungal growth occurs both prior to and after grain harvest. Ecological conditions such as drought or damage to seeds by insects or mechanical harvesting can enhance mycotoxin production during both cultivation and storage. Mycotoxin production also occurs over a wide range of moisture content, relative humidity, and temperature, and all major human food crops such as corn (maize), peanuts (groundnuts), cotton, wheat, rice, and the processed food derived from these commodities can be affected. Because the requirements for aflatoxin production are relatively nonspecific, commodities can become contaminated at final concentrations varying from <1 μg/kg (1 ppb) to >12000 μg/kg (12 ppm). In addition, contamination of, for example, maize kernels or a batch of peanuts is highly heterogeneous; consequently, measurement of human exposure to aflatoxin by sampling foodstuffs or by dietary questionnaires is extremely imprecise. Thus aflatoxin biomarkers have great potential for accurate assessment of exposure.

Following the discovery of the carcinogenic aflatoxins nearly 50 years ago, the search for mycotoxins has led to the identification of more than 100 toxigenic fungi and greater than 300 mycotoxins worldwide, and the metabolism and genetics of these fungal metabolites have been recently updated. Most of these mycotoxins have not been definitively linked to any toxic syndromes in animals or people, but in addition to aflatoxin, certain trichothecenes, fumonisins, and ochratoxins have been implicated in highly lethal episodic outbreaks of mold poisoning or chronic diseases in exposed animals and/or human populations. Mycotoxins with carcinogenic potency in experimental animal models include aflatoxins, sterigmatocystin, ochratoxin, fumonisin, patulin, and penicillic acid. Of these agents, aflatoxin B_1 has been classified as a Group 1 human carcinogen by the International Agency for Research on Cancer. Finally, because these mycotoxins co-occur in the developing world, the likelihood is high for additive and multiplicative toxic effects in people from this complex set of compounds.

14.2 BIOMARKERS FOR PROTECTIVE INTERVENTIONS

In wealthy grain-producing countries of the world, economic resources exist to ensure that regulations to limit aflatoxin exposure in the food supply are implemented. Furthermore, in the grain commodities markets, the prices of corn and groundnuts are often dictated by aflatoxin content, which contributes to much lower levels of exposure in wealthy countries. Thus, a result of these regulations and market forces is that people in economically developing countries are exposed to far higher levels of aflatoxins in the diet. This situation strongly motivates a number of groups to develop both effective and affordable intervention strategies. Biomarkers to assess the efficacy of these intervention strategies are becoming ever more important to validate and characterize these interventions in populations.

Biomarkers of exposure, internal dose, and the biologically effective dose of aflatoxin are the most commonly used for assessing intervention studies. These biomarker classifications are based on the classical paradigm for molecular epidemiology. A biomarker of exposure refers to measurement of the specific agent of interest, its metabolite(s), or its specific interactive products in a body compartment or fluid and indicates the presence and magnitude of current and past exposures. Biomarkers of an internal dose and a biologically effective dose for aflatoxins are generally the oxidative metabolites and DNA and protein adducts formed by the critical aflatoxin-epoxide derivative. These biomarkers are illustrated in **Fig. 1** .

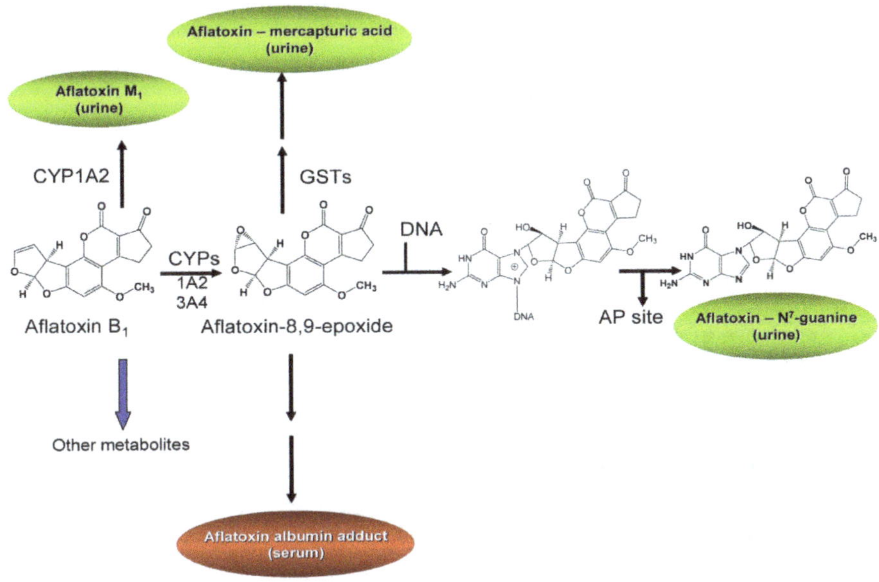

Groopman JD, et al. 2008.
Annu. Rev. Public Health. 29:187–203

Fig. 1. Biomarkers of aflatoxin exposure in an internal dose and a biologically effective dose. Biomarkers of exposure include aflatoxin M1, the internal dose includes the aflatoxin-mercapturic acid and aflatoxin-albumin adduct, and the biologically effective dose is reflected by the excretion of the aflatoxin-N7-guanine adduct formed by depurination leading to an apurinic (AP) site in DNA.

One can find a marked difference between the ability to measure a particular biomarker in a human biological sample employing high-quality analytical approaches and the ability to interpret that information on the basis of thorough validation of the biomarker. The validation step involves the careful characterization of the relationship between the biomarker and, for example, environmental exposure to the agent of interest or to the consequent progression of disease. This process of biomarker validation is well served by parallel experimental and human studies.

As shown in Fig.-2 (see color insert), a conceptually appropriate animal model is used to determine the associative or causal role of the biomarker on the disease pathway and to establish relations between dose (exposure) and response. The putative biomarker can then be validated in pilot human studies, where sensitivity, specificity, accuracy, and reliability parameters can be established. Data obtained in these studies are used to assess intra- or interindividual variability, background levels, relationships of the biomarker to external dose or to disease status, and feasibility for use in larger population-based studies. For a full interpretation of the information that the biomarker provides, prospective epidemiological studies may be necessary to demonstrate the role of the biomarker in the overall pathogenesis of the disease. Finally, these biomarkers can be employed as efficacy endpoints in interventions in both experimental models and high-risk human populations.

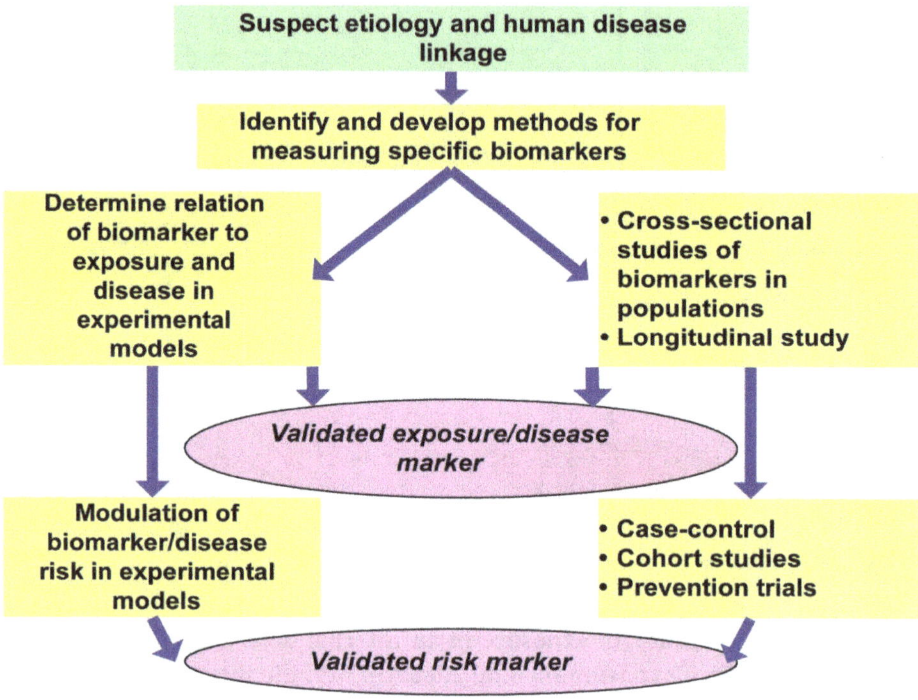

Groopman JD, et al. 2008.
Annu. Rev. Public Health. 29:187–203

Fig. 2 Biomarker validation strategy

14.3 AFLATOXINS: CHEMISTRY, OCCURRENCE, AND HUMAN TOXICOLOGY

The aflatoxins were discovered in the early 1960s as the causative agent of turkey X disease, which resulted in the death of thousands of turkey poults, ducklings, and chicks fed a contaminated peanut meal. Chemically, the aflatoxins are a highly substituted coumarin structure containing a fused dihydrofurofuran moiety. Four major aflatoxins designated B_1, B_2, G_1, and G_2 are produced by *A. flavus* and *A. parasiticus*. The B series were named because of their strong blue fluorescence under UV light, whereas the G series fluoresced greenish-yellow.

Widespread concern regarding the toxic effects of aflatoxins in humans and animals and possible transfer of residues from animal tissues and milk to humans has led to regulatory actions governing the interstate as well as global transport and consumption of aflatoxin-contaminated food and feed commodities. The United States Food and Drug Administration (FDA) has set action levels for aflatoxins in commodities. For feeding mature nonlactating animals, the action level is 100 ppb; for commodities destined for human consumption and interstate commerce, 20 ppb; and for milk, 0.5 ppb. The European Union has promulgated up to tenfold lower standards for aflatoxin content in foods.

The regulatory standards for aflatoxin originally promulgated in the 1960s were not based on toxicology data but were driven by the analytical method of the day: thin-layer chromatography. Thus, it was still unclear if ppb levels of aflatoxin posed a toxicological hazard. This issue was experimentally addressed by feeding aflatoxin to rats at levels of 1, 5, 15, 50, and 100 ppb. These levels induced liver tumors at incidences of 4.5%, 9%, 19%, 80%, and 100%, respectively. Therefore, even µg-per-day doses of aflatoxin could induce liver cancer, and this result provided a context for human cancer investigations exploring the linkage between aflatoxin and HCC.

HCC is a major cause of cancer morbidity and mortality in many parts of the world, including Asia and sub-Saharan Africa, where there are upwards of 600,000 new cases each year and more than 200,000 deaths annually in the People›s Republic of China alone. Furthermore, the median age of diagnosis and death from HCC is between 45 and 55 years of age. The map in Fig. 3 , based on the IARC cancer database, illustrates the unequal distribution of this disease. Upwards of 80% of the burden of HCC is manifest in the developing world. Because the occurrence of HCC is coincident with regions where aflatoxin exposure is high, efforts started in the 1960s to investigate this possible association. As in all ecologic investigations, the work was hindered by the lack of adequate data on aflatoxin intake, excretion, and metabolism in people; by the underlying susceptibility factors such as diet and viral exposure; and by the incomplete statistics on worldwide cancer morbidity and mortality. Despite these deficiencies, early studies did provide data illustrating that increasing HCC rates corresponded to increasing levels of dietary aflatoxin exposure.

Although investigators have focused extensively on the role of aflatoxin exposure in HCC, over the years a number of cases of acute aflatoxicosis in humans have been reported in regions of some economically developing countries. The clinical manifestations of aflatoxicosis were vomiting, abdominal pain, pulmonary edema, and fatty infiltration and necrosis of the liver. In the 1970s data showed putative aflatoxin poisoning in western India in conjunction with consumption of heavily molded corn. Officials reported at least 97 fatalities, and these deaths occurred only in households where the contaminated corn was consumed. Histopathology of liver specimens revealed extensive bile duct proliferation, a lesion often noted in experimental animals after acute aflatoxin exposure. An incident of acute aflatoxicosis in Kenya in the early 1980s was also associated with consumption of maize highly contaminated with aflatoxin. Reports showed 20 hospital admissions with

20% mortality. In a more recent report in 1995, the consumption of aflatoxin-contaminated noodles resulted in acute hepatic encephalopathy in children in Malaysia. Up to 3 mg of aflatoxin was present in a single serving of contaminated noodles.

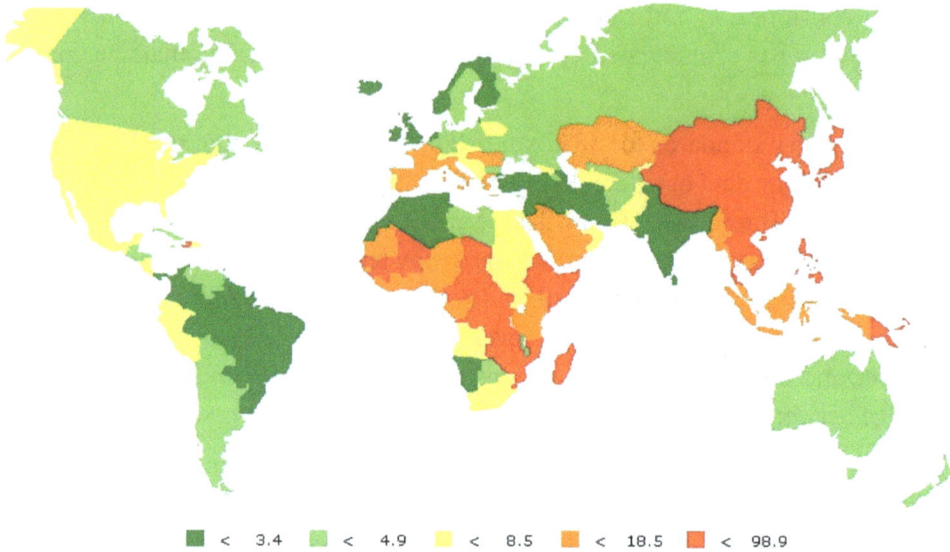

Groopman JD, et al. 2008.
Annu. Rev. Public Health. 29:187–203

Fig. 3. Age-standardized incidence rate of liver cancer per 100,000 men worldwide.

In April 2004, one of the largest documented aflatoxicosis outbreaks occurred in rural Kenya, resulting in 317 cases and 125 deaths. Aflatoxin-contaminated homegrown maize was the major source of the outbreak. In a survey of 65 markets and 243 maize vendors, 350 maize products were collected from the most affected districts. Fifty-five percent of maize products had aflatoxin levels greater than the Kenyan regulatory limit of 20 ppb, 35% had levels >100 ppb, and 7% had levels >1000 ppb. Makueni, the district with the most aflatoxicosis case-patients, had significantly higher market maize aflatoxin than did Thika, the study district with fewest case-patients (geometric mean aflatoxin = 52.91 ppb vs. 7.52 ppb, p = 0.0004). Maize obtained from local farms in the affected area was significantly more likely to have aflatoxin levels >20 ppb compared with maize bought from other regions of Kenya or from other countries (odds ratio = 2.71; 95% confidence interval, 1.12–6.59). In addition to the market survey for aflatoxin exposure, this outbreak in 2004 marked the first time that aflatoxin-albumin adducts independently confirmed the exposure in individuals.

Aflatoxin biomarkers of internal and biologically effective doses have been integral to establishing the etiologic role of this toxin in human HCC. These biomarker investigations revealed the multiplicative interaction of lifetime exposure to high dietary levels of aflatoxins and infection in early life with HBV in the promulgation of this disease.

The use of these biomarkers in etiologic investigations has been extended to preventive interventions in high-risk populations because experimental studies have established both quantitative and temporal linkages between modulation of aflatoxin biomarkers and changing disease risk.

14.4 METHODOLOGIES FOR MEASURING BIOMARKERS

The chemical structures of the major aflatoxin DNA and protein adducts were identified, and extensive research has been conducted to validate these structures for biomarker applications. Early experimental studies around 1980 demonstrated that the major aflatoxin-nucleic acid adduct, AFB_1-N^7-Gua, was excreted exclusively in the urine of exposed rats. The serum aflatoxin-albumin adduct was also examined as a biomarker of exposure because the longer half-life of albumin would be expected to integrate exposures over longer time periods, i.e., months instead of days. Studies in experimental models found that the formation of aflatoxin-DNA adducts in liver, urinary excretion of the aflatoxin-nucleic acid adduct, and formation of the serum albumin adduct were highly correlated events. These investigations provided the rationale for exploring the application of these biomarkers in human studies.

An immunoaffinity clean-up/HPLC procedure was developed for aflatoxin metabolites in urine samples. With this approach, initial validation studies investigated the dose-dependent excretion of urinary aflatoxin biomarkers in rats after a single aflatoxin B_1 (AFB_1) exposure. Investigators found a linear relationship between AFB_1 dose and excretion of the AFB_1-N^7-Gua adduct in urine over the initial 24-h period. Subsequent studies in rodents that assessed the formation of aflatoxin macromolecular adducts after chronic administration also supported the use of DNA and protein adducts as molecular measures of exposure. For example, in rats treated with relatively low doses of AFB_1 (3.5 μg) twice daily for 24 days there was an accumulation of aflatoxin binding to peripheral blood albumin followed by steady-state levels, which illustrated the potential for this biomarker (aflatoxin-albumin adduct) to integrate exposure over time.

Many different analytical methods are now available for quantitation of chemical adducts in biological samples, each with unique specificity and sensitivity. An area of considerable importance, which has received far less attention than it should, is internal standard development. All quantitative measurements require the use of an internal standard to account for sample-to-sample variations in the analyte recoveries. In the case of mass spectrometry, internal standards generally employ an isotopically labeled material that is identical to the chemical being measured, albeit of different molecular mass. In the case of immunoassays, internal standards pose a different challenge because the addition of an internal standard recognized by an antibody results in an incremental contribution to the positive value. The dynamic range is usually less than 100 in immunoassays, and therefore great care must be taken to spike a sample with an internal standard to obtain a valid result. In contrast, most chromatographic methods result in dynamic ranges of analyses that can be over a 10,000-fold range. Many of the aflatoxin studies used different analytical methods, and therefore the quantitative comparison of different data sets has been extremely problematic. However, a recent study compared methods of ELISA and mass spectrometry for aflatoxin-albumin adducts and found high correlation between these two ($r = 0.856$, $p < 0.0001$).

14.5 RELATIONSHIP OF AFLATOXIN BIOMARKERS TO EXPOSURE AND DISEASE RISK IN EXPERIMENTAL ANIMALS

Coincident with the development of aflatoxin biomarkers, studies were underway to identify effective chemoprevention strategies for aflatoxin carcinogenesis. The hypothesis tested in these investigations was that reduction of aflatoxin-DNA adduct levels by chemopreventive agents would be predictive of cancer-preventive efficacy. Preliminary data with a variety of established chemopreventive agents demonstrated that following a single dose of aflatoxin, levels of DNA adducts were reduced. A more comprehensive study using multiple doses of aflatoxin and the chemopreventive agent ethoxyquin reduced the area and volume of liver occupied by presumptive preneoplastic foci by >95% and dramatically reduced binding of AFB_1 to hepatic DNA, with a %90 initial reduction and a %70 reduction at the end of a -2week dosing period. An earlier study using a similar dosing protocol had established that ethoxyquin prevented HCC development in rats. The experiment was then repeated with several different chemopreventive agents, and in all cases, aflatoxin-derived DNA and protein adducts were reduced; however, even under optimal conditions, the reduction in the macromolecular adducts always underrepresented the effect on tumor burden. Therefore, these macromolecular adducts can track with disease outcome on a population level, but in the multistage process of cancer the absolute level of adduct provides only a necessary but insufficient measure of tumor risk.

Using the chemopreventive agent oltipraz, Roebuck et al. established correlations between reductions in levels of AFB_1-N^7-Gua excreted in urine and incidence of HCC in aflatoxin-exposed rats. Overall, reduction in biomarker levels reflected protection against carcinogenesis, but these studies did not address the quantitative relationship between biomarker levels and individual risk. Thus, in a follow-up study, rats dosed with AFB_1 daily for five weeks were randomized into three groups: no intervention, delayed-transient intervention with oltipraz during weeks two and three of exposure, and persistent intervention with oltipraz for all five weeks of dosing. Serial blood samples were collected from each animal at weekly intervals throughout aflatoxin exposure for measurement of aflatoxin-albumin adducts. The integrated level of aflatoxin-albumin adducts over the exposure period decreased 20% and 39% in the delayed-transient and persistent oltipraz intervention groups, respectively, as compared with no intervention. Similarly, the total incidence of HCC dropped significantly from 83% to 60% and 48% in these groups. Overall, integrated biomarker level and risk of HCC were significantly associated (p =

0.01). When the predictive value of aflatoxin-serum albumin adducts was assessed within treatment groups, however, there was no association between integrated biomarker levels and risk of HCC ($p = 0.56$). These data clearly demonstrated that levels of the aflatoxin-albumin adducts could predict population-based changes in disease risk but did not have the power to identify individual rats destined to develop HCC. Such limitations in the utility of these biomarkers must be recognized in human biomonitoring.

14.6 VALIDATION OF AFLATOXIN BIOMARKERS IN CROSS-SECTIONAL STUDIES

Initial studies of aflatoxin biomarkers in human populations began in the Philippines, where investigators demonstrated that an oxidative metabolite of aflatoxin, AFM_1, could be measured in urine as an internal dose marker. Subsequent work conducted in the People's Republic of China and the Gambia, West Africa, areas with high incidences of HCC, determined that the levels of urinary aflatoxin biomarkers followed a dose-dependent relationship with aflatoxin intake. However, as in the earlier experimental studies, this relationship was dependent on the specific urinary marker under study; for example, AFB_1-N^7-Gua and AFM_1 showed strong correlations with intake, whereas urinary AFP_1, a different oxidative metabolite, showed no such link. This type of study, to measure dietary aflatoxin intake and biomarkers at the individual level, is crucial to validate a biomarker for exposure assessment and is often overlooked in molecular epidemiology. Of particular interest in the Gambia was the observation that urinary aflatoxin metabolites reflected day-to-day variations in aflatoxin intake, whereas the aflatoxin-albumin adducts integrated exposure over the week-long study.

Data from these initial cross-sectional biomarker studies demonstrated short-term dose-response relationships for a number of the aflatoxin metabolites, including the major nucleic acid adduct, serum aflatoxin-albumin adduct, and AFM_1. This supported the validity of these exposure biomarkers for use in epidemiological studies, including investigations of intervention strategies and studies of the mechanisms underlying susceptibility.

14.7 INTERVENTION TRIALS FOR REDUCING AFLATOXIN EXPOSURE AND DOSE

Protective interventions against aflatoxin exposures in developing countries can take many forms (Table 1). It is axiomatic that as economic development increases, the ability of a population to afford a more diverse diet also increases. This dietary diversity usually leads to a reduction in aflatoxin exposures because the episodic events, such as those described in Kenya, become less frequent. Obviously, the economic development of a country can proceed at a very slow rate. and intervention strategies are needed during the interim times to lower risks associated with aflatoxin exposures. Any intervention strategies should be rigorously tested and validated using clinical trial designs with biomarkers serving as objective endpoints.

Table 1 Strategies for reducing exposure and risk from aflatoxin in developing countries

Intervention Type	Approach	Outcome
Primary	Post-harvest storage conditions	Reduction in aflatoxin exposure measured by decreased aflatoxin-albumin adduct levels
Primary	Trapping agents (NovaSil)	Reduction in aflatoxin exposure in animal models and clinical trials to begin
Primary	Chlorophyllin	Efficacy demonstrated in experimental models and 55% reduction in urinary aflatoxin-DNA adducts in human clinical trial
Secondary	Green tea polyphenols	Cancer prevention shown in rat models and reduction of oxidative DNA damage biomarkers found in clinical trial
Secondary	Dithiolethiones (Oltipraz)	Efficacy to prevent liver tumors in rat model demonstrated and modulation of aflatoxin biomarkers predictive of protection shown in human clinical trial
Secondary	Broccoli sprouts (Sulforaphane)	Cancer prevention shown in rodent models. Dose-related reduction of urinary aflatoxin-DNA adducts in human clinical trial

Clinical trials and other interventions are designed to translate findings from human and experimental investigations to public health prevention. Both primary (to reduce exposure) and secondary (to alter metabolism and deposition) interventions can use specific biomarkers as endpoints of efficacy. Such biomarkers can also be applied to the preselection of exposed individuals for study cohorts, thereby reducing study size requirements for validation trials. They can also serve as short-term modifiable endpoints. In a primary prevention trial, the goal is to reduce exposure to aflatoxins in the diet. A range of interventions includes planting pest-resistant varieties of staple crops, attempting to lower mold growth in harvested crops, improving storage methods following harvest, and using trapping agents that block the uptake of unavoidably ingested aflatoxins. In secondary prevention trials, one goal is to modulate the metabolism of ingested aflatoxin to enhance detoxification processes, thereby reducing internal dose and subsequent risk. These approaches have been previously summarized in terms of their application at the individual or community level.

There is active work in the development of mold-resistant strains of grains and in the genetic manipulation of molds to lower aflatoxin production. Detailed descriptions of these types of interventions are beyond the scope of this article, and the following sections will focus on the use of aflatoxin biomarkers to develop and validate both primary and secondary prevention strategies in developing countries.

14.8 PRIMARY PREVENTION

14.8.1 Post-Harvest Intervention

The use of aflatoxin biomarkers as efficacy endpoints in primary prevention trials has been recently reported. This work was built on earlier research that established high aflatoxin exposure in Guinea, West Africa, due to consumption of groundnuts as a dietary staple and revealed that postharvest storage was correlated with increases in exposure. On the basis of these observations, a community-based intervention study was conducted among subsistence farmers in the Kindia region of Guinea, West Africa. The intervention comprised a package of educational materials about postharvest measures to limit aflatoxin contamination of the groundnut crop delivered by local agricultural support workers. Farms from twenty villages were included in the design; ten were assigned to each of the intervention and control arms of the study. The intervention practices were all common in the region but were not systematically applied by the rural farmers. In the control villages, farmers were left to follow their normal postharvest practices, which therefore occasionally included one or more of the elements of the intervention strategy. Altered trajectories of aflatoxin-albumin adduct levels (**Fig. 1**) were used to assess the success of the intervention, and this biomarker was measured in 600 subjects over a 5-month period.

In the control villages, the aflatoxin-albumin level increased postharvest, whereas in the intervention villages the level after five months of storage was similar to that immediately postharvest. Mean levels at this time were 60% lower in intervention compared with control villages. The number of subjects with nondetectable aflatoxin-albumin at the time of harvest was approximately 30%, but this number decreased to 2% 5 months later in control villages. In contrast, almost 20% of individuals in the intervention group had nondetectable levels at the same time. The mean level of AFB_1 in groundnuts in household stores in intervention and control villages mirrored the pattern seen for aflatoxin-albumin adducts. The effectiveness of this intervention suggests that significant reductions in exposure can be achieved by using low-technology approaches at the subsistence farm level in sub-Saharan Africa.

14.8.2 Trapping Agents

For many years sodium calcium aluminosilicate marketed as NovaSil (NS) clay, a common anticaking agent in animal feeds, has worked to adsorb aflatoxins in the gastrointestinal tract of animals and diminish the bioavailability and adverse effects of these toxins. Animal model studies using male and female Sprague-Dawley rats fed diets containing 0, 0.25, 0.5, 1.0, or 2.0% (w/w) levels of NS for 28 weeks found no adverse health effects.

Thus clay-based enterosorption of aflatoxins may be a useful strategy for the prevention of aflatoxicosis in humans. These experimental findings, in combination with a larger body of literature, supported the exploration of this agent for intervention studies in human populations at high risk for aflatoxicosis.

An initial study reported on the safety and tolerance of NS in humans and established baseline protocols for long-term efficacy studies. In this work a randomized and double-blinded phase I clinical trial was conducted with 50 volunteers who were randomly divided into 2 groups: The low-dose group received 9 capsules containing 1.5 g NS/day, and the high-dose group received 9 capsules containing 3.0 g/day for a period of 2 weeks. Compliance was outstanding, and by the end of the study there were no statistically significant differences between the two groups for adverse health complaints nor were there differences in any of the clinical chemistries or any other health parameter measured. Thus these findings provide an initial step for the translation of NS from use in the protection of animals from the adverse health effects of aflatoxins to a potential application for use in human populations at high risk for aflatoxicosis.

Chlorophyllin.

The anticarcinogenic properties of chlorophyllin, a water-soluble derivative of chlorophyll, have been demonstrated in a number of animal models. The initial characterization of chlorophyllin as an anticarcinogen arose from inhibition of HCC development in aflatoxin-treated trout. A major mode of action of this agent is thought to be the complexation of aflatoxin by chlorophyllin in a 1:1 ratio. Although the primary mode of action of chlorophyllin appears to be sequestration of aflatoxin, experimental data have characterized enzyme-inducing properties that may also contribute to its mechanism of action. Most recently chlorophyll was protective against aflatoxin-induced cancer in the rat, and the aflatoxin biomarkers tracked with the protective efficacy.

Residents of Qidong, People's Republic of China, are at high risk for development of HCC, in part from consumption of foods contaminated with aflatoxins. In a randomized, double-blind, placebo-controlled chemoprevention trial, chlorophyllin was tested to determine if it could alter the disposition of aflatoxin. One hundred and eighty healthy adults were randomly assigned to ingest 100 mg of chlorophyllin or a placebo three times a day prior to each meal for 4 months. The primary endpoint was modulation of levels of urinary aflatoxin-N^7-guanine adducts collected three months into the intervention. Adherence to the study protocol was outstanding, and no adverse events were reported. Aflatoxin-N^7-guanine could be detected in 105 of 169 available samples. Chlorophyllin consumption at each meal led to an overall 55% reduction ($p = 0.036$) in median urinary levels of this aflatoxin biomarker compared with those subjects taking placebo. Thus, prophylactic interventions with chlorophyllin or supplementation of diets with foods rich in chlorophylls may represent practical means to prevent the development of HCC or other environmentally induced cancers.

14.9 SECONDARY PREVENTION

14.9.1 Green Tea Polyphenols

Many studies have demonstrated that green tea polyphenols **(GTP)** inhibit various chemically induced cancers in experimental animals and epidemiological studies also point to the potential benefit of these compounds. Qin et al. studied the effects of GTP in drinking water for two or four weeks to protect against the development of AFB_1-induced hepatocarcinogenesis in the rat. Data from this investigation revealed that AFB_1-DNA binding in the liver was significantly inhibited by ~20%–30% in animals pretreated with green tea and that the burden of preneoplastic lesions was significantly inhibited by 60%–70%.

The experimental data on GTP provided the impetus to translate this strategy to human clinical trials. In the only study reported to date in an aflatoxin-exposed high-risk group in Guangxi, People's Republic of China, the effects of GTP on the oxidative DNA damage biomarker, 8-hydroxydeoxyguanosine (8-OHdG), was assessed in urine samples collected from a randomized, double-blinded, and placebo-controlled phase IIa chemoprevention trial. 8-OHdG increased following aflatoxin exposure and is associated with HCC risk. All participants tested positive for aflatoxin-albumin adducts and took GTP capsules daily at doses of 500 mg, 1000 mg, or a placebo for 3 months. Twenty-four hour urine samples were collected before the intervention and at the first and third month of the study. At the end of the 3-month intervention, 8-OHdG levels decreased significantly in both GTP-treated groups, with medians of 2.02, 1.03, and 1.15 ng/mg-creatinine for the placebo, 500-mg, and 1000-mg groups, respectively ($p = 0.007$). Thus chemoprevention with GTP is effective in diminishing oxidative DNA damage; however, future studies will have to examine the efficacy of GTP to modulate aflatoxin-specific biomarkers.

14.9.2 Dithiolethiones (oltipraz)

One successful strategy for cancer chemoprevention is modulation of drug-metabolizing enzymes, leading to a facilitated elimination of endogenous and environmental carcinogens. Inducers of conjugating enzymes such as dithiolethiones and sulforaphane inhibit tumorigenesis of environmental carcinogens in various animal models. Increasing lines of evidence show that the Keap1-Nrf2 complex is a key molecular target of chemopreventive enzyme inducers. The transcription factor Nrf2 is a member of the basic leucine-zipper NF-E2 family and interacts with the antioxidant response element (ARE) in the promoter region of phase 2 detoxifying enzymes. A cytoplasmic actin-binding protein, Keap1, is an inhibitor of Nrf2 that sequesters it in the cytoplasm. Inducers dissociate this complex, allowing Nrf2 to translocate to the nucleus. Disruption of the *nrf2* gene in mice leads to enhanced sensitivity to carcinogens and the loss of chemopreventive efficacy by inducers.

Aflatoxin biomarkers were used as intermediate endpoints in a phase IIa chemoprevention trial of oltipraz in Qidong, People's Republic of China. This was a placebo-controlled, double-blind study in which participants were randomized to receive placebo or 125 mg oltipraz daily or 500 mg oltipraz weekly. Urinary AFM_1 levels were reduced by %51 compared with the placebo group in persons receiving the 500 mg weekly dose. No significant differences were seen in urinary AFM_1 levels in the -125mg group compared with placebo. This effect was thought to be due to inhibition of cytochrome P450 1A2 activity. Median levels of AFB_1-mercapturic acid (a glutathione conjugate derivative) were elevated sixfold in the 125-mg group, but were unchanged in the 500-mg group. Increased AFB_1-mercapturic acid reflects induction of aflatoxin conjugation through the actions of glutathione S-transferases. The apparent lack of induction in the 500-mg group probably reflects masking caused by diminished AFB,8,9-epoxide formation for conjugation through the inhibition of CYPlA2 seen in this group. This initial study demonstrated for the first time that aflatoxin biomarkers could be modulated in humans in a manner that would predict decreased disease risk by affecting the Nrf2-Keap1 signaling pathway.

14.9.3 Sulforaphane

Although the oltipraz clinical trial demonstrated the proof of principle for increasing pathways leading to aflatoxin detoxication in humans, the practicality of using a drug-based method for prevention in the economically developing world is limited. Not only is there a potential for adverse health effects from any long-term exposure to a drug, the expense of this type of intervention may make the intervention cost-prohibitive for these populations. Fortunately, oltipraz is not the only agent that affects enzyme changes through the Nrf2-Keap1 pathway. Many foods have high levels of these enzyme inducers.

Drinking hot water infusions of 3-day-old broccoli sprouts, containing defined concentrations of glucosinolates as a stable precursor of the anticarcinogen sulforaphane, was recently evaluated for its ability to alter the disposition of aflatoxin. Sulforphane has been extensively examined for its chemopreventive properties and is a potent activator of the Nrf2-Keap1 pathway. In this study, 200 healthy adults drank infusions containing either 400 or <3μmole glucoraphanin nightly for 2 weeks. Urinary levels of AFB_1-N^7-Gua were similar between the two intervention arms ($p = 0.68$). However, measurement of urinary levels of dithiocarbamates (sulforaphane metabolites) indicated striking interindividual differences in bioavailability. There were individual differences in the rates of hydrolysis of glucoraphan into sulforaphane by the intestinal microflora of the study participants. Accounting for this variability, an inverse association was observed for excretion of dithiocarbamates and AFB_1-N^7-Gua adducts ($r = 0.31$; $p = 0.002$) in individuals receiving broccoli sprout glucosinolates. This preliminary study illustrates the potential use of an inexpensive, easily implemented food-based method for secondary prevention in a population at high risk for aflatoxin exposures.

Over the past 25 years, the development and application of molecular biomarkers reflecting events from exposure to manifestation of clinical disease have rapidly expanded our knowledge of the mechanisms of disease pathogenesis. These biomarkers have increasing potential for early detection, treatment, interventions, and prevention.

14.10 PERSPECTIVES FOR THE FUTURE

The molecular epidemiology and prevention studies of aflatoxin and HCC represent probably one of the most extensive data sets in the field, and this work may serve as a template for future studies of the role of other environmental agents in human diseases with chronic, multifactorial etiologies. The development of these biomarkers was founded on knowledge of the biochemistry and toxicology of aflatoxins, which was gleaned from both experimental and human studies. The biomarkers have been subsequently utilized in experimental models to provide data on their modulation under different situations of disease risk. This systematic approach provides encouragement for design and successful implementation of preventive interventions in exposed human populations.

Over the next 50 years, much of the population growth in the world is going to occur in countries that have intrinsically high aflatoxin exposure, in addition to many other chemical and biological agents. The observation that aflatoxin exposure may affect child growth and susceptibility to infection, as well HCC risk, serves only to emphasize the public health need for development and implementation of interventions in these countries. Over the past 15 years, aflatoxin biomarkers have permitted the exploration of not just the etiology of aflatoxin-related disease but also the efficacy of various intervention strategies. One element in the appropriate application of prevention strategies is the targeting to various stages of the disease progression using knowledge of the underlying mechanisms of action. In addition, investigators need to consider the temporal relationship between exposure and disease. For example, it is now well established that aflatoxin exposure begins in utero and increases markedly as a child is weaned onto solid family foods. This early exposure, at the time the child is also first exposed to numerous infections including **HBV**, suggests that there may be particular merit in intervening to reduce exposure during this critical window. Finally, different interventions may be appropriate for different exposure scenarios. For example, the aflatoxicosis cases in Kenya discussed above occurred on a background of food insufficiency, where avoidance of exposure through primary prevention approaches was not feasible. However, some rapidly deployed secondary prevention may have provided benefits. Alternatively, in subsistence farming communities where dietary staples are frequently subject to aflatoxin contamination year after year, the introduction, through education of primary prevention approaches, may provide longer-term solutions of both health and economic benefit. Overall, the availability of a range of intervention approaches will help combat the adverse health effects of these common environmental contaminants. As the interventions are applied in longer-term studies the impact not only on exposure, but also on health endpoints will allow a more precise understanding of their benefits to public health.

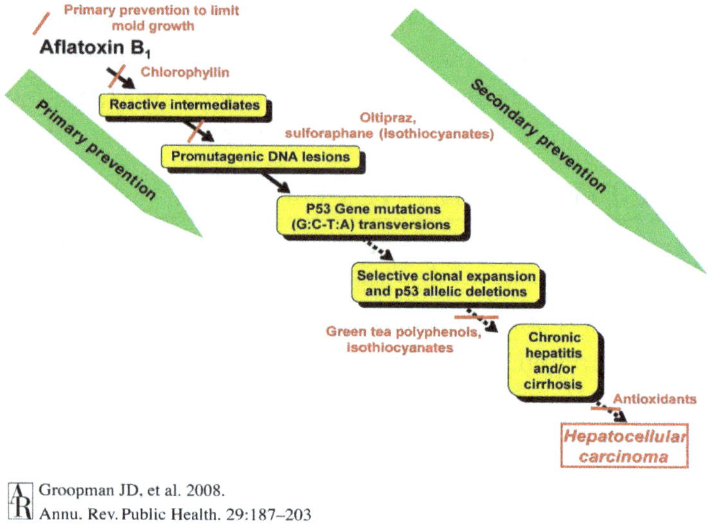

Fig. 4: Targets for intervention in populations at high risk for liver cancer

15

AFLATOXINS OCCURRENCE IN SPICES

15.1INTRODUCTION

The spices are produced mainly in developing countries where tropical and/or subtropical climate such as high temperature, heavy rainfall and humidity encourage fungal growth leading to increased occurrence of aflatoxins (AFs) in spices. Moreover, the inadequate implementation of good agricultural practice, good manufacturing practice and good hygienic practice in these countries are great alarming situation. AFs are considered as a carcinogenic, mutagenic, teratogenic and immunosuppressive to humans and are classified as hazardous food toxins. This chapter provides the worldwide production and regulations of spices, suitable conditions for the AFs production, worldwide occurrence of AFs, detection techniques and some aspect for the reduction of AFs in spices.

Nowadays, scientists are focusing on efficient control of the occurrence of xenogenous constituents in foodstuffs which might be risk for the public health. Spices native to India were grown as early as the 8th century BC in the gardens of Babylon. Spices are considered one of the valuable crops in the world due their important characteristic such as flavoring, colouring and aromatizing as well as antimicrobial and antioxidant effect. Spices are extensively used as a staple crop and cultivated in tropical and sub-tropical regions.

Spices are considered as the non-leafy fragments of plant such as seed, bud, bark, fruit, rhizome or bulp. However, the leaf and flower are designated as different group known as herbs. However, all parts of a plant should be considered to be spices if they possess the aforementioned properties for meal enhancement such as its color, flavor or even texture. Unfortunately, many spices are very susceptible to toxigenic fungal strains and are likely to produce aflatoxins (AFs) contamination. Fungal growth is also exaggerated by the landform, soil natures and properties along with interactions between the micro-fungus and micro- or macro-organisms in soil. In addition, harvesting, drying, handling, packing, carrying, due to probable physical rupturing, insect damage, growth and metabolic action of fungal are also responsible to propagate the fungal proliferation. Moreover, spices purchased in loose or open packing are proved to be considerably more contaminated than spices purchased in sealed or close packing.

AFs are naturally occurring metabolites mostly created by *Aspergillus flavus* and *A. parasiticus*. *A. bombysis, A. ochraceoroseus, A. nomius*, and *A. pseudotamari* are rarely AFs-producing species. From the mycological viewpoint, each strain shows different qualitative and quantitative abilities to produce AFs. For instance, only about half of *A. flavus* strains produced AFs-producing species more than 10^6 µg/kg. Aflatoxins B and G are produced by the *A. parasiticus* and more improved to a soil environment with limited spreading. Presently, 18 different types of AFs have been recognized, aflatoxin B_1, B_2, G_1 and G_2 (AFB_1, AFB_2, AFG_1 and AFG_2) is considered more toxic but AFB_1 is the most recurrently arising amongst all of them. The order of chronic and acute poisonousness is $AFB_1 > AFG_1 > AFB_2 > AFG_2$. AFs are among the most studied mycotoxins globally. AFs are linked with several diseases such as aflatoxicosis in humans, birds, fishes and livestock domestic animals. AFs are the most harmful and tremendously mutagenic mycotoxins. Moreover, AFs have been designated as the first class carcinogens by the International Agency for Research on Cancer (IARC). AFB_1 is deliberated as most toxic due to its extreme hepatotoxigenic and hepatocarcinogenic ability. The liver is recognized as the prime target organ. Furthermore, the cyclic nucleotide phosphodiesterase action in the heart, kidney, liver and brain tissues could be inhibited by AFB_1. This results in the malfunctioning of these body parts.

Due to above declared facts; many countries have enforced strict guidelines concerning AFs occurrence in food and their products. In addition, spices are frequently vulnerable to AFs contamination, as reported in assessments from various states. Spices are mainly produced in tropical climatic regions with higher range of humidity, temperature and rainfall]. Moreover, inadequate storage, prolonged drying periods and higher moisture contents may cause improvement of AFs in spices. The lacking of infrastructure against fungal attack on food commodities in developing countries (e.g. India, Sri Lanka and Pakistan) causes AFs problems. Limited execution of Good Hygienic Practice (GHP), Good Agricultural Practice (GAP), Good Manufacturing Practice (GMP), Good Storage Practice (GSP), improper storing and inadequate shipping could also liable to *Aspergillus* growth and proliferation the hazard of AFs contamination. The condition regarding AFs contamination in spices is well accepted. Still there is a need to collect information regarding AFs in different states around the world. Consequently, this chapter provides the comprehensive facts regarding occurrence of AFs in spices.

15.2. WORLDWIDE SPICES CULTIVATION

During the last 5 years (2015–2019), the average production of spices was c. 602126902 tonnes and 127787137 tonnes in 2019. These spices includes anise, badian, fennel, coriander, chilies and peppers, cinnamon, cloves, garlic, ginger, nutmeg, mace, cardamoms, mustard seed, pepper (piper spp.), peppermint and vanilla". Asia is considered the largest producer of spices in the world with the production share of 76.2% (197818212 tonnes in 2019). Whereas, India provides most to this segment (9508837 tonnes in 2019), followed by China (5307696 tonnes in 2019). Currently, the global spices and seasonings market is undergoing a healthy growth. Looking forward, the market is predicted to reveal a CAGR of around 4.7% during 2020–2025.

15.3 SUITABLE CLIMATIC CONDITIONS FOR AFS PRODUCTION

The climate is considerably influence on the accessibility and quality of the spices. The change in climate simultaneously impacts the complex of AF-producers to change its fungal community's structure. The temperature and water activity (aw) in the atmosphere alters by the climate which further impacts the gene appearance to produce AFs. The AF-producing genes are clustered on the genome and express the main regulatory genes as transcription activator aflatoxin (*aflR*), pathway regulator aflatoxin (*aflS*) as well as structural genes such as reductase aflatoxin (*aflD*) which are subjective by the contact of temperature and aw conditions. As revealed by the, the expression fraction of *aflR*/*aflS* significantly connects with the amount of AFB_1 produced. Most examinations in regions with warm climates have emphasized the occurrence of fungal species of the genus Aspergillus in spices. Generally, the greater contamination is found in warm, humid and even hot deserts and drought environments. However, ideal situations for the AFs formation is considered as moisture content between 18 to 20%, water activity >0.82, pH 3.0 to 8.5 and ambient temperature between 12 to 40 °C (54 to 104 °F) with an optimal at 25 to 30 °C (77 to 86 °F). Nutrition aspects such as carbohydrate, nitrogen sources, zinc, phosphates and other trace metals also influence the development of AFs.

15.4 REGULATIONS OF AFLATOXINS IN SPICES

Though large number of mycotoxins occur in nature however only few toxins (e.g. Aflatoxins) creates food safety and security problems. Therefore, it is necessary, to prevent dangerous outbreaks of these toxins in humans and animals, also to control them within tolerance limits assigned by international agencies. The international regulatory agencies & authorities establish maximum tolerated limit (MTL) for AFs in spices because of severe toxicity of these toxins. The MTL relating to AFs differ from country to country, as developed nations have set lower tolerance limits as compare to developing countries where these susceptible commodities are produced. In addition, the MTL differ from one country to another because of different agricultural practices and climatic conditions. The Food and Agriculture Organization (FAO) has stated that nearly 100 countries have been established MTL for mycotoxins or minimum only for AFs. While 13 countries are uncertain to provide specific regulations and almost 50 countries have no regulations or no data exist.

As it is evident from these data, at present, a fair number of the Codex member states have fixed the maximum limits on AFs in spices. These limits range from 1 μg/kg

(Honduras) to 30 µg/kg (India). The first tolerance level of 30 µg/kg for total AFs in all foods was legislated in 1965 by the USFDA. Later, it was reduced to 20 µg/kg due to the potent toxicity of AFs [26]. The European Union' Scientific Committee on Food (SCF) established the MTL in spices i.e. 5 µg/kg for AFB_1 and ١٠ µg/kg for total AFs. In 2017, the National Standard of China has been updated by the National Food Safety Standard for Maximum Levels of Mycotoxins in Foods (GB 2761–2017) and in January 2020, the public consultation on its revision was launched. While under the National Standard, the maximum level is set at 5 µg/kg for AFB_1 in spices. In India, the MTL prescribed for AFs in spices by the Food Safety and Standards Authority is 30 µg/kg. Also, MTL have also been established for AFs in different spices in many countries as listed in Table 1.

Table 1: MTL as established by various countries for AFB_1 and AFs (B_1, B_2, G_1 and G_2) in spices and foods.

Country	Commodities	Maximum acceptable limits (µg/kg)	
		AFB_1	$AFB_{1+}AFB_2 + AFG_1 + AFG_2$
United States (FDA & FAO)	All foods	—	20
EU	Spices	5	10
Bulgaria	Spices	2	5
Croatia	Spices	30	15
Brazil	Spices	—	20
China	Spices	5	—
Czech Republic	Spices	20	—
Finland	All Spices	—	20
Indonesia	Spices powder	15	5
Brazil	Spices	20	30
India	All foods	30	—
Turkey	Spices	10	—
Iceland	All Spices	30	—
Iran	Spices	5	10
Republic of South Africa	All foods	5	10
Australia	All foods	—	5
Uruguay	All Spices	5	20
Hong Kong	All foods	—	15
Malaysia	All foods	—	35
Japan	All foods	10	—
Singapore	All foods	—	0
New Zealand	All foods	—	5
Sri Lanka	All foods	—	30
Pakistan	Selected Spices	—	30

15.5 AFLATOXINS DETECTION TECHNIQUES IN SPICES

To develop effective and valid analytical methods extensive researches have been carried out for qualitative and quantitative detection or determination of AFs in spices. Generally, the determination of AFs is performed in two steps, (i) extraction or clean-up of samples and (ii) detection or quantification of AFs. The purpose for the use of different extraction and clean-up methods is to separate AFs from other matrix components and to minimize the impact of heterogeneous distribution of AFs. As a result, reduce the background signal during the instrumental analysis. Conventional extraction approaches are unable to precisely & accurately analyse AFs in spices. It is because of the presence of natural colour/pigment producing background interference in HPLC Analysis results masking of toxin. In addition, the complexity of natural constituents in spices matrix frequently makes it challenging to efficiently extract the AFs. Various extraction solvents are currently is in use such as Methanol: Water (80:20), Acetonitrile: Water, Acetone: Water and Methanol: Water. The reported maximum recovery was achieved by Methanol: Water. However, various clean-up methods have been proposed as mentioned below:

- Liquid–Liquid Extraction (LLE)
- Solid Phase Extraction (SPE)
- Immunoaffinity Column Clean-up (IAC)
- Supercritical Fluid Extraction (SFE)
- Energy-Assisted Extraction (EAE)

To clean up AFs in foodstuffs, SPE using a silica gel column, a florisil column, or multifunctional columns has been used. Recently, immunoaffinity (IA) chromatography utilizing immunological interaction which has a high clean-up effect has been employed as well. In addition, the Quick, Easy, Cheap, Effective, Rugged and Safe (QuEChERS) method is a simple and straightforward technique has also been utilized for AFs extraction. Furthermore, the detection methods are based on the emission and absorption characteristics such as thin layer chromatography (TLC), 2D-TLC (Two Dimensional), Enzyme linked immunosorbent assay (ELISA), Fluorometric determination, High performance liquid chromatography (HPLC), Gas chromatography (GC) and Liquid chromatography linked with tandem mass spectrometry (LC–MS–MS). Still, there are certain advantages and disadvantages linked with these developed techniques e.g. reduced sensitivity, unsatisfied accuracy, scope restriction, cost, laborious and sometimes extended analysis time.

Thin layer chromatographic technique is considered as preliminary analytical testing method for mycotoxins. From time to time, TLC is performed for the confirmation of other methods or techniques outcomes & effectively being practiced in developing countries because of low cost. The difficulties of TLC method of analyses are low sensitivity, sometimes poor separation and unsatisfactory accuracy. Special skills are also required to observe the separated spots of compounds (toxin) on TLC cards in presence of ultraviolet or fluorescent lamps. The analyte is difficult to separate from sample matrix in this technique. Also, the poor reproducibility or repeatability in TLC because of factors e.g. sample extraction in solvents, application of extracted sample as spots on TLC card or plate and observer visualization. The fluorometric system is relatively sensitive and accurate as compare to conventional spectrophotometric systems. More consideration and sensitive equipment technique is also prerequisite regarding environmental factors.

ELISA technique is widely applied since few years, because of its simplicity, sensitivity and fast quantification of AFs. More benefits of ELISA include simplicity of sample preparation and probably low analysis cost. However, ELISA is established on the immunological mechanism & it needed very specific monoclonal & polyclonal sera for sensitive and specific quantification of antigen. Also, it consume long incubation time, mixing & washing steps and it is unable to analyse AFs individually.

On the contrary to ELISA & TLC, the HPLC method has taken more attention because of its accuracy, sensitivity, specificity and reproducibility, also, the HPLC instrumental analysis is capable to measure single toxin exclusively. In other food matrices, HPLC has increased attraction where determination of AFB_1 is necessary requirement solely. HPLC determination is remarkably sensitive as compare to the TLC system and it can measure AFB_1 toxin even if less than 0.1 μg/kg. Conversely, the disadvantage of HPLC is the toxins need derivatization to improve their fluorescence properties. AFs have natural fluorescence characteristics due to the rigid and conjugated molecular structure. In these circumstances, small alterations in the molecular structure may increase fluorescence characteristic significantly. Hence, prerequisite derivatization with aid of chemicals boosts their fluorescence property. For instance, AFG_2 and AFB_2 are more fluorescent in comparison with AFB_1 and AFG_1.

Several derivatization methods of AFB_1 and AFG_1 are presently in practice. The frequent techniques include (i) pre-column treatment with trifluoroacetic acid (TFA) [41], (ii) post-column derivatization with iodine, (iii) cyclodextrins and (iv) pyridinium hydrobromide perbromide. The drawbacks related to the TFA based treatment method are the longer reaction time with elevated temperature, TFA toxicity and its corrosiveness. Similarly, iodine based method also does have prolonged reaction time (up to 2 min) at elevated temperatures along with chromatogram peak broadening and need of supplementary HPLC pump. Also, daily preparation of the iodine solution is not recommended because of its corrosiveness to avoid capillary blockage and draining of the reagent pump seals. All of the above four methods are prolonged and linked with the some weaknesses. In these circumstances, an exciting methodology of photochemical derivatization is known as Kobra Cell™ (R-Biopharm, Rhone Ltd., UK). This derivatization does not need everyday preparation of any reagent. There is no instability or corrosiveness issues and photochemical derivatization time is 4 seconds only at ambient temperature.

Likewise the method with a low maintenance cost no supplementary pump is needed.

The Gas Chromatography methodology is also operated for the determination of particular AFs which are challengingly measured by HPLC. Unfortunately, this technique is time-consuming and expensive. Sometimes it is required to transform analyte into the volatile compound compared to other methods. Sometimes, pre-derivatization is prerequisite for the AFs before GC injection. Thermal stability of the AFs extract is also an issue because AFs decomposes at high temperature. In addition, the GC is not widespread using for the commercial testing of AFs because of greater running cost than HPLC analysis.

Nowadays, the advanced version of Liquid Chromatography Mass Spectrometry i.e. LC/MS/MS system has achieved more attention for AFs determination. The LC/MS/MS benefits over other techniques are of extra ordinary sensitivity (lower level detection), greatly specific and the confirmation support of mass spectral fragments and identification of interfering impurities. A single run of LC/MS/MS can support the quantification and determination of multiple mycotoxins. Conversely, to get desire selectivity and sensitivity, there is need of additional accuracy in the sample preparation steps. The LC/MS/MS is an expensive equipment unit and it requires more skill for analytics hence it is more recommended for R & D. The extraction specific solvents for LC/MS/MS sample preparation is an additional requirement.

The variation in these approaches could be established because of the effect of matrix and matrix parameters, the intensity of AFs contamination in product. A validated method is desirable in these situations, a method which avoid maximum matrix effect and close by the tolerable limits as obligatory by the international legislative authorities.

15.6 WORLDWIDE CONTAMINATION OF AFLATOXINS IN SPICES

This section of the chapter describes the studies regarding AFs in spices over the last few years. A total of 27 studies altogether covering 19 spices were included. AFs incidence in spices differs place to place due to temperature & moisture differences, microflora and agronomics variations. The worldwide occurrence of AFB_1 and total AFs with respect to each unique spice are shown in Table 2. In terms of AFs, studies are most often concerned with red chilli, black pepper, caraway, cinnamon, aniseed, cumin, ginger, red pepper, clove, fenugreek, coriander, cardamom, turmeric, paprika, curry, garlic and mix spices. The occurrence of total AFs in the above-mentioned spices is usually high to very high.

Composite spices are common in South Asian countries (Bangladesh, India, Pakistan and Sri Lanka) due to enrich aroma, color and flavor in variety of cooked meals. Yet no survey is accessible on AFs incidence in composite spices. Some studies have informed that spices are considerably more contaminated with AFs than other foods. Furthermore, chilli, red and black peppers were found to be more contaminated as compare to other spices as reported from India, Pakistan, Turkey, Iran, Bangladesh, Indonesia and Nigeria. The samples of ground chilli were found more exposed to AFs occurrence than whole or uncut red chilli. It is because of the possibility of inferior raw material usage for powder chilli production. Also, the tendency of powder to be hygroscopic makes it susceptible for high AFs contamination. The inappropriate handling, processing and high amount of fat content in black pepper and the solubility of AFs in fat is most likely the reason of high level of AFs contamination. In addition, the presence of essential oils may prevent the occurrence of AFs in crude spices.]

The antifungal and anti-toxicogenic properties of ginger, turmeric, clove, cinnamon, garlic and cumin can maintain the AFs occurrence at lower level. Some spices show antifungal activity and disrupt the integrity of fungi cell wall which creates AFs. Also, some stated that the extract of turmeric can downregulate the gene expression in the biosynthesis of AFs in *A. flavus* . The lack of AFs in cinnamon and cumin is due to the inhibition of the aflatoxigenic fungi by essential oils and aromatic constituents in these plants. The cinnamon is likely not to be a good substrate for aflatoxigenic fungi development and accumulation. Additionally, the effect of 09 oils studied against the growth and the toxicity of *A. parasiticus* . The clove oil was found capable to inhibit the development of fungi to limit the formation of discontinue the AFs biosynthesis.

Considering highly reported contamination with AFs in spices particular chillis and peppers regular monitoring of the imported spices are highly recommended to maintain the food quality. Advance studies are still needed to address the source of occurrence and to control the AFs level in the spice from pre-harvesting to post-harvesting and from packaging, storage and shipment stages. A worldwide potential risk for AFs contamination may occur during prolonged storing of spices in poor temperature and moisture control. The factor of storage environmental conditions plays a major role in the occurrence of secondary metabolites such as AFs.

Table 2: The worldwide occurrence of AFB_1 and total AFs with respect to each unique spice. AFB_1 = Aflatoxin B1, Total AFs = Total Aflatoxins, n = number of tested samples.

Country	Category of spices	n	Positive samples (%)	AFB_1		Total AFs		Study Year
				Range	Mean	Range	Mean	
Pakistan	1	331	97	—	—	1.3–93.7	17.2	2006–2011
Turkey	2,3,5,6,8,9,19	93	58	0.5–52.5	20.7	—	—	2010–2011
Pakistan	1	69	67	—	—	1.2–600.0	57.5	2012
Turkey	9	42	90	—	—	0.4–86.0	17.1	2013
Turkey	1	182	82	0.2–165.0	—	—	—	2013
India	1,3	18	18	31.1–174.7	—	—	—	2013
Iran	3,9	76	54	0.88–28.6	15.5	1.4–30.2	15.5	2014
Algeria	2,3,4,5,6,7,8,9,11	36	63.9	0.1–26.5	—	—	—	2015
Iran	3,5,9,13	80	40	0.8–17.9	—	0.8–24.1		2015
South Africa	1,8,15,19	70	40	3.0–19.0	—	—	—	2015
Iran	3,7, 8,9,13	120	31	0.2–57.5	—	0.7–57.5		2015
Malaysia	1,3,5,6,7,13	25	88	0.3–28.4	7.3	0.32–31.17	8.4	2015
Pakistan	19	75	77	—	—	0.7–25.7	4.6	2015
Saudi Arabia	1	60	57	—	—	0–200	16.0	2015
USA	1	169	64	ND–94.9	4.8	—	—	2015–2016
Nigeria	1	55	93	ND–156.0	13.5	—	—	
Italy	12	45	31	LOQ–155.7	12.8	LOQ–529.1	13.9	2017
Tanzania	5,8,10,16	120	57	0.15–11.2	0.8	0.1–11.9	1.4	2017–2018
Pakistan	1,3	50	30	23.9–75.8	—	—	—	2018

Country	Category of spices	n	Positive samples (%)	AFB_1		Total AFs		Study Year
				Range	Mean	Range	Mean	
Lebanon	1,2,5,6,7, 8,10,11,13,15,16,17,18,19	94	19	2.2–1118.2	193.4	2 . 2 – 1118.2	168.1	2018
Pakistan	1,3,13,19	120	100	1.0–30.4	6.54	1.5–44.3	9.7	2018
Bangladesh	1	50	75	—	—	12.0–68.7		2018
Iran	3,5,9,13	80	50	—	—	1.2–77.3	8.6	2019
Malaysia	1,20	20	40	4.7–16.9	—	—	—	2019
Nigeria	1	70	69	—	—	0–97.0	8.9	2019
Greece	2,5,6,7,8,9,13,14,15,16,18,19	29	69	LOD–132.7	9.9	—	—	2020
Indonesia	1	6	50	39.3–139.5	—	—	—	2020

Number of spice category: Red chilli; Aniseed; Black pepper; Caraway; Cinnamon; Coriander; Cumin; Ginger; Red pepper; Clove; Sweet cumin; Sweet pepper; Turmeric; Curry; Paprika; Cardamom; Garlic; Fenugreek; Mix spices.

15.7 MITIGATION OF AFLATOXINS IN SPICES

The widespread elimination or inhibition of AFs contamination during pre- and post-harvest steps is not an easy work however strategies to control fungal growth are essential to minimize the exposure to humans. Numerous methods for the detoxification or elimination of AFs by means of physical, chemical and biological approaches have been proposed]. The product safety outcomes of these methods and reducing agents are not clear.

Approaches to address AFs fall under two main areas. The first includes reducing AFs occurrence in the growing cycle by applying good agricultural practices and the other is mitigating the accelerated toxin growth in the post-harvest supply chain both approaches reduce AFs levels in food commodities. Farmers need to use those crop varieties robust to native growing environment, mainly drought, insects and pests, also show resistance to fungal contamination. Postharvest control of humidity is a key to reducing chances of AFs contamination. Irrigation and fungicides can develop plant health to resist the AFs-producing fungus. Solar dryer is also a solution to control moistness in spices earlier to storage.

AFs occurrence could be reduced by different inhibition methods. The toxins toxins levels could be minimized when the defected chilies like midget, dwarf, damaged and broken are physically sorted from achieving an average of 78% reduction in toxin content. The use of gamma radiation can be helpful for the protection of chilies with respect to the production of AFs during storage. In addition, some novel detoxification technologies including a microwave, ultraviolet, electrolyzed water, ozone, pulsed light, cold plasma and gamma irradiation in a support with biological, physical, chemical or genetic engineering methods have the potential to detoxify. The application of each technique has its benefits and drawbacks. Consequently, biocontrol processes in synchrony with other physical and chemical methods with improved packaging materials should be implemented to attain spices safety and security.

By the execution of advanced agricultural technologies, good agricultural practices, good manufacturing practices and good storage practices can mitigate the AFs occurrence or contamination. AFs contamination occurs in pockets of high concentration which are not randomly dispersed throughout the commodity. Thus, sampling is of key importance before sample preparation in laboratory. The sampling, sample handling and analyses are not yet standardized at growers and farmers eventually the users are at risk. Therefore, attention must be taken in the determination of laboratory results and quality testing should be performed from ISO-17025 accredited or similar laboratories.

Furthermore, the unpacked composite spices are susceptible to the AFs occurrence because of direct exposure to climate. The higher levels of AFs presence could be credited to tropical condition which may favor the spread of toxigenic fungi. Organization of American States and Mayan Reserve Foundation jointly reported the corrective measures to reduce

AFs contamination in chilies such as storage at low relative humidity and temperature, shorten the drying time and quick supplying to the user. And last but not least, the skilled personnel to involve in these processes. Hence, consumers are guided to take measures such as procure from reliable retailers, store food in cool conditions and avoid unpacked products.

16

MECHANISMS OF MYCOTOXICITY

16.1 INTRODUCTION

- Mechanisms of mycotoxicity were recently reviewed. Many potential pitfalls are encountered when interpreting the results of in vitro studies and extrapolating them to in vivo studies or to diseases associated with mycotoxin consumption.
- The factors that tend to confound in vitro mechanistic studies include (1) failure to differentiate secondary effects from the primary biochemical lesion (Fig. 1), (2) failure to relate the effective intracellular concentration in vitro to the tissue concentration of toxin that induces disease in vivo, (3) an in vitro model that is either deficient in the biochemical
- target or unresponsive due to other inadequacies of the model system, and (4) failure to adequately model the complexity of the in vivo exposure with regard to potential interactions with other toxins, drugs, environmental, and/or nutritional factors.

16.2 MECHANISM OF ACTION

The study of mechanism of action of toxic compounds is founded in the belief that cells are molecular machines. To understand how mycotoxins interfere with cellular machinery, it is necessary to understand how toxic compounds alter the behavior of the molecules of life. Examples of how mycotoxins interact directly with the molecules of life or prevent their biosynthesis are common (Table 1).

Investigating mechanism of action is an important area of mycotoxin research for several reasons:

- to reveal the initial biochemical lesion leading to the onset and progression of diseases,
- to differentiate between biological effects that only occur at high dosages and those more likely to occur at environmentally relevant ones,

- to predict downstream biochemical effects that will develop as a consequence of the initial biochemical lesion, and
- to predict potential chronic toxicity and potential for interaction with other mycotoxins or bioactive agents, e.g., toxins, drugs.

Because biochemical targets can be similar in all eucaryotic cells, e.g., AAL-toxin understanding the mechanism of action in animal cells may provide insight into strategies to decrease the phytotoxicity of mycotoxins that are virulence factors. This information could be used to engineer biochemical target molecules so that they do not interact with the toxin or so that bioactive metabolites are rapidly metabolized or eliminated.

Mechanism of action studies can provide insight into why fungi produce chemicals that appear to functionally mimic bioactive animal and plant molecules. Understanding the mechanism of action can engender testable hypotheses concerning the role of mycotoxins in pathogenesis.

Unfortunately, mycotoxins as a group cannot be classified according to their mechanism of action. This is not surprising when one considers the diversity of chemical structures that are encompassed by the group of secondary fungal metabolites called mycotoxins.

To grasp the potential diversity of mycotoxin mechanisms of action, it is necessary to understand the potential number of mycotoxins in nature. New mycotoxins are being discovered at a high rate and, while few human diseases are definitively caused by mycotoxins, there is considerable evidence to support mycotoxins being agents of animal disease.

There is little doubt that mycotoxins kill farm animals; however, perhaps of greater importance is the fact that consumption of mold-contaminated feeds by farm animals has been associated with clinical cases involving reproductive effects, feed refusal, emesis, decline in productivity, lameness, bruising, decreased resistance to infectious agents, and other nonlethal effects Thus, mycotoxins are potentially contributing factors in chronic disease and may confound therapeutic and/ or other interventions.

Given the possibly large number of mycotoxins and the diversity of the mechanisms of action, the potential for extremely complex toxin interactions is also great. Mycotoxins with similar modes of action would be expected to have at least additive effects. Conversely, some interactions could have subtractive effects.

For example, cyclopiazonic acid prevents the lipid peroxidation induced by patulin and fungal serine palmitoyltrans ferase inhibitors such as sphingofungins prevent the fumonisin-induced accumulation of free sphingoid bases.

16.3 QMYCOTOXINS OCCUR NATURALLY IN FEEDS AND FOODS

Mycotoxins occur naturally in feeds and foods in low concentrations, with several mycotoxins present. For example, Fusarium proliferatum on corn might produce several fumonisins (e.g., B1, B2, B3, B4, etc.), fusaric acid, beauvericin, fusarins, moniliformin, and fusaproliferin.

Primary cellular biochemical lesion = the molecular interaction between the parent toxin and target biomolecules, which occur within the cell and from which the cascade of events (secondary effects) leading to cellular deregulation and/or toxic cell injury are initiated. It should be recognized that there is often more than one mechanism of action with different dose dependencies and time courses and that these mechanisms may run parallel paths both leading to toxic cell injury (see Fig.1). Where there is more than one proposed mechanism of action the alternative mechanism is presented in parentheses. Question marks indicate uncertainty, or events not documented in vivo, or repeated in more than one study or model system.

Co-occurrence of other *Fusarium* spp. and *Aspergillus* spp. can co-contaminate corn with fumonisins, T-2 toxin, DON, nivalenol, aflatoxin B1, cyclopiazonic acid, zearalenone, and nitrosamines. Thus, with a diverse diet, humans will be exposed to multiple mycotoxins, intermittently and at low concentrations, over a long period of time. Acute toxicity studies cannot reveal the risks associated with this type of exposure; chronic toxicity studies of all mycotoxins and their combinations are logistically impossible. Understanding the mechanism of action in simple in vitro systems can provide a rational basis for predicting possible toxin, drug, and/or nutritional interactions. Coupling this approach with the known co-occurrence of mycotoxins will make discovery of those interactions that pose a health risk from consumption of contaminated foods or feeds more likely.

For example, fumonisins and aflatoxins are hepatotoxic, co-occur on corn, and have quite different mechanisms of action (Table1). Similarly, DON, fumonisins, and ochratoxins are nephrotoxic, are common contaminants of human foods, and have quite dissimilar mechanisms of action. Because the mechanisms are different, is there a possibility for synergy or additive risk from multiple mycotoxin exposure? Are there potential adverse interactions with infectious agents, drugs, pesticides, stress, or nutritional deficiencies that might be aggravated by exposure to specific mycotoxins or their combinations (Fig.1).

Mycotoxin	Initial Lesion → Cascade of Events
Aflatoxin	Metabolic activation → DNA modification → cell deregulation → cell death/ transformation (metabolic activation) → disruption of macromolecular synthesis → cell deregulation → cell death (apoptotic)
Adenophostins	ER IP3 receptor → Ca^{2+} release → cell deregulation → ?
Anthraquinones	Mitochondrial un-coupler → loss of respiratory control → cell death (apoptotic?)
Beauvericin	K^+ ionophore → K^+ loss → cell deregulation → cell death/apoptosis (inhibition of cholesterol acytranferase → disruption of cholesterol metabolism →?
Citrinin	Loss of selective membrane permeability → cell deregulation → cell death (apoptotic?) disruption of macro-molecular synthesis →?
Cycloiazonic	ER and SR C_2^{+-}ATPase → disruption of C_a^{2+} homeostasis → cell deregulation → cell death
Cytochalasisn	Cytoskeleton → disruption of endocytosis → cell deregulation → cell death
Deoxynivalenon	Inhibition of protein synthesis → disruption of cytokine regulation → altered cell proliferation → cell death/apoptosis ?
Fumonisins	Sphinganine N-acultransferase → disrupted lipid metabolism → cell deregulation → cell death/apoptosis (disrupted delta-6 desaturase activity → disrupted fatty acid and arachidonic acid metabolism cell death)
Gliotoxin	Calcium homeostasis → zinc homeostatis → endonuclease activation → apoptosis (radical mediated damage → oxidative stress → cell death) (inhibition of protein synthesis → apoptosis?)
Griseofulvin	Cytoskeleton → deregulation of cytoskeletal control → cell death
Luteoskyrin	Radical mediated damage → oxidative stress → lipid peroxidation → cell death
Monilliformin	Pyruvate and a-ketoglutarate decarboxylation → loss of respiratory control → cell death
Ochratoxin	Disruption of phenylallanine metablism → reduced PEPCK → reduced gluconeogenesis →cell death (metabolic activation → inhibition protein/ DNA synthesis → apoptosis ?) (altered membrane permeability → disrupt calcium homeostatis → cell deregulation → cell death)
Patulin	Non-protein sulfhydryl depletion → altered ion permeability and/or altered intercellular communication → oxidative stress cell death (inhibition of macromolecular biosysthesis → cell death)
Sphingofungin and ISP 1	Serine palmitoyltransferase → decrease sphingolipids → cell deregulation → cell death/apoptosis
Sporidesmin	Radical mediated damage → oxidative stress → ithiol depletion → cell death (disrupt calcium homeostatis → zinc homeostatis → endonuclease activation → apoptosis)
Swainsonine	Manosidase II → disrupted glycoprotein processing → cell deregulation → cell death

Mycotoxin	Initial Lesion → Cascade of Events
Tremorgens	GABA receptors → altered Cl^{+} permeability → disrupted neuromuscular control →?
T-2 toxin	Inhibition of protein synthesis → ? → cell death (apoptotic?) transient Ca^{2+} elevation → endonuclease activation → apoptosis (altered membrane structure → disruption of membrane functions → deregulation cell death)
Wortmannin	Inhibition of phosphatidylinositon 3-kinase → responsiveness to insulin/ growth factors/ apoptosis (inibition of myosin light chain kinase → inhibition of IP3 signaling pathway →?) (Inhibition of phospholipase A-2 →?)
Zearalenone	Cytosolic estrogen receptor → estrogenic response → disruption of hormonal control →?

Many potential mycotoxin interactions can be studied. Unfortunately, resources to conduct these stud-ies are limited. Understanding the mechanism of action can help select those interactions relevant to human, animal, and plant safety. Because many mycotoxins remain to be discovered, the actual combined health risk from mycotoxin exposure is un-known. Therefore, it is proposed that chemicals be classified by their function in vitro as a guide for toxicological studies. This same concept, which has been used extensively in drug discovery efforts, is also appropriate for studying feeds and foods naturally contaminated with fungi and suspected of being involved in human or animal disease outbreaks. Functional "food toxicology" would screen suspected feeds and foods for mechanisms known to be under lying causes of chronic disease.

For example, cereal grains, milling fractions, or fungal culture material could be screened for ability to inhibit sphinganine N-acyltransferase (fumonisin-like activity), inhibit IP3 binding to the IP3 receptor (adenophostin A-like ac-tivity), produce G → T transversion mutations at codon 249 (aflatoxin-like activity), bind to the SERCA-ATPase (cyclopiazonic acid-like activity), or bind to the cyto-solic estrogen receptor (zearalenone-like activity). The number of possible biochemical targets is great and unraveling the source of the biological activity would be difficult but not impossible. Screening fungal isolates would pinpoint those biological activities of fungal origin. The possibility of nonfungal sources of the toxic principal would also need to be assessed.

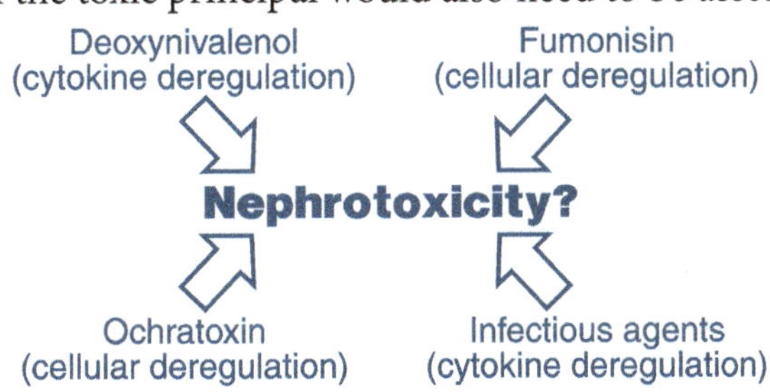

Fig. 1. A theoretical example of mycotoxins and an infectious agent acting in concert to initiate or exacerbate renal dysfunction.

17

FUNGAL GROWTH AND MYCOTOXIN DEVELOPMENT

17.1 INTRODUCTION

- It is not possible to describe a single set of conditions that is favorable for the growth and production of mycotoxins by fungi. Mycotoxins are produced by a diverse group of fungi that differ in their morphology, biochemistry, and ecological niches. Even a single fungal genus, *e.g. Aspergillus*, contains species that differ greatly in their optimum temperature forgrowth and their parasitic abilities.
- In general, mycotoxigenic fungi are not aggressive pathogens but often are well adapted to growing on substrates with low moisture, and they can readily colonize improperly stored grain. Species from a number of genera, however, can colonize and produce mycotoxins in developing plant tissue. This section focuses on species from six genera, *Aspergillus, Fusarium, Penicillium, Claviceps, Stachybotrys*, and *Neotyphodium* (sexual stage, *Epichloe*). These genera were chosen because they contain fungal species that produce mycotoxins of major concern and because they are representative of mycotoxigenic fungi that inhabit diverse ecological niches. The genera *Aspergillus, Fusarium*, and *Penicillium* represent the largest number of mycotoxigenic fungi and are often found on stored grain. Certain species of *Fusarium* and *Aspergillus* also can attack grain in the field. In contrast, infection by *Claviceps* spp. and *Neotyphodium* spp. occurs only in the field. *Neotyphodium* spp. also differ from other mycotoxigenic fungi in that they are endophytes and colonize vegetative tissue as well as reproductive tissue. *Stachybotrys* spp. represent yet another type of mycotoxinproducing fungus. Unlike most mycotoxin-producing fungi, *Stachybotrys* rarely colonizes grain but is of concern because it colonizes wet building material, particularly in houses.

- The fungus also differs because animals and humans apparently are exposed to the mycotoxin it produces by inhalation of mycotoxin- laden conidia rather than by ingestion. It is difficult to make general statements about control strategies for mycotoxins that will apply universally because mycotoxins are produced pre- and postharvest by a diverse group of fungi on numerous crop species. Presently, no absolute controls are available for eliminating mycotoxin contamination. Mycotoxin occurrence in the field can be decreased by good agronomic practices and planting resistant varieties. Although heritable resistance to accumulation of some mycotoxins has been dentified, adequate resistance is not yet available in commercial genotypes.
- Harvesting and storage techniques also can be used to control the occurrence of mycotoxins in crops. The two overriding factors that influence mycotoxin formation in storage are moisture or aW and temperature. Other procedures such as cleaning, insect control, use of antifungal agents, and maintaining the integrity of the seed coat are important in decreasing or keeping mycotoxin levels to a minimum.
- Mycotoxins are produced by a diverse number of fungal species. In general, these fungi are not aggressive pathogens, but some species can invade and colonize plant tissue before harvest and in storage. The two primary factors influencing growth and mycotoxin production, both preharvest and postharvest, are temperature and moisture. These two factors have direct effects on the ecology and pathogenicity of the fungus and are often important in predisposing the plant to infection. While the most common exposure to mycotoxins is from the consumption of plant seeds, there are exceptions. Neotyphodium spp. establish an endophytic relationship with the plant and may produce indole alkaloids in all above-ground tissue. Stachybotrys, a fungus that has re-emerged as a health concern, is associated with damp building material in houses. Exposure to mycotoxins produced by this fungus presumably comes from inhalation of Stachybotrys conidia. It is not uncommon for a substrate to be contaminated with more than one fungus at the same time or for fungi to colonize a substrate in succession as the conditions favor one species over another.No absolute controls are available for eliminating mycotoxins. Also, it is difficult to make general statements about control strategies because of the diversity of fungi producing mycotoxins, both pre- and postharvest, in a wide variety of crop species. Preharvest control strategies are based on use of good agronomic practices and resistant varieties. Harvesting and storage techniques are important in decreasing mycotoxin contamination of crops as well.

17.2 ASPERGILLUS INFECTION AND MYCOTOXIN DEVELOPMENT

- The genus *Aspergillus* represents a large family of fungi that occupies very diverse ecological niches. Although members are distributed worldwide, *Aspergillus* spp. appear most abundant between latitudes 26o to 35o north or south of the equator. Thus, these fungi are more common in subtropical and warm temperate climates. Generally regarded as saprophytes, *Aspergillus* spp. grow on a large number of substrates and are very important in nutrient cycling. Their ability to thrive in high temperatures and with relatively low aW makes them well suited to colonize a number of grain and nut crops. Under favorable conditions, some species have limited parasitic abilities and can colonize crops in the field.
- Within this genus are some of the most important fermentation fungi, e.g., *A. niger, A. sojae, A. oryzae*, cultured for their ability to produce industrial enzymes and metabolites and to impart flavor to foods. Other members, however, are notorious for the mycotoxins they produce. Mycotoxins associated with *Aspergillus* species include aflatoxins, ochratoxins, versicolorins, sterigmatocystin, gliotoxin, citrinin, CPA, patulin, citreoviridin, and tremorgenic mycotoxins.

17.2.1 *Aspergillus flavus*

- Aflatoxins represent the major class of mycotoxins produced by *Aspergillus* spp. Only four species of fungi are known to produce aflatoxins and each belongs to *Aspergillus* section *Flavi*. These species are *A. flavus, A. parasiticus, A. nomius*, and *A. pseudotamarii*. Of these four species, only *A. flavus* and *A. Parasiticus* are economically important. These two fungi have overlapping niches and can produce aflatoxin in developing seeds of corn, peanut, cotton, almond, and pistachio. Other tree nuts such as walnuts and Brazil nuts also are affected. Figs also may be infected, but the incidence is low. These fungi also can produce aflatoxin on most any poorly stored substrate. *Aspergillus Aspergillus flavus* is the predominant species on all commodities , although *A. Parasiticus* is common on peanuts. Because these two species are similar, *A. flavus* will be used to refer to either *A. flavus* or *A. parasiticus*.
- *Aspergillus flavus* was reported to cause an ear mold of corn as early as 1920, but the fungus was of little concern until the 1960s when it was shown to produce the factor (later identified as aflatoxin) associated with Turkey X disease. The significance of preharvest infection of corn by *A. Flavus* was largely discounted prior to 1971 because aflatoxin contamination was considered to be a storage problem only. The report of aflatoxin contamination in the southern and midwestern United States in the 1970s awakened the research community to the significance of preharvest contamination. The occurrence of aflatoxin contamination is sporadic and

highly dependent on environmental conditions. Large populations of *A. flavus* and aflatoxin contamination occur each year somewhere in the southern United States, but serious outbreaks are associated with above-average temperature and below- average rainfall. These two environmental conditions were associated with a high incidence of aflatoxin contamination in the U.S. corn belt in 1983 and 1988. A high incidence of aflatoxin has been reported also in southern China, southeast Asia, and Africa. The infection process of *A. flavus* is best characterized in corn. *Aspergillus flavus* is a soil-inhabiting fungus that reproduces by asexual conidia. Conidia carried to the corn silks by wind or insects can grow into the ear shortly after pollination and colonize kernel surfaces. If environmental conditions are favorable, the fungus may directly invade the seeds and cobs or it may enter through wounds created by insects. In either case, significant infection and aflatoxin contamination do not occur until the kernel moisture is below 32%. Aflatoxin can continue to be produced in kernels until the moisture reaches 15%. Although insects are not required for aflatoxin contamination, their presence increases the level of contamination and high levels of aflatoxin are almost always associated with insect injury,especially by the European corn borer, *Ostrinia nubilalis*. There is evidence that peanut flowers also may be infected with *A. flavus*, although this route of infection appears minor compared to infection of the pods. The exact route of infection into pods is not known but insects appear to play a major role. Both mites and lesser stalk borer larvae (*Elsmopalpus lingosellus*) are known vectors of the fungus. Even microscopic damage to the pods increases infection by the fungus

- Even though there is evidence for direct infection of cotton by *A. flavus*, insects always are associated with high levels of aflatoxin in the field. Exit holes made by the pink boll worm larvae (*Pectinophora gossypiella* Saunders) appear to be the entry point for the fungus.
- Infection of pistachios by *A. flavus* is associated with *early splits*, a condition in which the hull splits before the nut is mature. In both pistachios and almonds, high aflatoxin contamination is associated with damage by the navel orange worm larvae. The two overriding conditions that influence aflatoxin contamination are temperature and moisture. In corn and peanuts, high temperatures and drought stress lead to high levels of aflatoxin contamination. Under field conditions where soil moisture and temperature were controlled, showed that neither by itself is sufficient. The researchers found that peanuts grown with adequate moisture had no aflatoxin. Similarly, peanuts grown under prolonged drought with temperatures less than 25°C or greater than 32°C were free of aflatoxin. Colonization by *A. flavus* and aflatoxin contamination were maximized at 30.5°C.

- High temperature and drought conditions increase the airborne inoculum of the fungus . The increased growth and reproduction of the fungus at higher temperatures is presumably related to its relatively high optimum growth temperature. The fungus can grow over a wide range of temperatures (12 to 48°C), but its optimum for growth is 37°C . The higher temperatures and drought conditions also may favor *A. flavus* over other fungi because of its ability to grow on substrates with low water activity. The fungus can grow at an aW as low as –35 megapascals (MPa). Interestingly, the optimum temperature for aflatoxin production is 25 to 30°C. Temperature and drought stress also likely predispose the plant to increased infection; however, little is known about the mechanisms.
- The effect of temperature on aflatoxin contamination of cottonseed seems more complex and poorly understood . Although aflatoxin contamination of cottonseed is rarely a problem in the southern United States, it can be a serious problem in western-grown cotton.
- High day and night temperatures also have been associated with higher aflatoxin levels in almonds.
- The source of inoculum for *A. flavus* is the soil but the predominant survival structure is not known. The fungus produces sclerotia in culture and in cornfields in the southern United States however, sclerotia have not been reported in the Midwest. The fungus probably survives as mycelium and, to some extent, as conidia and sclerotia. Soil temperature and moisture greatly influence the number of conidia in the soil and the air.

17.2.2 Preharvest Control Strategies

- Any management practice to maximize plant performance and decrease plant stress will decrease aflatoxin contamination. This includes planting adapted varieties, proper fertilization, weed control, and necessary irrigation. Even the best management strategies will not eliminate aflatoxin contamination in years favorable for disease development. Previous cropping history has been shown to influence soil populations of the fungus but the importance of initial inoculum has not been established. Further, *Aspergillus* molds grow on litter from pistachio trees but it is not known if practices to decrease this litter would decrease infection and aflatoxin contamination. Decreasing damage by the naval orange worm can decrease aflatoxin contamination. Compared to plant stress, cropping history and plant debris likely play a minor role in aflatoxin contamination. Breeding programs are underway for all major crops affected by aflatoxin contamination however, no genotypes are available commercially with adequate resistance to aflatoxin accumulation.

- Inbred lines of corn and genotypes of peanut with some resistance to aflatoxin accumulation have been identified. Evidence exists for resistance in corn and peanuts to aflatoxin production per se. Breeding for resistance to aflatoxin contamination also is underway in almonds.
- Because chemical control procedures for mycotoxin contamination are not economically feasible for most grain crops, interest exists for developing effective biocontrol agents to decrease mycotoxin contamination. Recent research indicates the potential for a biocontrol agent to decrease aflatoxin contamination of cotton, peanuts, and corn. Researchers have treated these crops with nonaflatoxigenic isolates of *A. flavus* or *A. parasiticus*. Because they occupy the same or similar ecological niche as the aflatoxigenic strains, the rationale for using nonaflatoxigenic isolates of the two fungi is that they are likely the best biocompetitors. So far, there is no evidence that the ability to produce aflatoxin confers a competitive advantage to *A. flavus* or *A. parasiticus*. The most extensive testing of a biocontrol agent for prevention of aflatoxin contamination has been performed. They obtained a permit from the U.S. Environmental Protection Agency (EPA) to treat fields of cotton in Arizona with wheat seed colonized by a naturally occurring nonaflatoxigenic strain (AF36) of *A. flavus*. This treatment has increased the population of AF36 and decreased that of the toxigenic strains and the concentrations of aflatoxin in the cottonseed.

17.2.3 Aspergillus ochraceus and Other Aspergillus Species

In addition to the *A. flavus* group, a number of other *Aspergillus* spp. can be found on commodities before harvest or, most commonly, in storage. One of the more common *Aspergillus* species is *A. ochraceus*. Recently, this species has been renamed as *A. Alutaceus* var. *alutaceus* Berkley and Curtis 1875, and both names appear in the literature. This fungus is best known for its ability to produce ocratoxin, and *A. ochraceus* and *Penicillium verrucosum* are the major producers of ochratoxin. However, other *Aspergillus* and *Penicillium* spp. also produce ochratoxin . The *Penicillium* spp. are probably a more important source of ochratoxin contamination of grain than the *Aspergillus* spp. Controversy exists over which species of *Penicillium* is most important . Ochratoxins are considered a potential cause of the human disease, Balkan endemic nephropathy. In Canada, the predominant mycotoxin produced by *Aspergillus* spp. is sterigmatocystin. It is associated with grain contaminated with *A. Versicolor* and *Eurotium* spp. In warmer climates, grain may be contaminated preharvest with *A. fumigatus, A. nidulans, A. sydowii, A. terreus*, and *A. versicolor*, species more commonly associated with storage. Control strategies for *A. ochraceus* and other *Aspergillus* species are covered in the harvest and storage section later in this chapter.

17.3 FUSARIUM INFECTION AND MYCOTOXIN DEVELOPMENT

17.3.1 Fusarium

Fusarium is a large and complex genus with species that are adapted to a wide range of habitats throughout the world. *Fusarium* populations in agricultural field soils can be very high and include saprophytes that break down plant residues in the soil as well as pathogens that can cause rots, wilts, and other diseases. Because of its importance as a plant pathogen, a great deal is known about *Fusarium*. Present-day nomenclature has developed largely as a working tool of plant pathologists and others seeking usable, practical systems for correctly identifying *Fusarium* species. Practical identification has been based largely on morphological characters, such as colony structure and color and spore size and shape of strains of these fungi grown under carefully controlled conditions in the laboratory.

But because some species are morphologically very similar, misidentifications occur, especially by those with limited experience in fungal identification. Recent advances in identifying and classifying *Fusarium* species include biochemical and genetic markers and vegetative and sexual compatibility among strains.

Although there are dozens of *Fusarium* species, a rather limited number are responsible for most mycotoxin contamination of the food and feed supply. The *Fusarium* species of most concern are those that produce mycotoxins in wheat, corn, rice, barley, oats, and other cereal grains that are human dietary staples and animal feed ingredients Many *Fusarium* species can infect heads of wheat and other small grain cereals under field conditions causing head scab or blight . In most regions of the world, *F. graminearum* (the sexual stage is called *Gibberella zeae*) is the major causal agent of head scab of small grains, including wheat, barley, rye, and oats . Other species, e.g., *F. avenaceum, F. culmorum, F. poae, F. sporotrichioides*, can predominate in small grains in some regions and environmental conditions.

Fusarium graminearum is the world's major causal agent of red ear rot of corn. *Fusarium graminearum* appears as a pink to red mold of the kernels and, in severe infections, of the cob, husks, and shank. Although *F. graminearum, F. fujikuroi*, and related species can cause ear blights of rice, the grain within the ears is not usually infected. Pink and white ear rots of corn can be caused by a group of closely related *Fusarium* species, predominately *F. verticillioides* (syn., *moniliforme*), *F. proliferatum*, and *F. subglutinans*. *Gibberella fujikuroi* is the teleomorph for these three anamorph species.

On suitable culture media, *F. graminearum* rapidly produces profuse mycelial growth, which usually becomes tinged with yellow or pink as the culture ages. The undersurface of the culture is often a brilliant carmine red. This species can be distinguished also by the long, straight, multicelled spores (macroconidia) produced in the mycelium. Many strains of *F. graminearum* are self-fertile and produce blueblack fruiting bodies (perithecia) of the *Gibberella* sexual stage on the surface of culture media and on infected plant tissues. Although cultures of *F. Culmorum* and *F. crookwellense* may also exhibit *F. graminearum* in rapid mycelial growth and a red undersurface, only *F. graminearum* is sexually selffertile. These *G. zeae* perithecia contain large numbersof sexual spores (ascospores), which are an important inoculum source in epidemics of wheat head scab. *Fusarium graminearum* produces trichothecenes, mainly DON and its acetylated derivatives, and zearalenone. Strains of *F. graminearum* can be distinguished using deoxyribonucleic acid (DNA) markers.

Epidemics of wheat and barley head scab, also known as head blight, can occur worldwide wherever rainy or humid weather coincides with host flowering and early kernel-fill stages. In the United States, wheat head scab was first described described in the 1890s in Delaware, Indiana, and Ohio. Throughout the twentieth century, head scab epidemics occurred sporadically in most wheat- and barley growing regions of the United States and Canada. From 1991 through 1997, head scab was a serious problem for wheat and barley growers in the eastern and midwestern United States and in the provinces of Ontario and Manitoba. For example, wheat and barley growers in the Red River Valley region of Minnesota, North Dakota, South Dakota, and Manitoba lost an estimated $1 billion to scab in 1993 alone. Recent epidemics in North America have been attributed to increases in conservation tillage practices and to cropping systems in which corn and wheat are rotated or wheat is grown each year. Wheat scab affects more than 7 million hectares in the People's Republic of China, with yield losses of 20 to 40% in severe epidemics. All European wheat- and barley-growing regions are affected by scab, with recent severe epidemics in southern Russia, Romania, and Hungary during the 1990s. Scab incidence and severity also have increased in Argentina and Uruguay, with yield losses reaching 50% in some regions in 1993.

Wheat and barley head scab have been thoroughly studied in the greenhouse and other controlled environments as well as in the field. Premature loss of green color of the spikelets in the head is the first symptom of the disease and, under persistent wet conditions, bleached spikelets can be visible a few days after infection. This mixture of bright green and pale yellow spikelets makes infected heads easy to detect in the field. Infection of a spikelet during the early flowering stages can prevent kernel development. Spikelet infection during kernel fill yields kernels with various degrees of fungal growth, ranging from chalky "tombstone" kernels to kernels with varying degrees of shriveling and discoloration. If humid weather persists, *F. graminearum* can infect the entire head and produce blue-black perithecia and a pinkish mass of mycelium and macroconidia on the surface of the head. This pinkish crust on the grain's surface is the source of the name *scab* for this disease.

The major scab pathogen, *F. graminearum*, is a widespread colonizer of crop residues, where the pathogen survives during the winter. Wheat stubble and corn stalks are major inoculum sources in North America, while rice stubble is the major source of inoculums in the People's Republic of China. As temperatures increase in the spring, large numbers of perithecia form on the surface of infested residues. Spore trapping indicates that, at the time of wheat flowering, the majority of airborne *Fusarium* spores are *F. graminearum* ascospores, which can easily be distinguished by their curved shape and four-cell structure from the longer, straighter macroconidia. Ascospore release peaks each night, correlated with a rise in relative humidity. Airborne release of ascospores from perithecia peaks during and after rainy periods. Recent scab epidemics in the midwestern United States have been associated with above-average July rainfall, while wheat heads are in flower.

17.3.2 Ear Rot of Corn Caused by Fusarium graminearum

Fusarium graminearum has a broad host range and can cause an ear disease of corn often called red ear rot or *ibberella* ear rot. Red ear rot has been reported in corn-growing areas worldwide but is especially prevalent in temperate climates when relatively cool temperatures and wet weather coincide with silk emergence. *Fusarium graminearum* ear rot epidemics have occurred sporadically throughout the Midwestern United States, *e.g.*, in 1965 and 1972 in Indiana, in 1986 in Minnesota, and region wide during the unusually cool, wet summers of 1992 and 1993 . Localized epidemics of *F. graminearum* ear rot are common in Ontario and eastern provinces, with outbreaks reported in years with wetter summers, *e.g.*, 1972, 1975, 1976, 1977, 1986, 1987, 1990. Red ear rot also affects corn grown in northern Italy, eastern Europe and the former Soviet Union, the People's Republic of China, and central and southern. In some epidemics associated with persistent cool, wet weather, *F. sporotrichioides* has been recovered from corn with ear rot symptoms. *Fusarium graminearum* ear rot of corn is a more complex and challenging disease than wheat head scab. Controlled environmental studies have been limited because mature corn plants are so large and because it has been difficult to consistently establish infection in the greenhouse. Natural epidemics are often localized and sporadic and thus difficult to predict. During early stages of symptom development, infected ears are not easily detected in farm fields because husks may show few symptoms. In severe infections, however, red or pink mold can progress throughout the entire ear, colonizing kernels, cobs, and husks. If wet weather persists, blue-black perithecia and a pinkish mass of mycelium and macroconidia can form on the husk surface.

There are two routes of entry for *F. Graminearum* infection of corn. First, spores landing on the emerged silks can infect the ear by the silk channel. Controlled field inoculations into the silk channel indicate that corn ears are most susceptible to *F. Graminearum* within 4 to 7 days after silk emergence. Second, wounds caused by birds, insects, or extreme weather can provide an opportunity for fungal invasion; such ear damage is positively correlated with ear rot. Infection through wounding must occur before the kernels have hardened significantly and have ceased to be an attractive substrate for spore germination. Ascospores and macroconidia produced by *F. graminearum* growing on corn stalks and other crop residues are the major inocula for both wheat head scab and corn ear rot. In addition, studies in Ontario have shown a positive correlation between *F. graminearum* ear rot and rainfall in July and August, when corn is silking. Epidemics also appear to be associated with wet weather late in the growing season.

Both deoxynivalenol and zearalenone can be produced by many strains of *F. graminearum* on autoclaved grains under laboratory conditions. However, the technical difficulty and expense of mycotoxin analysis have limited quantitative analysis of these mycotoxins in large-scale field experiments. Antibodies raised against DON and other analytical tests have become increasingly available, which should facilitate field analyses. Studies of wheat and corn infected with *F. graminearum* in the field indicate that wheat can have high levels of DON but usually has little or no zearalenone, whereas corn can be contaminated with both classes of mycotoxins. Corn cob tissue often contains the highest concentrations of DON in the ear. Similarly, the rachis of fieldinoculated wheat often was found to contain the highest average DON concentration. DON levels in kernels of wheat and corn are highly correlated with disease severity ratings and with various measures of fungal load, including colony-forming units and ergosterol levels. However, even visibly healthy grain can contain significant amounts of DON. In an Ontario study, 52% of normal-appearing corn kernels collected from an infected field contained 0.28 to 5 micrograms (μg)/gram (g) DON by ELISA (enzyme-linked immunosorbent assay) detection.

17.3.3 Ear Rot of Corn Caused by Other Fusarium Species

Farm animal disease outbreaks at the end of the nineteenth century drew the attention of agricultural scientists in Nebraska, who determined that "moldy corn poisoning" was associated with consumption of corn ears, stalks, and fermented silage heavily contaminated with a fungus described as *F. Moniliforme*. The fungi originally described as *F. moniliforme* currently are considered to comprise at least six different biological species, three of which are now recognized as *F. verticillioides*, *F. proliferatum*, and *F. subglutinans*. To help clarify the taxonomy of this complex genus and to develop a common taxonomic nomenclature, *G. fujikuroi* was organized into eight mating populations (A-H). Mating population A is most often associated with corn, and this prolific fumonisin producer has been renamed from *F. Moniliforme* to *F. verticillioides*. Interestingly, members of the D population, which includes

F. proliferatum, also are able to produce significant amounts of fumonisins. In contrast to mating population A, members of mating population F (*Fusarium* anamorph *verticillioides*, syn., *moniliforme*) do not produce fumonisins and primarily are limited to sorghum as a host. In scientific literature and disease reports from 1904 through the 1970s, it is often unclear whether fungi referred to as *F. verticillioides* are actually *F. verticillioides*,

F. proliferatum, or *F. subglutinans*. During this period, *F. verticillioides* was considered to be the most common pathogen affecting ears of corn (Kommedahl and Windels 1981) but it is likely that many of these observations involved all three species collectively. Several reports from the northern corn belt states and from Canada indicate that *F. Subglutinans* can be more prevalent than *F. Verticillioides* and that *F. proliferatum* is also very common. Disease symptoms caused by these three species are indistinguishable and are commonly called *Fusarium* ear rot. Other *Fusarium* species are isolated infrequently from corn ears with ear rot symptoms. *Fusarium verticillioides, F. proliferatum, F. subglutinans*,and related species produce white mycelia growth, which may become tinged with purple as the culture ages. The undersurface can vary from colorless to orange or dark purple. All these species are characterized by the production of copious amounts of one-celled spores (microconidia) as well as long, multicelled macroconidia in the mycelium. Unfortunately, for the nonspecialist, some of these species can be difficult to distinguish based on colony color, spore size, and shape. *Fusarium verticillioides* and closely related species can usually form *Gibberella* sexual stages but, unlike *F. graminearum*, these species are not self-fertile. To obtain the sexual stage, strains must be interbred or mated with another strain of the same species. The ability of strains to interbreed and produce perithecia-containing ascospores is a useful method to identify species of these morphologically similar fungi. Biochemical and genetic markers are being developed for this challenging group of *Fusarium*. Current evidence indicates that trichothecenes and zearalenone are not produced by *F. verticillioides* and related species. Fumonisins can be produced by all of these species but, in general, strains of *F. subglutinans* are poor fumonisin producers and strains of *F. proliferatum* vary from low to high production.

Fusarium verticillioides, a consistent, highlevel fumonisin producer, is the predominant fumonisin- producing *Fusarium* species in corn grown in the United States and many other regions of the world . Throughout the twentieth century, *Fusarium* ear rot has occurred widely in the United States and is now the most common ear disease of corn in the corn belt states. *Fusarium verticillioides* infects corn worldwide; the distributions of *F. proliferatum* and *F.subglutinans* are less well documented but these species have been reported from temperate regions on every continent. *Fusarium* ear rot is common in Canada but fumonisin levels generally have been low in naturally infected corn. High levels of *Fusarium* ear rot and fumonisins, the major mycotoxin produced by *F. verticillioides* and *F. proliferatum*, have been reported in corn in Europe, the People's Republic of China, central and southern Africa, and South America. *Fusarium verticillioides, F. proliferatum*, and *F. subglutinans* survive in corn residue, which is probably the most important source of inoculum for kernel infection. Macroconidia and microconidia from residue are splashed or carried by wind to above-ground infection sites on the plants. It is not certain whether root infection is caused by conidia or by mycelia growth from infested residue. These *Fusarium* species can be associated with every part of the corn plant: roots, stalks, leaves, cobs, and kernels. From 50 to 100% incidence of kernel infection is not uncommon, with the majority of kernels showing no visible damage. When ear rot appears, kernels can be visibly moldy, darkened, chalky, blistered, cracked, or streaked with white. In severe infections, white orpinkish mycelia can progress throughout the entire ear, colonizing husks and gluing them to the surface of the ear.

Because these *Fusarium* species are ubiquitous in corn plants and colonize corn residues in the soil, it is difficult to establish uninfected plants for controlled experiments in the field or greenhouse. Using straintracking approaches, several infection pathways have been established for *F. verticillioides* and related species. These species can be common in corn seed and can be transmitted to seedlings. There are many additional infection pathways, including root, leaf, and stalk infection, some of which can lead to systemic infection. The contribution of systemically infecting strains to mycotoxin contamination in kernels is not clear. The most important pathways for kernel infection are silk infection by airborne or rain-splashed microconidia and macroconidia and kernel damage by insects or birds. Larvae of the European corn borer (*Ostrinia nubilalis*) have been shown to act as vectors of *F. verticillioides*, transmitting the fungus from corn leaves to kernels. Other insects may act as vectors. In recent field tests, both visible ear rot symptoms and symptomless kernel infection were decreased in corn genetically engineered for resistance to the European corn borer.

The weather conditions that favor *Fusarium* ear rot are not well understood. The three species in question may differ in optimum temperatures and moisture conditions for infection. There is evidence that *F. ubglutinans* is more prevalent than *F. Verticillioides* under cooler conditions. Some studies have associated severe ear rot with dry weather during June and July early in the growing season, followed by wet weather during silking and later in the growing season. Silks are most susceptible to infection during the first week of silking and moisture on the silks favors infection.

Fumonisins can be produced by almost all strains of *F. verticillioides* (*G. fujikori* mating population A) and by many strains of *F. proliferatum* on autoclaved grains under laboratory conditions. Analyses of shelled kernels have found higher fumonisin levels in visibly moldy corn than in good-quality corn. Due to the technical difficulty and expense of fumonisin analysis, the distribution of fumonisins in individual ears of corn infected with *F. verticillioides* has received little study. In one recent study, distribution of fumonisin B1 in symptomatic and symptomless kernels from 116 individual ears of corn was determined. Over a wide range of fumonisin levels, 82 to 100% of the fumonisin B1 in each ear was localized in symptomatic kernels.

17.4 PREHARVEST CONTROL STRATEGIES

The major source of inoculum for the *Fusarium* spp. is colonized plant debris. Large numbers of perithecia are produced by *F. graminearum* on infested residues. Ascospores released from these perithecia in the spring infect wheat at flowering. Recent epidemics of wheat and barley scab in North America have been attributed to increases in conservation tillage practices and to cropping systems in which corn is rotated with wheat or in which wheat is grown each year in the same field. Thus, avoiding rotation of corn with wheat may afford some control for wheat and barley head scab. *Fusarium verticillioides, F. proliferatum,* and *F. subglutinans* survive in corn residue, which is probably the most important source of inoculum for kernel infection. Tillage, however, has not been shown to decrease *Fusarium* ear rot intensity, possibly because these fungi can survive for 21 months or more in infested residue. Dry weather during grain fill and late-season rains have been associated with fumonisin production. Therefore irrigation to avoid water stress during critical periods for the plant may decrease the risks of fumonisin contamination. For head scab of wheat and ear rot of corn, very few sources of resistance have been found. Scab resistance breeding programs in China have identified wheat lines with moderately high resistance to the spread of *F. graminearum* within the head. Chinese wheat germplasm is being used for scab resistance breeding programs worldwide. Corn breeding programs in Canada have identified lines with *F. graminearum* ear rot resistance expressed in silks and in kernels . Corn breeding programs have made little effort to screen for germplasm highly resistant to *F. verticillioides*. Furthermore, screening based on visible ear rot symptoms may not reflect symptomless kernel infection with *F. verticillioides* or the potential for fumonisin accumulation. Enhanced breeding strategies are exploiting advances in understanding the biosynthesis of *Fusarium* mycotoxins and their interactions with plant hosts. Efforts are underway to express mycotoxindetoxifying enzymes in plants, with the goal of preventing mycotoxin accumulation and ear rot symptoms. A gene encoding a fumonisin ester hydrolase cloned from a saphrophytic microbe present in corn ears is being expressed for fumonisin detoxification. A *F. graminearum* gene encoding a trichothecene 3-O-acetyltransferase significantly decreases toxicity of trichothecenes and has potential for expression in corn and wheat.

Another genetic engineering approach is altering the ribosomal protein that is the trichothecene target binding site in plant tissues. Harris and Gleddie (2001) have modified a rice gene to produce a trichothecene- tolerant ribosomal protein. When transformed into tobacco, the modified rice gene provided transgenic cell cultures with tolerance to the trichothecene DON. Research is underway to express this gene in corn and wheat plants, to evaluate its effect on disease symptoms and trichothecene accumulation. Plants genetically modified for enhanced insect resistance have shown decreased contamination with fumonisins. A significant decrease was found in each year of the three-year study. However, infection of corn by *F. Verticillioides* seems complex. Although insect damage clearly increases contamination by fumonisin, there appear to be other modes of fungal entry. Thus, fumonisins can occur in Bt corn as a result of infection that is not related to insect injury. Penicillium Infection and Mycotoxin Development *Penicillium* species are more commonly associated with storage than with preharvest contamination of grain. Preharvest contamination of corn can occur, with *P. funiculosum* and *P. Oxalicum* being the most common in the Midwestern United States; *P. purpurogenum*, *P. funiculosum*, and *P. citrinum*, the most common in the southeastern United States; and *P. chrysogenum*, *P. steckii*, and *P. purpurogenum*, the most common in Spain. *Penicillium oxalicum* can cause lesions on husks and kernels and may cause ear rot. *Penicillium funiculosum* can cause streaking of the pericarp but usually does not affect seed viability. *Penicillium* spp. can occasionally be found on small grains, with infection favored by prolonged wet weather or lodging of grain. Control strategies for *Penicillium* species will be discussed in the harvest and storage section later inthis chapter.

17.5 ERGOT AND ENDOPHYTIC FUNGAL INFECTION AND MYCOTOXIN DEVELOPMENT

17.5.1 Claviceps Species

Claviceps spp. produce a variety of alkaloids. Ergot alkaloids refer specifically to those with the clavinet or ergoline ring system. These include lysergic acid, lysergic acid amide (precursor for the illegal nacotic, lysergic acid diethylamide), and ergopeptines such as ergotamine. In addition to ergot alkaloids, some *Claviceps* spp. produce tremorgenic alkaloids of the indolediterpene class. The malady, paspalum staggers, is attributable to *Claviceps paspali* sclerotia on florets of *Paspalum* spp. grasses. This species also produces ergot alkaloids, so that animals grazing on *Paspalum* that is seeding or in flower may suffer the combined effects of both toxic alkaloid classes.

Outbreaks of *Claviceps africanus* and/or *C. Sorghi* in North America and Australia have become a major concern. Toxicoses in pigs and cattle resulting from the ingestionof contaminated feed have been reported. Ergotism is one of the oldest known mycotoxicoses, and is the result of consuming grain contaminated with sclerotia of *Claviceps* spp. The *Claviceps* spp.are "replacement parasites" in that they replace the developing plant structures with fungal structures called ergots, or *sclerotia*. The early stage of infection involves production of honeydew, a sweet liquid suspension of asexual spores that are transferred by insects from floret to floret. The growing mycelium infects grass florets very close to the time of pollination, grows into the ovary, and surrounds (largely killing) the female structures of the florets. Fungal growth continues until the fungus produces a resting structure termed a sclerotium, or ergot. Ergots fall from the plant, overwinter in the ground, and then fruit sexually the following spring when grass hosts flower. Ergots give rise to ascospores, which are forcefully ejected so that they land on stigma (female part of the flower) of grass florets. These ascospores give rise to hyphae that invade the ovary, thus completing the infection cycle.

17.5.2 Epichloe and Neotyphodium Species

Toxic alkaloids are also produced by group of fungi that establish an endophytic relationship with the host. Many cool-season grasses possess endophytic symbionts of the fungal genera *Neotyphodium* (if asexual) and *Epichloe* (if sexual). *Neotyphodium* was previously known as *Acremonium* until the name was changed in 1996–1997. *Neotyphodium* spp. and *Epichloe* spp. colonize most of the aerial tissues of the host plant, the floral meristems, the ovule, the embryo, and continue infection of new seedlings. *Neotyphodium* and *Epichloe* species produce a diverse array of alkaloids, some with antimammalian activity and some with little or no antimammalian activity. All of these alkaloid classes are active against insects. The important antimammalian alkaloids are the ergot alkaloids and indolediterpenes. These alkaloid classes also are present in sclerotia produced by *Claviceps purpurea* and *C. fusiformis* (in the case of ergot alkaloids) as well as *C. paspali* (which has both ergot and indolediterpene alkaloids). Furthermore, the endophytes vary similarly to *Claviceps* spp. in the levels and chemical structures of ergot alkaloids and indolediterpenes expressed. Some grass-endophyte associations accumulate ergopeptines, e.g., ergovaline, to levels that can cause symptoms to grazing livestock. In these instances, the amount of infected plant material that must be ingested to receive a toxic dose is much higher than for *Claviceps* spp. because the endophyte biomass is relatively low and the levels of ergot alkaloids in infected plants are much lower than in *Claviceps*-infected seed heads.

Neotyphodium and *Epichloe* species producing various levels of antiherbivore alkaloids are known from many cool-season grasses . Reports of associated toxic effects are rare. Episodes of staggerslike toxicoses have been reported from the native New Zealand grass, *Echinopogon ovatus*, which was recently shown to possess a *Neotyphodium* spp. Endophyte . The stupefying effects of lysergic acid amide and related alkaloids have been reported in horses grazing *Stipa robusta* in the southwestern United States and a related grass *Achnatherum inebrians* in northwestern China.

In the United States, most of the widely grown forage grass, tall fescue (*Festuca arundinacea*), possesses endophytic *Neotyphodium coenophialum*. This endophyte produces ergovaline and is implicated in episodic ergotism-like toxicosis of animals grazed on this grass. Estimated losses to the U.S. beef cattle industry in 1990 were $600 billion and substantial losses (undetermined) undoubtedly occur in the dairy and equine industries as well.

Some endophytes produce sufficiently high levels of indolediterpenes, e.g., lolitrem B, to cause tremors in grazing livestock. An example of this malady is ryegrass staggers, suffered by livestock that graze perennial ryegrass (*Lolium perenne*) bearing the endophyte *Neotyphodium lolii*. The malady is usually not considered let al. Case reports from South America, however, suggest an association of staggers and lethal toxicosis with certain endophyte-infected grasses.

Notably, not all endophytes of grasses produce potent antimammalian toxins. Many produce only loline alkaloids (saturated 1-aminopyrrolizidines), peramine (a pyrrolopyrazine alkaloid), or both. These two alkaloid classes exhibit anti-insectan activity but little or no antimammalian activity at physiological concentrations. The protection afforded by these anti-insectan alkaloids — as well as endophyte enhancements of drought tolerance, growth, nematode resistance, and resistance to some fungal diseases —makes endophyte infections of grasses desirable in many circumstances, including pasture and forage.

The *Neotypodium* spp. have a simple life cycle that involves the colonization of most of the aerial tissues of the host plant, the floral meristems, the ovule, the embryo, and continued infection of new seedlings. In this manner, *Neotyphodium* spp. are transmitted vertically through the maternal lineages of infected grassplants. Typically, such seed transmission is nearly100% efficient, meaning that virtually all seeds set by an endophyte-infected mother plant give rise to new infected plants in the next generation.

Epichloe spp. often exhibit the same life cycle of seed transmission as their *Neotyphodium* spp. relatives. However, the *Epichloe* spp. also can be transmitted horizontally by spores arising on infected plants to the florets of uninfected plants nearby .

Table 1. Values reported for some endophyte-grass associationsa

Host grass	Symbiont	Alkaloids (µg g-1 dry wt)			
		EV	SAP	LM	PM
Lollum perenne	*Epichloe typhina*	0	0	0	53
Festuca gigantea	*Elestucae*	0	300	0	4
Festuca longifolia	*E. festucae*	0.9	0	4.0	22
Festuca rubra	*E festucae*	1.2	0	0	0
Achnatherum inebrians	*Neotyphodium spp.*	2,900	ND	ND	ND
Echinopogon ovatus	*Neotyphodium spp.*	0	1,060	0	0
Festuca arundinacea	*N. coenophalum*	0.5	1.100	0	2
Festuca pratensis	*N. uncinatum*	0	5,600	0	0
L. perenne	*N. lolii*	1.3	0	4.7	19

A Abbreviations: EV= ergobaline , SAP= Saturated 1- amino- pyrrolizidines (lolines), LM= lolitrems, PM= peramine , ND = not determined. Date are from Miles et.al. (1996,1998) and Bush et.al. (1997).

This process is coordinated with host flowering. In an infected plant, some or all of the flowering tillers may have fungal proliferation around the leaf sheaths surrounding the immature inflorescence. The proliferating fungal structure is termed a *stroma* and produces abundant spores of the type (conidiospores) that can serve as spermatia. A symbiotic fly, *Botanophila* spp. (formerly *Phorbia phrenione*), transfers spermatia between stromata, causing fertilization of each mating type with its opposite and thus initiating ascospore development. The ascospores are forcibly ejected, land on open florets of neighboring plants, and initiate new infections of the developing seeds. Those seeds then give rise to infected plants.

17.6 PREHARVEST CONTROL STRATEGIES

17.6.1 Ergot

Ergot poisoning is an ever present danger in livestock agriculture. Wild grasses in pasture, if permitted to produce seed heads, often will become infected with *Claviceps* spp. A partial solution is to maintain adequate grazing pressure to keep the pasture grasses in a vegetative state. However, grasses growing in fence rows may become accessible to the livestock after flowering, thereby serving as a source of ergot. Also, hay cut too late may bear ergot or *Claviceps* honeydew. Because the pasture grass, tall fescue, is both an excellent host of *Claviceps purpurea* and the host of an ergot-alkaloid producing endophyte (*Neotyphodium coenophialum*), potential confusion exists regarding the specific fungus responsible for episodes of ergot-like poisoning on tall fescue pastures. It is important to recognize that "fescue toxicosis," if caused mainly by *Claviceps* rather than the endophyte, cannot be ameliorated by replacing the pasture with endophyte-free tall fescue or with other pasture grasses that can host *Claviceps* spp. Thus, pasture management systems should be geared toward minimizing toxicosis of livestock from both endophytes and ergot fungi. Farmers should be watchful for ergotism-like signs in livestock, regardless of the endophyte status of their pasture grasses.

The nature of the infection process makes open-pollinated grains such as rye much more prone to *Claviceps* infection than close-pollinated grains such as wheat. However, the increased use of male-sterile wheat for breeding purposes increases ergot infections. Although this is not a major concern for the food supply, yields of hybrid wheat seed can be greatly decreased by the pathogen. With the advent of more hybrid wheat, the fungus could become a problem.

17.6.2 Stachybotrys

About 15 species of *Stachybotrys* are known. *Stachybotrys chartarum* can be commonly found in food, and *S. alternans* is a cellulosedegrading saprophyte that commonly colonizes straw and hay. The fungus produces a sooty dark layer, especially around the nodes, on hay or straw, which are substrates most commonly associated with stachybotryotoxicosis of animals. *Stachybotrys alternans* can be isolated from wheat, oats, barley, rye, peas, cotton, sugar cane roots, and soil. *Stachybotrys atra* requires a minimum aW of 0.94 for growth and toxin production. Signs of disease in certain types of stachybotryotoxicoses resemble some diseases produced by certain *Fusarium* species. Stachybotryotoxicosis was first reported about 1931 in the Ukraine, and for a considerable time, it was considered a disease only of horses. The disease was found to occur in other animals, however, and became recognized in many parts of the world. In recent years, there has been concern for the association with *Stachybotrys* and illness in humans occupying buildings where there is proliferation of this fungus. Subsequent to a report that *S. Chartarum* apparently was the cause of an illness that occurred for several years in a home in Chicago, Illinois, there was a report of 10 cases of acute idiopathic pulmonary hemorrhage in infants in Cleveland, Ohio. *Stachybotrys chartarum* was isolated from the homes of some of the infants and was considered a potential cause of the syndrome. The association of growth of the fungus with water damage in the homes was evident. No true cause-and effect relationship has been completely established in these cases, however. Further investigation is warranted relative to the toxins from the fungus and the disease occurring in individuals exposed to airborne fungal conidia.

17.7 HARVEST AND STORAGE TECHNIQUES

17.7.1 Time of Harvest

Many species of mycotoxigenic fungi that colonize grain and oil crops are well adapted to grow on substrates with low moisture. For this reason, delayed harvest can result in increased contamination with mycotoxins. For example, aflatoxin concentrations in corn left in the field increases until the kernel moisture is 16 to 18%. Delayed harvest of contaminated grain can result in greater amounts of aflatoxin. Thus, early harvest of contaminated corn, followed by drying, may help avoid increased aflatoxin contamination. Such a strategy is not always economically feasible because of the added cost of drying the grain. The kinetics of fumonisin production in corn have not been determined but *F. verticillioides* can grow in grain until the moisture content reaches 18 to 20%. Late-season rains after crop maturity have been associated with increased aflatoxin and fumonisin contamination.

Peanuts are indeterminate plants that should be harvested when the greatest percentage of peanuts is mature. If peanuts grown under drought stress are left in the field and subjected to precipitation, they become increasingly contaminated. In general, following the recommended practices of crop management and harvesting crops when they are mature helps to decrease aflatoxin contamination.

17.7.2 Cleaning and Drying at Harvest

During harvest, it is important to properly adjust combines to prevent excess damage to kernels, which may predispose them to infection during storage. Furthermore, as the highest levels of mycotoxins often are associated with broken and insect-damaged kernels, careful adjustment of the combine may eliminate these contaminated kernels in the field, with minimum loss of sound kernels. Notably, sound kernels may contain high concentrations of fumonisin and aflatoxin, so removal of only damaged kernels will not eliminate mycotoxin contamination. An overall decrease of total fumonisins (FB1 and FB2) of 60% was obtained by screening and gravity separating corn being discharged from a storage silo.

Sclerotia of *C. purpurea* and *C. fusiformis* tend to be larger than host seeds, while those of *C. Paspali* tend to be approximately the same size. The use of sieving techniques to rid rye and wheat seed of the larger ergots apparently has selected strains whose ergots more closely approximate the size and shape of the seeds. Harvested grain, coffee beans, fruits, and oilseed crops should be dried immediately. The final safe moisture content depends on the crop and the climatic conditions where the commodity is stored. Allowing corn contaminated with *A. flavus* to be held at kernel moistures above 18% for more than 4 to 6 hours can rapidly increase aflatoxin contamination. Care also should be taken to clean storage bins and auger pits and to maintain clean trucks, trailers, and combines, to minimize future contamination of the crop. It is well accepted that grains should be dried to 15% or below before storing. Drying of coffee seems to be critical as well to avoid ochratoxin occurrence. Contamination of coffee with the ochratoxin-producing organism from the soil seems to occur in the drying process. Ochratoxin production seems to occur when the beans have an aW activity of 0.94 to 0.80. If the time the beans are within this window of is longer than three days, there is increased likelihood of the occurrence of ochratoxin in the coffee.

17.7.3 Temperature and Moisture Control in Storage

Growth of mycotoxigenic fungi on grain is influenced by, substrate temperature, seed damage, aeration, fungal inoculum, microbial interactions, and insects. Avoiding mycotoxin accumulation in stored grains and oilseeds depends primarily on moisture control. If the product is too dry to allow fungal growth and it is kept dry, no further deterioration will occur. However, if there is insect or rodent activity, moisture migration, condensation, or water leaks, fungal growth that could lead to mycotoxin contamination will occur. If storage conditions are favorable for *A. flavus*, aflatoxin concentrations will increase in stored grain previously infected in the field. In addition, the fungus can directly infect stored grain. Germination and growth of *A. flavus* require aW greater than 0.85 and temperatures greater than 10°C. The lower limits for growth of *A. flavus* on seeds for a number of crops have been determined and are shown in Table 3.3. *Aspergillus ochraceus*, like *A. flavus*, can infect seeds in the field and in storage. *Aspergillus ochraceus* and *A. flavus*, however, differ in their limits for temperature and aW (Table 3.3). *Aspergillus ochraceus* can germinate and grow at 10oC with an aW of 0.85 to 0.87. *Penicillium* spp. can grow and produce mycotoxins over a wider range of temperatures than *Aspergillus* spp. *Penicillium* species are more abundant in temperate climatic zones, whereas the *Aspergillus* spp. predominate in warmer cli mates. Patulin production by *P. expansum* can occur between 0oC and 24oC. Moisture requirements for *Penicillium* spp. vary widely among species. Some species can colonize substrates with no free water, whereas others can invade and destroy grains and their products with moisture contents in equilibrium with relative humidities of 90% or above at temperatures of –2% to +5%C (Table 3.). Thus, at low temperatures and high moisture, some *Penicillium* species are very competitive with other fungi.

Table. 2 Moisture requirements for *A. ochraceus*, *A.flavus* & *Penicillium* spp.

	A. Ochraceus	*A. Flavus*	*Penicillim* spp.
Plant	**Moisture content of grain(%)**		
Starch cereal grains	15.5-16	17-18	16.5-20
Soyabeans	14.5-15	17-17.5	17-20
Sunflower, Safflower, Peanuts and copra	9.0-9.5	10-10.5	10-15

Penicillium aurantiogriseum Dierckx and *P. Verrucosum* var. *verrucosum* require an aW of only 0.80 to 0.82 to start growing found that germination of conidia and growth of *P. aurantiogriseum* could occur at 5oC at an aW of 0.9 to 0.95. At temperatures between 15oC and 25oC, it could grow at an aW of 0.85. As the fungi continue to grow, moisture released during metabolism allows the substrate to be colonized by other species. *Penicillium* spp. are known to produce a number of mycotoxins on different commodities but only a few are found in naturally contaminated cereals. These include ochratoxins, citrinin, penicillic acid, and xanthomegnin. Although these mycotoxins are attributed to *Penicillium* spp., most are also produced by the *Aspergillus* spp. (Wilson and Abramson 1992). The disease of corn known as blue eye can be caused by species of either *Penicillium* or *Aspergillus*. If the disease, whose symptoms are caused by the accumulation of fungal spores on the germ just under the pericarp, occurs in the field or shortly after storage, it is usually due to *Penicillium* spp. If the disease appears in corn stored for several months to a year at 14 to 14.5% moisture, it is most likely caused by *A. restrictus* or *A. Glaucus*. Over 24 species of *Fusarium* have been associated with human or animal health problems and more than 100 active secondary metabolites are known to be produced by *Fusarium*. Species of *Fusarium* that may be found on stored grain include *F. culmorum, F. graminearum, F. verticillioides, F. tricinctum, F. poae, F. sporotrichioides, F. avenaceum, F.acuminatum*, and *F. sambucinum*. Most of the contamination in storage comes from infections that began in the field. (Colonization of corn and wheat in the field with *Fusarium* species was described in an earlier section.) The fungus can continue to grow in stored grain if the moisture content is high. There have been no reports of fumonisin accumulation during storage.

Zearalenone concentrations appear to be low in grain contaminated in the field but can increase in storage if the moisture is greater than 34%. Zearalenone is not produced under anaerobic silage conditions but zearalenone present in corn before ensiling is not destroyed by the silage treatment. In general, *Fusarium* species require a high aW to colonize grain, usually above 0.90. *Fusarium culmorum* requires at least 0.90 aW for growth.

Temperature and moisture environments can be altered by other factors such as rodent and insect activity. Therefore, insect and rodent control are important because their activity in stored products creates favorable microclimates for fungal growth.

Once fungal growth starts, the water of metabolism is sufficient for further growth and mycotoxin development.

17.7.4 Antifungal Agents

Antifungal agents of low toxicity should be used to supplement good management practices, rather than as a substitute for safe, clean handling practices of mixed feeds. Fungal growth and mycotoxin contamination of high-moisture grains can be prevented with propionic acid or mixtures of propionic and acetic acids. Grain treated with these acids is somewhat corrosive, which can create problems when handling treated grains or feeds. Liquid sprays generally are more effective than dry formulations of propionic acid but the dry formulations are easier to use.

17.7.5 Interaction Among Fungal Species

When conditions are favorable for fungal contamination of substrates it is not uncommon for more than one fungus to be involved. In storage, grain is often colonized by a succession of fungi, the sequence of which is determined in part by the temperature and moisture of the grain. The most xerophytic fungi in stored grain are members of the *A. glaucus* group and *A. restrictus*, which can grow when the aW is as low as 0.7. As the moisture of the grain increases through microbial activity, moisture migration, or insect activity, other fungal species begin to grow. Some species may colonize grain at conditions suboptimal for growth because they can out compete other species under these conditions. For example, *P. aurantiogriseum* and *P. viridicatum* are most abundant at 0oC and 1.0 aW, although they grow best in pure culture at 25oC and 1.0 aW. Likewise, members of the *A. glaucus* group are most abundant in grain at 30oC and at 0.7 aW but grow best at 30oC and 0.90 aW. Marin et al. (1998b) compared the effect of temperature (10 to 30oC) and water activity (0.92 to 0.994) on in vitro microbial interactions of *F. verticillioides, F. proliferatum, F. graminearum, A. flavus, A. niger, A. ochraceus, P. aurantiogriseum, P. griseofulvum*, and *P. citrinum*. Both temperature and aW affected the relative competitiveness of each strain. The researchers found, however, that under the conditions examined, *F. verticillioides* and *F. proliferatum* were able to dominate several corn-contaminating fungi. *F. proliferatum* was more competitive than *F. verticillioides*; however, at 15oC, *F. graminearum* had a competitive advantage over the other two species. *Aspergillus* and *Penicillium* species became more competitive at lower aW. *Aspergillus flavus*, for example, was dominant over *F. verticillioides* and *F. Proliferatum* at 30oC and an aW of 0.94. Similarly, *A. ochraceus*, and *A. niger* were dominant over *F. Verticillioides* and *F. proliferatum* at 15oC and an aW of 0.98. Because of the possible interaction of several fungal species, grain may be contaminated with a number of different mycotoxins.

17.8 GENETIC AND MOLECULAR ASPECTS OF MYCOTOXIN BIOSYNTHESIS

17.8.1 Aflatoxins

Aflatoxins are the most thoroughly studied mycotoxins. Their biosynthesis is the best characterized of any fungal secondary metabolite, with the possible exception of penicillin. Aflatoxins are polyketides derived from acetate as the starter unit. Most of the biosynthetic steps in the pathway are known. Genetic analysis of the biosynthesis of aflatoxin by *Aspergillus flavus* and *A. parasiticus*, and sterigmatocystin production by *A. nidulans*, has led to the cloning of 17 genes responsible for 12 enzymatic conversions in the aflatoxin/sterigmatocystin pathway. Sterigmatocystin seems to be synthesized by a pathway similar to that for aflatoxin. The tractable genetic system of *A. Nidulans* is adding to our understanding of the biosynthesis and regulation of these two mycotoxins. The genes for aflatoxin and sterigmatocystin biosynthesis and the pathway specific regulatory gene, *aflR*, are clustered in a 75 kb region of DNA.

17.8.2 Trichothecenes

Trichothecene biosynthesis begins with the formation of the sesquiterpene trichodiene, which is then oxygenated, esterified, and cyclized to yield a large family of sesquiterpene epoxides. The entire trichothecene biosynthetic pathway has been established for several *Fusarium* species. More than eleven pathway genes have been cloned for both *F. sporotrichioides* and *F. graminearum*. All of these genes, which encode enzymes, a transcription factor, and a transporter, are clustered in a region of DNA.

17.8.3 Fumonisins

The fumonisin biosynthetic pathway is being investigated. Precursor feeding experiments have established that fumonisins originate by condensation of the amino acid alanine with an intermediate derived from acetate. Sexual genetic analysis has shown that several genes involved in fumonisin biosynthesis are clustered on chromosome 1 of *F. verticillioides.* Recent studies have identified the first fumonisin biosynthetic gene, a polyketide synthase that is located in this gene cluster (Proctor et al. 1999). Research also is focusing on using mycotoxin pathway genes to study *Fusarium* biology and the factors that regulate fungal growth and mycotoxin production in agricultural commodities. Using gene disruption to block trichothecene biosynthesis has been a powerful tool for investigating the role of these mycotoxins in plant disease. In-field trials in the United States and Canada, trichothecene-nonproducing mutants of *F. graminearum* demonstrated decreased abiity to cause wheat scab and corn ear rot.

17.8.4 Ergot Alkaloids

Considerable effort is underway to identify or generate genotypes of grass and endophyte that minimize expression of the antimammalian alkaloids while retaining beneficial characteristics. One approach is touse tall fescue genotypes associated with decreased ergot alkaloid levels in symbioses with *Neotyphodium coenophialum*. Another approach is to identify endophyte genotypes that produce few or none of the major antimammalian alkaloids but still exhibit other plant protective characteristics. An advantage to these approaches is that they make use of alreadyavailable biological materials. A disadvantage is that the benefits already well documented for *N. Coenophialum* and *N. lolii* in existing grass cultivars need to be field tested for the new grass and endophyte genotypes.

Another approach to address tall fescue toxicosis is to specifically eliminate genes for biosynthesis of antimammalian alkaloids in the well-characterized endophytes. To this end, the gene for the first determinant step in ergot alkaloid biosynthesis has been cloned from the related fungus, *Claviceps purpurea*. By eliminating the gene for the first determinant step in the pathway, the endophyte should be rendered incapable of producing ergot alkaloids. However, the endophyte should retain the capability to produce lolines and peramine because these anti-insect alkaloids appear chemically unrelated to ergot alkaloids. Thus, the potential exists to improve *N. coenophialum* for agricultural use through genetic engineering. Speculative Function of Mycotoxins to the Fungus Many mycotoxin-producing fungal species can cause plant diseases under field conditions. Thus, it becomes logical to ask whether mycotoxins themselves play a role in plant diseases in addition to their role in animal diseases. Although it has been easy to demonstrate that some mycotoxins are highly toxic to plants, it has proven far more difficult to establish a causal role for mycotoxins in plant disease. Critical analysis of the role of mycotoxins in plant disease awaited the development of laboratory methods to specifically eliminate a toxin from a biological system. DNA-mediated transformation and other molecular biological methods are providing the essential tools to rigorously test the role of mycotoxins in plant disease. In principle, if production of a mycotoxin increases the ability of a fungus to cause a plant disease, then increasing plant resistance to the toxin should increase plant resistance to the disease itself. Various strategies have been used to isolate and disrupt mycotoxin biosynthetic genes in plant pathogenic fungi. As of this writing, mycotoxin biosynthetic pathways that have been blocked by gene disruption include aflatoxin biosynthesis in *Aspergillus* species, ergot alkaloid biosynthesis in *Neotyphodium* species, fumonisin biosynthesis in *F. verticillioides*, and trichothecene biosynthesis in *Fusarium* species. Laboratory and field tests of mycotoxinnonproducing transformants have demonstrated that trichothecenes play an important role in wheat head blight and corn ear rot caused by *F. Graminearum*, but that fumonisins are not required for corn ear rot caused by *F. verticillioides*. Similar efforts are underway to test the roles of aflatoxins, ergot alkaloids, and other mycotoxins in plant disease in the laboratory and in the field.

18

MYCOTOXINS AND HUMAN DISEASE

18.1 INTRODUCTION

Humans likely are exposed to mycotoxins through several routes such as ingestion (the most prominent means of exposure), contact, and inhalation. Historical accounts reveal that ergotism was one of the oldest recognized mycotoxicoses and, potentially, moldy grains were involved in other mycotoxicoses as well. The aflatoxins are known causes of acute aflatoxicosis in humans. But chronic forms of aflatoxicosis, especially carcinomas, are ore problematic because epidemiological evidence is not as clearly defined due to other factors such as hepatitis B that may be interactive in the disease process. Ochratoxin has been conjecturally associated with a disease known as Balkan endemic nephropathy.

Although ergotism in humans likely occurred in medieval times (see cover photograph) as being associated with ingestion of the plant parasitic fungus, *Claviceps purpurea*, little concern for toxic compounds produced by fungi was evident outside the veterinary and agricultural communities. Today, there is an apparent growing concern within the medical community regarding mycotoxin involvement in human diseases. Animal studies and epidemiological evidence indicate definite involvement of fungal metabolites as etiological agents of human disease. Furthermore, the potential for agents of mycoses to produce toxic metabolites that could be virulence factors involved in the pathogenesis of the mycotic disease is largely ignored. One must be careful, however, when blaming mycotoxins for human disease, as the literature includes incriminating statements without complete scientific evidence. As in animal mycotoxicoses, the disease in the human recipient is the result of exposure by ingestion of contaminated food. However, other routes such as inhalation, contact, and passive exposure resulting from a mycotic infection by a toxigenic fungus must be considered.

The rural populations in the Balkans have a high incidence of chronic kidney problems and tumors of the excretory organ system. Russian literature contains some evidence that Fusarium spp. were involved in human disease. The most likely toxic products of this genus, the trichothecenes, are conjecturally associated with the most prominent disease described, alimental toxic aleukia. The trichothecene, DON, is capable of producing a disease in mice that is similar in histological descriptions to human glomerulonephropathy. Stachybotryotoxicosis is a disease that occurs in humans and other animals and is suspected to be caused by toxins of the organism Stachybotrys chartarum. The exact toxic product has not been identified in a cause-and-effect relationship in this disease. Some epidemiological evidence exists, however, for the involvement of the organism in disease. The fumonisins have been associated with esophageal cancer in certain human populations, but again, no exact cause-and-effect relationship has been established. Epidemiological evidence indicates, however, that fumonisin-producing species of Fusarium on corn are involved. Of interest to many investigators is that a number of mycotoxins are immunosuppressive and likely could be involved in human disease. These are difficult to prove, however, because the underlying involvement of the mycotoxin probably is overshadowed overshadowed by disease resulting from the immunosuppressed state of the host.

18.1.1Mycotoxicosis

Mycotoxicosis, the disease resulting from exposure to a mycotoxin, may be manifested as acute to chronic, and ranges from rapid death to tumor formation. More occult disease may occur when the mycotoxin interferes with immune processes, rendering the patient more susceptible to infectious diseases. The underlying mycotoxic event may be overshadowed by the infectious disease and thus not be considered in the overall syndrome. This may explain why little information is available on the natural occurrence of mycotoxins having immunosuppressive activities. Therefore, we are largely dependent on results from experimental studies in animals.

Preharvest contamination of the commodity by toxigenic fungi can be carried over into and persist during storage. Historical accounts suggest that moldy grains and their toxic products have been responsible for major outbreaks of disease and are even considered events of the *Ten Plagues of Egypt*. Stored grains may be damaged by insects or moisture to the extent that a portal of entry is provided to the often-present toxigenic fungi in the storage environment.

Inhalation exposure can result from handling contaminated materials or from airborne fungal constituents, e.g., conidia. Industrial processes using fungi and the milling and other processing of grain for foods should be reviewed. Certainly, some environmental foci provide potential sources for airborne fungal inoculums of humans. Some mycotoxins are considerably dermanecrotic and exposure of skin or mucous membranes by contact can result in disease. Exposure of individuals to mycotoxins due to infection by a toxigenic fungus is problematic but is being investigated for the pathogenesis of certain infectious diseases. The following section discusses mycotoxicoses for which there is considerable evidence for involvement of a specific mycotoxin(s). These and other human diseases where mycotoxin involvement is likely are presented in Table .1.

Table 1: Some human diseases in which analytic and/or epidemiologic data suggest or implicate mycotoxin involvement

Diseases	Species	Substrate	Etiologic agent
Akakabio-byo	Human	Wheat, barley, oats, rice	*Fusarium Spp.*
Alimentary toxic aleukia (ATA or septic angina)	Human	Cereals grains	*Fusarium Spp.*
Balkan nephropathy	Human	Cereal grains	*Penicillilum*
Cardiac befiberi	Human	Rice	*Aspergillus Spp., Penicillilum Spp.*
Celery harveter's disease	Human	Celery (Pink rot)	*Scierotinia*
Dendrodochiotoxicosis	Human, horse	Fodder	*Dendrodochium toxicum*
Ergotism	Human	Rye, cereal grains	*Claviceps purpurea*
Esophageal tumors	Human	Corn	*Fusarium Monliforne*
Hepatocarcinoma (acute aflatoxicosis)	Human	Cereal grains, peanuts	*Aspergillus flavus, A parasiticus*
Kashin Beck diseases "Urov diseases"	Human	Cereal grains	*Fusarium*
Kwashiorkor	Human	Cereal grains	*Aspergillus itavus, A parasiticus*
Onyalai	Human	Millet	*Phoma sorghina*
Rye's syndrome	Human	Cereal grains	*Aspergillus*
Stachyboryotoxicosis	Human, horse and other livestock	Hay, cereal grains, fodder (skin contact, inhated haydust)	*Stachybotrys altra*

18.1.2 Aflatoxicosis

The relationship of aflatoxins to disease in animals has been extensively studied. Although acute aflatoxicosis is well documented in humans, the relationship of aflatoxins to hepatocellular carcinoma and other human maladies is still being investigated. Several epidemiological studies have implicated aflatoxins in the increased incidence of human gastrointestinal (GI) and hepatic neoplasms in Africa, the Philippines, and China. Aflatoxin B1 also has been implicated in human liver cell carcinoma.

- **Acute Aflatoxicosis:** Acute disease in humans resulting from aflatoxin ingestion has been manifested as an acute hepatitis usually associated with highly contaminated foodstuffs, especially corn. In some cases, exposure was sufficient to find aflatoxins in selected tissues and histopathologic evidence was convincingly adequate to allow for a diagnosis of aflatoxicosis. Typical but nonspecific changes in patients with acute aflatoxicosis include jaundice, low-grade fever, depression, anorexia, and diarrhea, with fatty degenerative changes in the liver evident, upon histopathologic examination, as centrolobular necrosis and fatty infiltration. Tenderness near the liver was evident in patients with acute, aflatoxin-caused hepatitis in Kenya; ascites may develop. Mortality reached 25% in outbreaks in India. Samples of liver obtained from patients that died contained detectable levels of aflatoxin B1. Two human diseases of undefined etiology have been linked to the consumption of aflatoxin- contaminated foods: kwashiorkor and Reye's syndrome. Kwashiorkor has been geographically associated with the seasonal occurrence and distribution of aflatoxin in food (Hendrickse et a. Animals given dietary aflatoxin possessed some of the same attributes of kwashiorkor, namely, hypoalbuminemia, fatty liver, and immunosuppression. Aflatoxins were detected in liver tissue taken at autopsy from 36 children with kwashiorkor, which has added to the credibility of establishing aflatoxin as the cause of this human disease with no other known etiology.
- Malnutrition may change the metabolism of dietary aflatoxin, resulting in its detection in some patients with kwashiorkor. The etiology of Reye's syndrome, however, is more problematic. This disease, which includes an acute encephalopathy with fatty degeneration of the viscera, has been associated with aflatoxin because this mycotoxin has been found in Reye's syndrome patients in Thailand, New Zealand, the former Czechoslovakia, and the United States. Furthermore, aflatoxin B1 produced a disease similar to Reye's syndrome in macaque monkeys. However, found no significant differences between matched controls and patients with Reye's syndrome relative to aflatoxins in serum and urine. Similar inconsistencies were found regarding the occurrence of aflatoxins in tissues and Reye's syndrome in patients. Also, the U.S. cases seem to lack any geographical relationship to aflatoxin exposure. Thus, the cause-and-effect relationship of aflatoxins with Reye's syndrome has not been adequately established. Again, Reye's syndrome, which involves the liver, may change the metabolism of dietary aflatoxin in some patients.
- **Chronic Aflatoxicosis:** Chronic aflatoxicosis in humans usually implies the association of this mycotoxin with hepatocellular carcinoma. Several epidemiological studies in countries or localities with a high incidence of liver cancer examined the relevance of dietary aflatoxin and other factors with this disease. Most of the studies, which occurred mainly before 1980, sought to determine dietary levels

of aflatoxin B1 and correlate them with occurrence of hepatocellular carcinoma. After determined that aflatoxin B1 can form a DNA adduct whose excision product, aflatoxin B_1-guanine, occurs in urine, found this product in urine from individuals from high liver cancer risk areas who were presumably exposed to aflatoxin. Discovered that aflatoxin B1 binds to proteins and forms a lysine adduct that can be found in serum of inividuals for a considerable period of time (half-life = 20 days) after consuming aflatoxin B1. Monoclonal antibodies developed to quantify aflatoxin B1 adducts in humans were used in attempts to measure exposure of populations to aflatoxins. More recently, biomarkers have been used to understand the etiology of tumors, such as mutations of the tumor suppressor gene, p53, which is commonly mutated in human cancers.

18.1.3 Ochratoxicosis

Ochratoxin A, produced primarily by *Aspergillus ochraceous* or *Penicillium verrucosum*, occurs on several commodities prevalent in human diets, including barley and green coffee beans. Known for its nephrotoxic effects, ochratoxin A can also impact the liver. A major renal disease of swine known as porcine nephropathy occurs in some European countries, especially Denmark, and is associated with consumption of ochratoxin-contaminated barley. Most effects of ochratoxin A were discovered during experiments with swine that, when intoxicated, exhibited pain near the kidneys, drank excessive water, appeared depressed, urinated almost continuously, and ate less. The impaired renal function, which is characterized histopathologically as a tubular degeneration and atrophy with interstitial fibrosis and often hyalinization of the glomeruli, results in glucosuria and proteinuria, with casts evident in the urine.

In 1956, the first clinical description of a human kidney disease of unknown etiology known as Balkan endemic nephropathy was published. With the recognition that mycotoxins can cause nephropathies and the epidemiological evidence of ochratoxin A in food of patients in the Balkan countries, ochratoxin A became a prime suspect in the causation of this disease. Furthermore, because ochratoxin A is carcinogenic in rats and mice and patients with Balkan endemic nephropathy frequently have kidney tumors, ochratoxin has been more strongly linked to this disease. However, researchers also have found the fungus (*P. aurantiogriseum*) to be common in food from endemic areas in the former Yugoslavia, Bulgaria, and Romania. While this fungus caused nephrotoxicity in rats, ochratoxin A was not a known metabolite of one of the major isolates examined. This has led to the search for another mycotoxin as a potential cause for Balkan endemic nephropathy. Epidemiological evidence suggests that approximately half of the European population is exposed to ochratoxin A, and estimates of human exposure of Europeans have been published. Table 2 compiles data illustrating levels of ochratoxin that can occur in human tissues and fluids from various countries.

The airborne nature of either ochratoxin or particles laden with this mycotoxin is now considered a potential risk for human exposure. Acute renal failure occurred in a female agriculture worker who had been exposed to granary and grain dust from wheat contaminated with *Aspergillus ochraceus* and its metabolite ochratoxin A. Although ochratoxin A was not demonstrated in granary air samples, guinea pigs and rabbits experimentally exposed to air passed through a layer of the ochratoxincontaminated wheat experienced renal failure. Interestingly, high levels of ochratoxin A (≥1,500g/kg) were found in dust samples obtained from heating ducts in a household in which occupants (humans and animal) had complained or exhibited signs of recurring health problems. Subsequently ochratoxin A was found in dust collected from cow barns in Norway.

Ochratoxin A biotransformation is cytochrome P450 dependent in animals and humans and results in the formation of metabolic intermediates active in the carcinogenic and other toxic activities of ochratoxin A. Phenylalanine, a structural component of ochratoxin A, is likely involved in the complex toxicological activities of the parent compound. The dihydroisocoumarin moiety is likely involved in as-yet-undetermined toxic activities of ochratoxin in humans and animals. Animal studies indicate that ochratoxin is absorbed in the GI tract and in the proximal and distal tubules. It enters the enterohepatic circulation and can be excreted and reabsorbed. It can bind to the albumin fraction in blood and thus can persist in animal tissues for extended periods of time. Intestinal microflora can convert ochratoxin A into ochratoxin alpha, a nontoxic metabolite that can be measured in the urine and feces; small amounts of the parent compound are usually present also.While most evidence for the causal relationship of Balkan endemic nephropathy points toward ochratoxin A, data are not conclusive and further work is needed to substantiate claims in this nephrotoxic disease.

18.1.4 Trichothecene Toxicoses

The trichothecenes are the largest family of known mycotoxins, chemically called sesquiterpenoids. *Fusarium* species are the most notable fungi that produce trichothecenes, although species of other genera, e.g., *Trichothecium, Trichoderma, Myrothecium, Stachybotrys*, also are important producers of these compounds. The trichothecenes are potent protein inhibitors, a basic mechanism of their toxicity, and have been shown to produce a wide variety of effects in experimental animal studies.

18.2 ALIMENTARY TOXIC ALEUKIA

In Russia during the first half of the twentieth century, a disease known as alimentary toxic aleukia (ATA) occurred — particularly in the Orenburg District in 1944 — that was characterized by total atrophy of the bone marrow, agranulocytosis, necrotic angina, sepsis, hemorrhagic diathesis, and mortality reaching 80%. Clinical stages of the disease have been described, with progressive severity of signs and symptoms in the patients.

Patients experienced vomiting, diarrhea, abdominal pain, and burning in the upper GI tract; onset could occur shortly after consuming contaminated food. Later, petechial hemorrhages developed on the skin, often accompanied by hemorrhages in the oral cavity, followed by development of necrotic lesions and enlargement of the local lymph nodes. Joffe (1986) noted that treatments employed for ATA victims were blood transfusion, administering nucleic acid and calcium preparations, antibiotics, vitamins C and K, and dietary improvement. This disease was determined to be caused by consuming overwintered cereal grains or their products. Two common organisms isolated from the grains after over 20 years of storage were *Fusarium poae* and *Fusarium sporotrichioides*, which subsequently were shown capable ofproducing trichothecenes such as T-2 toxin, neosolaniol, HT-2 toxin, and T-2 tetraol. Some signs of ATA could be reproduced in cats given purified T-2 toxin orally. Due to a lack of proof of the involvement of trichothecenes in the original outbreak, they remain conjecturally associated with the disease. However, most signs of ATA have been well documented in animals given T-2 toxin, the major toxic component from the two fungi isolated from the overwintered grain.

18.2.1 Stachybotryotoxicoses

While this disease has been known for some time to occur in horses and cattle consuming *Stachybotrys*contaminated hay (Forgacs 1972), more recently, involvement of toxins (most notably, macrocyclic trichothecenes) from this genus of fungi has become suspect in illnesses of humans occupying *Stachybotrys*- contaminated buildings. This cellulolytic fungus can grow on moist cellulose products in buildings. The consumption of contaminated hay by horses and cattle causes an acute manifestation of disease characterized by neurological signs such as tremors, incoordination, impairment or loss of vision, or, more chronically, dermonecrosis, leukopenia, and GI ulceration and hemorrhage. Humans handling contaminated hay exhibited dermatitis, while inhalation of hay dust caused inflammation of the nose, fever, chest pain, and leukopenia. Although macrocyclic trichothecenes were isolated from the hay in the human and animal intoxications, illnesses in humans occupying *Stachybotrys*-contaminated buildings have not been proved to be causally related to these toxic products. Individuals living in and personnel working in a house contaminated with *S. atra* developed pulmonary irritation and headaches, fatigue, malaise, and diarrhea. Samples of air duct dust and ceiling fiber board covered with *S. chartarum* from a problem residence yielded the trichothecenes verrucarol, verrucarins, satratoxin H, and trichoverrins. Airborne particles of contaminated materials can be collected on membrane filters (polycarbonate) and trichothecenes can be identified from analysis of membrane-collected materials.

18.2.2 Glomerulonephritis

Among the most frequently isolated trichothecenes from grains and foods is DON, a metabolite of the plant pathogenic fungi *Fusarium graminearum* and *F. culmorum*. This compound was shown to be indirectly associated with a nephropathy in a mouse model. Investigators found that DON increased immunoglobulin A (IgA) levels in the sera of mice, resulting in mesangial accumulation of this immunoglobulin and a disease similar to human glomerulonephropathy. Involvement of DON in the latter disease is speculative. However, relative to this toxin and not to glomerulonephritis, an outbreak of intoxication in humans who consumed bread made from mold-damaged wheat in the Kashmir Valley, India, implicated DON as the etiologic agent. Symptoms of the disease included nausea, GI distress or pain, vomiting, and throat irritation. Some patients had blood in their stools or developed a rash.

Symptoms in humans — including headaches, vomiting, and diarrhea — have been attributed to a variety of trichothecenes from cereal grains with redmold disease (akakabi-byo) and black-spot disease (kokuten-byo). Multiple causation, including mycotoxins, has been suggested for these sporadic episodes of human disease.

18.2.3 Citreoviridin Toxicosis

Acute cardiac beriberi or "shoshin kakke," a disease that occurred for centuries, including the early twentieth century in Japan and other Asian countries, was characterized by palpitations, nausea, vomiting, rapid and difficult breathing, rapid pulse, abnormal heart sounds, low blood pressure, restlessness, and violent mania leading to respiratory failure and death.

The disease was regarded as either an infection or avitaminosis until a fungus was isolated, identified, an extract component identified, and its structure elucidated. This dark yellow compound was called citreoviridin and the neurologic syndrome and respiratory failure were reproduced in laboratory animals. The Rice Act of 1921 passed by the Japanese government decreased the availability of moldy rice in the markets and resulted in rapid decline of the disease. Improved diet and inspection have made cardiac beriberi of little importance in modern times. However, it should be recognized that substantial quantities of citreoviridin can occur in corn infected with *Eupenicillium ochrosalmoneum.*

18.2.4 Ergotism

The oldest recognized mycotoxicosis of humans is ergotism. After periodic outbreaks in central Europe, the disease became epidemic in the Middle Ages, where it was known as St. Anthony's fire. Gangrenous symptoms were described in medieval episodes of ergotism in humans, where early symptoms were swollen limbs with burning sensations, with subsequent necrosis leading to loss of appendages (see cover photo). Ergotism results from consumption of products made with grains contaminated with ergots. The ergots or sclerotia are often larger than the normal grain, are typically black in color, and may replace several grains in one spike or head of the respective grain. Because the sclerotia may be inadvertently included in grains processed for human consumption, there is a potential threat for human disease and the need for continued surveillance of grains used for such purposes.

Two different manifestations of ergotism are likely due to different modes of action of the various ergot alkaloids produced by the different species of *Claviceps*. Convulsive and gangrenous ergotism are described in the literature. The gangrenous form likely results from the vasoconstrictive action of certain alkaloids belonging primarily to the ergotamine group, associated primarily with wheat and rye. Edema, pruritis, necrotic extremities, prickling sensations, and severe muscular pain are associated with the gangrenous form of ergotism. The ergot of pearl millet involved in an outbreak of the convulsive form in India in 1975 contained alkaloids of the clavinet group. No manifestations of circulatory effects were noted in the latter outbreak. Clinical evidence of convulsive ergotism includes tingling under the skin, pruritis, numbness of extremities, muscle cramps, convulsions, and hallucinations. Wheat and rye infections are usually caused by *C. purpurea*, while the 1975 outbreak in Indian pearl millet was caused by *C. fusiformis*. Apparently, different species produce different alkaloids.

Attempts to associate disease severity with quantity of sclerotia were unsuccessful, as alkaloid concentration varies among sclerotia and only a chemical quantitation of total alkaloids is meaningful relative to form and severity of ergotism.

Two human outbreaks have occurred in recent times. In Africa, 93 cases of gangrenous ergotism were reported involving grain infected with *C. Purpurea*, while 78 cases of GI ergotism involving millet infected with *C. Fusiformis* occurred in India.

18.2.5 Zearalenone Toxicosis

Little conclusive information is available regarding the effects of zearalenone on humans. However, concern for ingestion of this estrogenic compound has increased due to popular press citations of studies where premature puberty occurred in 7- and 8-yearold children. This precocious development included premature thelarche, premature pubarche, prepubertal breast enlargement in boys, and pseudopuberty in girls in epidemic proportions in Puerto Rico. Researchers examined local food and found high concentrations of an estradiol equivalent in some of the meat. There was concern that exposure could have been from mothers who ingested contaminated food during pregnancy. Exogenous food contamination was also suspected and considered the most likely source of exposure to the estrogen-like substance. Zearalanol , a processed anabolic product made from zearalenone and approved for use as an anabolic agent in cattle and sheep, is used in Puerto Rico. It was discovered that FDA recommendations that animals implanted with this agent be kept from slaughter for at least 60 days postimplantation were not observed because farmers did not consider it as a hormonal product. Earlier, suggested that zearalenone was possibly involved in human cervical cancer and premature thelarche. More recently measured zearalenone in 5 of 36 patients exhibiting thelarche in Hungary and found high concentrations of zearalenol in the cereal foods of the patients. They discovered that both Zearalenone zearalenone and zearalenol bind to estrogen receptors of human myometrial tissue. Evidence exists for the lasting effects of zearalenone on the endocrine system. Furthermore, toxicological responses to natural toxins such as zearalenone, aflatoxins, or certain trichothecenes can be modulated by endocrine metabolites, leading to increased responses to the toxin or perhaps protection against toxicity. Glucocorticosteroids can induce cytochrome P-450 enzymes, which are potentially active in certain mycotoxin metabolic schemes.

18.2.6 Fumonisin Toxicosis

An association between consumption of moldy corn and the high human esophageal cancer rates found in parts of southern Africa and the Linxian area of China has been known for some time. Esophageal cancer rates and the consumption of corn products, which likely contain fumonisins, have simultaneously increased in northeastern Italy in recent years. Various *Fusarium* species are common on corn worldwide, but it is the consumption of food prepared from "home-grown" *F. Verticillioides* (syn., *F. moniliforme*)-contaminated corn that has been correlated with esophageal dysplasia (suggesting a precancerous condition) and esophageal cancer in southern Africa and China. Marasas found esophageal hyperplasia in rats fed diets supplemented with *F. Verticillioides* culture material (ground, moldy corn produced in the laboratory under controlled conditions), while van Rensburg showed that *F. Verticillioides* culture material enhanced nitrosamine-induced tumorigenicity in the esophagus of rats. Others found epithelial hyperplasia, papillomas, and carcinomas in the forestomach of rats fed *F. verticillioides* culture material or *F. Verticillioides* inoculated corn bread containing nitrosamines, including N-methyl-N-benzylnitrosamine. Interestingly, nitrosamines and nitrosamine precursors are also present in foods from a high esophageal cancer area of China, further suggesting that nitrosamines may contribute to esophageal cancer, perhaps as cancer initiators. *Fusarium verticillioides* culture materials of strains MRC 602 and 826 are hepatocarcinogenic in rats. Not all fungi identified as *F. verticillioides*, however, elicit the same effects, as illustrated by strain MRC 1069 culture material, which was not hepatocarcinogenic in rat feeding studies. *Fusarium* spp. produce a number of mycotoxins. One of these, the highly mutagenic compound fusarin C has received attention as a possible carcinogen. However, when *F. verticillioides* cultures containing high or low levels of fusarin C were fed to rats, only the "low fusarin C" culture material was hepatocarcinogenic. Fusarin C's instability during cooking under neutral or slightly basic conditions and its rapid conjugation and excretion by the liver further argue against fusarin C being a carcinogen. On the other hand, fusarin C is stable at the low pH encountered under some cooking conditions or in the stomach and, therefore, until the factor(s) causing esophageal cancer in southern Africa and China are positively identified, fusarin C should not be dismissed as a possible contributing factor.

Fumonisin B1 and other fumonisins with a primary amino group are liver cancer promoters in rats, and fumonisin B1 (50 ppm, ≥90% purity) caused hepatocarcinomas when fed to male BD IX rats for 26 months. However, when fed to Fischer 344/N/Netr rats for two years, fumonisin B1 (≥92% purity) did not cause liver tumors but rather was a kidney carcinogen carcinogen, causing renal tubule adenomas and carcinomas in males fed 50 or 150 ppm. Focal tubular hyperplasia also was increased in males fed at least 50 ppm for two years. Apoptosis, a type of programmed cell death, was increased in males fed ≥15 ppm for up to 26 weeks. At that time, cell proliferation in the renal tubules was increased in males fed 50 or 150 ppm and was marginally increased in females fed 100 ppm (National Toxicology Program 1999). Fumonisin B1 was not carcinogenic to female rats. In mice, the incidence of hepatocellular adenoma and carcinoma was increased in females fed 50 or 80 ppm fumonisin B1. Increased apoptosis was also seen in these groups and hepatocellular hypertrophy was found at two years.

The latter occurred in males fed 15, 80, or 150 ppm and in females fed 50 or 80 ppm (National Toxicology Program 1999). Although fumonisin B1 was clearly carcinogenic to rodents when fed at levels of 50 ppm or more, the reasons for the differences in sex-related response and target organs found in the two studies are unknown. Resolving this issue is important and may contribute significantly to assessing the relationship of fumonisins to human health. There were no significant esophageal changes in either rats or mice after two years, and efforts to enhance nitrosamine-induced esophageal neoplasia in rats with fumonisin B1 were unsuccessful. In liver and kidney, apoptosis is the initial, and sometimes only, effect seen in fumonisin-exposed animals. Only at higher doses or after longer exposures do the other features of fumonisin toxicity, e.g., overt necrosis, mitosis and regeneration, fibrosis, become evident. This is important because apoptosis may play a key role in cancer promotion or carcinogenesis and is induced by alterations in cellular sphingoid bases or sphingolipids. Because they inhibit the enzyme ceramide synthase, fumonisins have pronounced effects on cellular sphingolipid metabolism that, in turn, may alter sphingolipid-mediated regulatory processes related to cell survival and replication. Tolleson found that fumonisin B1 (100 ìM) inhibited growth and induced morphological features consistent with apoptosis in HET-1A (an SV-40 T-antigen-immortalized human esophageal epithelial cell line) and other human cells in vitro, suggesting that fumonisins may cause esophageal epithelial apoptosis in humans. It has been suggested that fumonisins play a role in human diseases other than esophageal cancer. Fumonisins in *F. verticillioides* culture material disrupt sphingolipid metabolism in nonhuman primates and, as in other species, this may be the mechanistic trigger for hepatotoxicity in vervet monkeys fed *F. verticillioides* culture material. Interestingly, atherogenic serum lipid profiles, including increased cholesterol, were found in vervets fed *F. verticillioides* culture material, suggesting that fumonisins may, under some conditions, contribute to human atherosclerotic vascular disease, either directly or as a secondary effect of hepatotoxicity. Whether fumonisins play a role in human esophageal cancer or other diseases is far from being resolved. A number of factors must be considered, including the possibility that fumonisin derivatives may also be toxic. Under alkaline conditions used in nixtamalization,i.e., the process used to make masa flour from corn, fumonisins may be converted to their hydrolyzed forms. Nixtamalized culture materials containing hydrolyzed fumonisin B1 but no measurable fumonisin B1 elicited the same liver cancer-promoting activity, liver and kidney toxicities, and in vivo sphingolipid effects in rats as did untreated culture materials or purified fumonisin B1. Fumonisins also may exert effects by other, indirect mechanisms. It has been proposed that neural tube defects in newborns from southern Texas may have been mediated by a fumonisin or hydrolyzed fumonisin-induced deficiency in a critical folate receptor. Fumonisin B1 was nevertheless not teratogenic to rats, mice, or rabbits. As with *F. verticillioides*, epidemiological surveys of fumonisin B1 suggest but do not prove a relationship with human cancer. Populations from "high-risk" areas in southern Africa and China have a relatively high fumonisin exposure, compared to populations from nearby "lower-risk" areas in which corn is also a dietary staple. In Italy, some corn-based foodstuffs have been found to contain relatively high fumonisin levels, suggesting that the ongoing exposure to fumonisins in polenta, a popular food in the northeastern part of the country, may contribute to the rising esophageal cancer rates in this region. In response to these concerns about known and suspected health effects of fumonisins, the U.S. FDA recently issued a draft guidance for fumonisin levels (total FB1 + FB2 + FB3) in foods and feeds (U.S. Food and Drug Administration 2000a). The recommendations for human foods vary by product type and range from 2 ppm in some degermed, dry-milled products to 4 ppm in other dry-milled products and corn intended for masa production. In conclusion, fumonisin B1 is toxic and carcinogenic to rodents and there are data suggesting that fumonisins or *F. verticillioides* cause esophageal cancer or other human health problems. However, an International Agency for Research on Cancer (IARC) working group on fumonisins concluded that there is "inadequate evidence" for carcinogenicity in humans from oral exposure to fumonisin B1 (International Agency for Research on Cancer 1993) and a role for fumonisins in any other human disease has likewise not been proven. Additional research on the relationships between fumonisins, sphingolipid metabolism disruption, and apoptosis is critical for understanding the experimental carcinogenicity of fumonisins and determining the relevance of these mycotoxins to human health.

18.2.7 Gliotoxin Toxicosis

Gliotoxin is an immunosuppressive mycotoxin that first attracted the interest of investigators searching for new antibiotics. However, its toxicity precluded clinical use as an antibiotic. Interest in gliotoxin was revived when it was discovered to be produced by *Aspergillus fumigatus* and other fungi such as *Candida albicans* and that it had unusual immunosuppressive activities. Of significance to the medical and veterinary mycologist is gliotoxin's possible involvement in the pathogenesis of diseases such as aspergillosis. Gliotoxin is produced in mice experimentally infected with *A. Fumigates*, in natural bovine uder infection, and both experimentally and naturally infected urkeys. This toxin has been found in vaginal secretions of women with *Candida vaginitis*. The involvement of gliotoxin in the pathogenesis of aspergillosis and candidiasis could be important because the immunosuppressive nature of this toxin could exacerbate the infection and possibly be a virulence factor. This suggestion has been made regarding human infections caused by *Fusarium* spp., as these are known to produce several immunosuppressive compounds. However, their production has not been demonstrated during the pathogenic state in humans or animals. The medical mycologist should consider this aspect of infections caused by any toxigenic fungus, especially those that produce immunosuppressive compounds. The significance of gliotoxin in edible tissue is unknown, although a recent report found that contaminated hay caused intoxication in camels, gliotoxin does not appear to be important as an ingested mycotoxin.

18.3 IMMUNOMODULATION

Immunosuppression is a likely major economic effect of mycotoxins. Mycotoxins known to have this effect are aflatoxins, certain trichothecenes, ochratoxin A, and gliotoxin. Immunosuppression by mycotoxins, studied primarily in animal experiments, may involve specific classes of immunoglobulins, or antibodies if the amount of mycotoxin consumed is sufficient. However, the major effects appear to involve cellular immune phenomena and nonspecific humoral factors associated with immunity (Richard 1991). These mycotoxins can cause a variety of immune- related changes, including thymic aplasia and inhibition of phagocytosis by macrophages, delayed cutaneous hypersensitivity, lymphocyte proliferation, and leukocyte migration. The involvement of specific mycotoxins in infectious diseases depends on the agent of disease, the toxin dose and constitution, the animal species, and perhaps the sensitivity of the test employed.

One of the most consistent features of the effects of the aflatoxins on immunity is the decrease of cellmediated immunity. Aflatoxins have been shown to affect the production of cytokines important in relating certain immune processes by a number of cell types. While aflatoxins are not notable in impacting antibody production, they have an important effect on levels of nonspecific humoral factors such as complement, interferon, and some bactericidal serum components. Increased susceptibility to yeast infections, pasteurellosis, and salmonellosis has been demonstrated in poultry. Aflatoxins can move transplacentally in swine, immunocompromising neonatal piglets. Chicken embryos can exhibit depressed immune responses when exposed to aflatoxin B1. Trichothecenes generally decrease the serum proteins (other than the increase in IgA levels discussed earlier with deoxynivalenol in the mouse). The trichothecenes usually decrease phagocytosis and delayed hypersensitivity; exceptions do occur, as has been found with delayed hypersensitivity where small amounts of T- 2 toxin were given to mice subcutaneously and in phagocytosis by macrophages if T-2 toxin is given prior to immunization.

Trichothecenes administered in vivo and in vitro can decrease the response of lymphocytes to mitogens. However, in some cases, low levels of certain toxins appear to function as mitogens, evident with T-2 toxin and deoxynivalenol in that at low concentrations in *in vitro* assays, they increased lymphocyte proliferation. The trichothecenes, especially T-2 toxin, have resulted in decreased resistance to a variety of organisms. However, when intraperitoneal injections of T- 2 toxin were given to mice a few days prior to inoculation with *Listeria monocytogenes*, an enhanced resistance to disease resulted. This did not occur when the toxin was given after inoculation with the organism. Notable among the effects of ochratoxin was decreased natural killer cell activity. This decrease was overcome if production of interferon, a component decreased by ochratoxin A, was stimulated with Poly I:C treatment. Ochratoxin A is an inhibitor of protein synthesis and likely contributes to the overall effect of decreased humoral factors, especially immunoglobulins, by this mycotoxin. Some discrepancies regarding the effect of ochratoxin on complement activity may be related to species differences or dietary factors such as phenylalanine..

19

MYCOTOXICOSES OF ANIMALS

19.1 INTRODUCTION

Fungal toxins produce a wide range of injurious effects in animals, in addition to posing foodborne hazards to humans. The economic impact of decreased productivity, subtle but chronic damage to vital organs and tissues, increased disease incidence because of immune suppression, and interference with reproductive capacity is many times greater than that of acute livestock death. Mycotoxins can be classified according to the organ systems they affect. At least one mycotoxin affects each system in the animal body as a direct or indirect mechanism of toxicity. Several important mycotoxins affect the same system, e.g., the immune system, and a given mycotoxin may affect several systems concurrently.

Mycotoxins cause illness and lethality in domestic animals fed moldy feedstuffs. These acute intoxications can have devastating effects and are difficult to diagnose and treat because the suspect feed may be consumed before it can be tested. Because of the large number of structurally unrelated mycotoxins produced by the various fungi, it is difficult to pinpoint which toxin(s) is responsible for a particular outbreak, even if a mycotoxicosis is strongly suspected. Some indication of mycotoxin potencies can be obtained from LD50 (the single dose of a substance required to cause 50% mortality in a population) studies, such as those shown in Table 1. However, the LD50 has limited value in assessing toxicity of a compound or its potential risks. Of more practical use are reports of natural cases of intoxication caused by consumption of mycotoxin-contaminated food or feed. Whereas it is difficult to prove that a particular outbreak was caused by a mycotoxin, a number of reports in the literature are reasonably well documented. Selected examples are presented in Table 2.

Aflatoxins cause liver damage including cancer, decreased milk and egg production, and suppression of immunity in animals consuming low dietary concentrations. While the young are most susceptible, all ages are affected; clinical signs include GI dysfunction, decreased reproductivity, decreased feed utilization and efficiency, anemia, and jaundice. Nursing animals may be affected by exposure to aflatoxin metabolites secreted in the milk. The trichothecene mycotoxins cause necrosis and hemorrhage throughout the digestive tract, depress blood regenerative processes in the bone marrow and spleen, and cause changes in reproductive organs. Affected animals show signs of anorexia, weight loss, poor feed utilization, vomiting, bloody diarrhea, abortion, and death. Dysregulated immune and neuroendocrine function are significant features of certain trichothecene intoxications.

Table 1: LD$_{50}$[a] of selected mycotoxins

Toxin	Animal	Sex	Age/Size	Route		LD$_{50}$
Aflatoxin B_1	Duckling	M	Day-old	PO[b]	0.37	mg/kg
	Rat	M-F	Day-old	PO	1.0	mg/kg
	Rat	M	21 days	PO	5.5	mg/kg
	Rat	F	21 days	PO	7.4	mg/kg
	Rat	M	100 g	IP[c]	6.0	mg/kg
	Rat	F	150 g	PO	17.9	mg/kg
	Dog	M-F	Adult	IP	ca. 1.0	mg/kg[d]
	Dog	M-F	Adult	PO	ca. 0.5	mg/kg
	Hamster	M	30 days	PO	10.2	mg/kg
Aflatoxin B_2	?					
Aflatoxin G_1	Duckling	M-F	Day-old	PO	0.79	mg/kg
Aflatoxin G_2	Duckling	M-F	Day-old	PO	172.5	mg/duckling
Aflatoxin M_1	Duckling	M-F	Day-old	PO	16.6	mg/duckling
Aflatoxicol	?					
Aflatrem	?					
Citreoviridin	ddys[e] mice	M	—	SC[f]	11.0	mg/kg
	ddys mice	M	—	IP	7.2	mg/kg
	ddys mice	M	—	PO	29.0	mg/kg
Citrinin	Rat	—	—	SC/IP	67.0	mg/kg
	Mice	—	—	SC/IP	35.0	mg/kg
	Guinea pig	—	—	SC	37.0	mg/kg
	Rabbit	—	—	IV[g]	19.0	mg/kg
Cyclopiazonic	Rat	M	—	IP	2.3	mg/kg

Toxin	Animal	Sex	Age/Size	Route		LD50
acid	Rat	M	—	PO	36.0	mg/kg
	Rat	F	—	PO	63.0	mg/kg
	Chick	M	Day-old	PO	12.0	mg/kg
	Chick	F	Day-old	PO	12.1	mg/kg
	Poult	M	Day-old	PO	19.0	mg/kg
	Poult	F	Day-old	PO	17.9	mg/kg
	Duckling	M-F	Day-old	PO	38.6	mg/kg
	Quail	M	6 wk-old	PO	69.6	mg/kg
Cytochalasin H	Chick	M	Day-old	PO	12.5	mg/kg
Moniliformin	Chick	M	Day-old	PO	4.0	mg/kg
	Mice	M	25 g	IP	29.1	mg/kg
	Mice	F	25 g	IP	20.9	mg/kg
Ochratoxin A	Rat	—	Weanling	PO	22.0	mg/kg
	Trout	—	6 mos	IP	5.53	mg/kg
Penitrem A	Mice	—	—	IP	1.05	mg/kg
PR toxin	Rat	—	Weanling	IP	11.0	mg/kg
	Rat	—	Weanling	PO	115.0	mg/kg
	Mice	—	—	IP	5.8	mg/kg
Roquefortine A	Mice	—	—	IP	340.0	mg/kg
Slaframine	Chick	—	Day-old	PO	81.6	mg/kg
Sterigmatocystin	Rat	M	—	PO	166.0	mg/kg
Rat		F	—	PO	120.0	mg/kg
Rat		M	—	IP	60.0	mg/kg
Monkey		M	—	IP	32.0	mg/kg
Tenuazonic acid	?					
Trichothecene toxins: Deoxynivalenol ddys mice		M	—	IP	70.0	mg/kg
ddys mice		F	—	IP	76.7	mg/kg
Diacetoxyscirpenol	Rat	M-F	—IP		0.75	mg/kg
	Rat	M-F	— PO		7.3	mg/kg
	Mice	M-F	— IP		23.0	mg/kg
	Mice	M-F	—IV		10.0	mg/kg
Fusarenon-X	ddys mice	M	—IP		3.3	mg/kg
HT-2	ddys mice	M-F	—	IP	9.0	mg/kg

Monoacetoxy-	Rat	F	20 days	SC	0.75	mg/kg
scirpenol Neosolaniol	ddys mice	M-F	—	IP	14.5	mg/kg
Scirpenetriol	Rat	—	—	IP	0.81	mg/kg
T-2	Rat	M-F	21 days	IP	3.0	mg/kg
	ddys mice	M-F	—	IP	5.2	mg/kg
	Swine	M-F	—	PO	4.0	mg/kg
T-2 tetraol	Chick ?	M	Day-old	PO	1.84	mg/kg
Zearalenone Verruculogen	? Mice	F	20 g	IP	2.4	mg/kg
	Mice	F	20 g	PO	126.7	mg/kg
	Chick	M	Day-old	IP	15.2	mg/kg
	Chick	M	Day-old	PO	265.5	mg/kg

aLD = lethal dose
bPO = per os (by mouth)
cIP = intraperitoneal
dca. = approximately
eddys = recombinant congenic strain of mice used in the study of the major histocompatibility complex region of chromosomes
fSC = subcutaneous

gIV = intravenous

The diagnosis of mycotoxicoses is based on knowledge gained in experimental studies with specific toxins and specific animal species. Signs of disease caused by natural occurring mycotoxins can be changed by environmental and other factors such as nutrition, sex, and breed. Thus, diagnosis of mycotoxicoses is difficult but can be achieved by good observations of the signs in affected animals and analysis of the feed involved in the intoxication.

Fig. 1. Effects of dietary T-2 toxin on growth of turkey poults. The turkey poult on the left was given a normal ration, and the poult on the right was given a ration containing 10 ppm of T-2 toxin for three weeks.

The impact of fungal toxins on animals extends far beyond the obvious effect of causing death. The economic impact of lowered productivity, decreased weight gain (see Fig. 6.1), decreased feed efficiency, decreased meat and egg production, increased disease incidence because of immune system suppression, subtle damage to vital body organs, and interferences with reproduction is many times greater than that of immediate morbidity and lethality. Potential threats of cancer from mycotoxins in feeds and human foods, along with the unknown subtle effects induced by these mycotoxins, are coupled to universal concerns about health risks.

Most mycotoxin toxicity data have come from experimental studies with purified compounds in otherwise healthy animals. In this section, toxicities will be characterized by a "systems approach" where the emphasis is on the body system(s) affected by the mycotoxin. Specifically highlighted are the effects of important mycotoxins on productivity, liver, kidney, hematopoiesis, immune system, reproduction, birth defects, neuroendocrine system, and cancer. The diagnosis of animal mycotoxicoses is based on experimental studies with specific toxins and specific animals. Diagnosis of naturally occurring mycotoxicoses, however, is difficult because there are a multiplicity of factors such as breed, sex, environment, nutritional status, and other toxic entities that can affect the intoxication. Diagnosis is dependent on receiving a sample of the food or feed that was involved in the intoxication, collecting data from the animals relative to the illness, and conducting a thorough postmortem examination if death is a result. Therefore, the best diagnosis is achieved when a toxin, which is referable to the kind of disease observed, is found in the sample of feed involved in the intoxication.

19.2 GENERAL EFFECTS OF MYCOTOXINS ON HEALTH AND PRODUCTIVITY

19.2.1 Aflatoxins

Aflatoxins are potent liver toxins. Their effects in animals vary with dose, length of exposure, species, breed, and diet or nutritional status. These toxins may be lethal when consumed in large doses. Sublethal doses produce a chronic toxicity and low levels of chronic exposure can result in cancer, primarily of the liver, in some species. Generally, young animals are more susceptible than older ones to the toxic effects of aflatoxins. Aflatoxins have elicited the greatest public health concern because of their widespread occurrence in several dietary staples, e.g., peanuts, tree nuts, milk, corn, dried fruits, and their potential as human carcinogens.

One of the experimental animals used in the early studies of aflatoxicosis was trout. They seem to be quite sensitive to the effects of the aflatoxins and the LD50 was calculated to be equivalent to 0.5 to 1.0 mg/ kg crystalline aflatoxins B1 and G1 in the same proportions. Eighty parts per billion total dietary aflatoxins produced hepatomas in the trout at a very high incidence. Rainbow trout in the early stages of development are quite sensitive to hepatogenicity. Immersion of fry or embryos in 0.5 ppm aflatoxin B1 for 0.5 hr resulted in 30 to 40% incidence of hepatocellular carcinoma nine months later. As epizootics of aflatoxicosis still occur in fish and were likely the cause of a trout liver cancer epizootic that occurred in California in hatcheries from 1939 to 1942. Aflatoxin-contaminated cottonseed meal was implicated as the causative agent in this outbreak. Trout fed aflatoxins develop hepatic cancer. Aflatoxin toxicity has been reported in suckling piglets, growing and finishing swine, and breeder stock. Clinical and pathological signs include decreased rate of weight gain, decreased feed conversion efficiency, toxic hepatitis, nephrosis, and systemic hemorrhages. The effects of aflatoxins in pigs vary, depending on age, diet, concentration, and length of exposure. Swine appear to be resistant to dietary levels of aflatoxins up to 300 ppb fed from time of weaning to marketing.

Fig. 2. Livers from guinea pigs given increasing doses of aflatoxins over the same period of time. From left to right beginning in the upper left corner with the liver from a guinea pig given no aflatoxins, to the lower right corner with a liver from a guinea pig given the greatest dose of aflatoxins. Note the increasingly pale livers with increasing doses of aflatoxins.

Acute aflatoxicosis in cattle has been thoroughly described. Clinical signs consist of decreased feed consumption, dramatic drops in milk production, weight loss, and liver damage.

However, chronic exposure of dairy and beef cattle to naturally occurring levels of aflatoxins may have an even greater economic impact due to decreased feed efficiency, immunosuppression, and lower reproductivity. Aflatoxins affect rumen function in vitro and in vivo by decreasing cellulose digestion, volatile fatty acid formation, and proteolysis. Chronic exposure of a dairy herd to aflatoxin-contaminated corn (120 ppb) resulted in severe health problems. Also, breeding efficiency decreased 2% for a five-month period after exposure, while milk production increased 28% after the aflatoxin- contaminated corn was removed from the diet. Other problems include the birth of smaller and unhealthy calves, diarrhea, acute mastitis, respiratory disorders, prolapsed rectum, hair loss, and decreased feed consumption.

Another characteristic of aflatoxin exposure in dairy cattle is the conversion of aflatoxin B1 to the hydroxylated metabolite, aflatoxin M1, which is excreted in milk. Milk from Holstein cows given aflatoxin B1 for seven days contains aflatoxin M1, although aflatoxin M1 is not detected in milk four days after termination of aflatoxin B1 administration. The excreted amounts of aflatoxin M1 as a percentage of aflatoxin B1 average 1 to 2%, but values as high as 6% have been reported at daily intake levels of aflatoxin B1.

Aflatoxicoses have produced severe economic losses in the poultry industry, affecting ducklings, broilers, layers, turkeys, and quail. Clinical signs of intoxication include anorexia, decreased weight gain, decreased egg production, hemorrhage, embryotoxicity, and increased susceptibility to environmental and microbial stressors. Histopathologic changes, including fatty liver, necrosis, and bile duct hyperplasia, are found in chickens given a high level (1.5 ppm) of dietary aflatoxins. In chickens given half of this dose, clinical responses include hypoproteinemia; decreased hemoglobin; and decreased serum triglycerides, phospholipids, and cholesterol. Aflatoxins can decrease activities of several enzymes important to digestion of starches, proteins, lipids, and nucleic acids in broiler chickens. The decreased activities of pancreatic amylase, trypsin, lipase, ribonucleic acid

(RNA) enzyme RNAse, and DNAse could contribute to the malabsorption of nutrients associated with aflatoxicoses.

Aflatoxin-contaminated feed (up to 10 ppm) consumed by layers for 4 weeks decreases egg production and size. Total yolk weight and yolk as a percentage of total egg weight decreased, accompanied by greater yolk and plasma carotenoid concentrations.

19.2.2 Trichothecenes

General signs of trichothecene toxicity in animals include weight loss, decreased feed conversion, feed refusal, vomiting, bloody diarrhea, severe dermatitis (Fig. 3), hemorrhage, decreased egg production, abortion, and death. Histologic lesions consist of necrosis and hemorrhage in proliferating tissues of the intestinal mucosa, bone marrow, spleen, testis, and ovary. Trichothecenes inhibit protein and DNA synthesis; many of their toxic effects stem from this mechanism. Because T-2 toxin was one of the first trichothecenes discovered, its effects on animals were characterized early.

Deoxynivalenol is common in North American and European cereal grains, and of the trichothecenes, poses the greatest problems to animal health. Al- though DON can be acutely lethal when ingested in large quantities, moderate- to low-level ingestion of the toxin can cause poor performance and altered immune function. Monogastric animals, particularly swine, exhibit the greatest sensitivity to DON, while chickens and turkeys, followed by ruminants, appear to have higher tolerance.

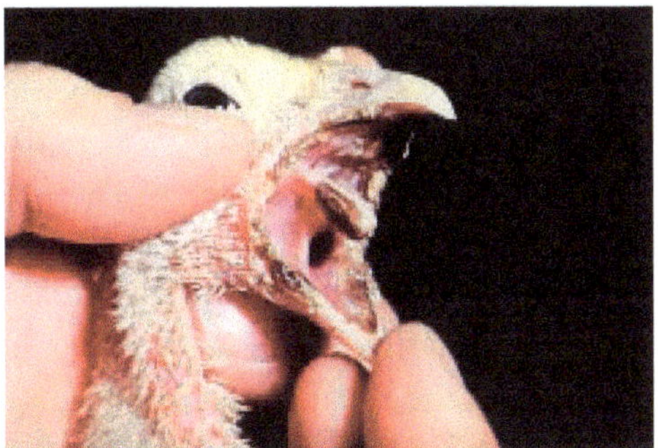

Fig.3. Severe oral lesion in a turkey poult that consumed a ration containing 10 ppm T-2 toxin for three weeks.

Diminished feed consumption and lower weight gain are the principal clinical effects seen in pigs that have eaten DON in naturally contaminated feeds (≤ 2 ppm feed). At 1.3 ppm DON in diet, feed intake by growing pigs is significantly decreased, followed by complete feed refusal at 12 ppm and vomiting at 20 ppm. The most common signs of acute DON exposure are abdominal distress, increased salivation, and malaise; however, vomiting has been reported at higher dietary concentrations. Extensive lesions are not typically documented in field cases, because pigs regulate toxin ingestion by adjusting their feed intake. Although pigs fed DON exhibit altered blood parameters, these effects cannot be easily separated from nutritional status, i.e., weight loss as a result of significantly decreased feed intake. Nevertheless, altered stomach condition and serum protein status do indicate a specific effect of *Fusarium* toxins/DON.

The extent of DON effects in pigs relates to age and sex as well as the contamination source. Initial studies reported that detrimental effects can be observed when purified

DON is added at a level of 5 ppm; however, the situation with naturally contaminated diets is more complex. Because *F. graminearum* produces many metabolites besides DON , mycotoxicosis may be caused by multiple toxins. Unidentified/bound toxins, conjugated mycotoxins, or toxic agents of other origin might contribute substantially to the animal response.

Poultry are much less sensitive than pigs to DON. Levels as high as 8 ppm in feeds have not been associated with impaired productivity. Rapidly growing broilers are generally more sensitive than laying hens to feed refusal. In day-old broiler chicks given DON, the seven-day oral LD50 is approximately 140 mg/kg body weight (bw), with widespread hemorrhaging, urate deposition, neural toxicity, and upper GI tract irritation being the primary signs.

Ruminants are relatively insensitive to DON because rumen microorganisms are able to metabolize/ detoxify this toxin. Dairy cows fed 6.4 ppm for 6 weeks or 66 ppm for 5 days showed no signs of impaired performance or illness. Suggested, however, from the results of epidemiological studies of milk production and DON-contaminated feed that there should be an investigation of the presence of other mycotoxins in DON-contaminated feeds.

19.2.3 Ochratoxin A

Ochratoxin A, a nephrotoxic mycotoxin produced byseveral *Aspergillus* and *Penicillium* species, primarily affects the kidneys in animals exposed to naturally occurring levels. Changes in the renal function of pigs exposed to ochratoxin A include impairment of proximal tubular function, altered urine excretion, and increased excretion of urine glucose. Extrarenal effects may have occurred in animals exposed to levels of ochratoxin A in feed greater than 5 to 10 ppm. These effects included enteritis, necrosis of lymphoid tissue, and fatty change in liver.

Turkeys and chickens both exhibit decreased productivity during field outbreaks of ochratoxicosis. Symptoms include retarded growth, decreased feed conversion, nephropathy, and mortality. There is an apparent increased susceptibility to air sacculitis caused by *Escherichia coli* among affected birds. While feed refusal was demonstrated in turkeys exposed to ochratoxin A, it was not observed in chickens. Ochratoxin A also has detrimental effects on egg production in laying hens. There is a dose-related decrease in egg production in layers experimentally fed ochratoxin A (0.25 to 2.0 ppm) for 12 weeks. Higher levels of ochratoxin A decreased eggshell quality and increased the percentage of eggs with blood and meat spots.

19.2.4 Cyclopiazonic Acid

The biological effects of CPA in rats, dogs, pigs, sheep, laying hens, and chickens have been thoroughly described. Clinical signs of intoxication include anorexia, diarrhea, pyrexia, dehydration, weight loss, ataxia, immobility, and extensor spasm at death.

Histologic examination of tissues from CPA-exposed animals revealed alimentary tract hyperemia, hemorrhage, and focal ulceration. There was widespread focal necrosis in most tissues, including liver, spleen, kidneys, pancreas, and myocardium.

Cyclopiazonic acid has the ability to chelate metal cations due to the structure of the tetramic acid moiety. The chelation of cations such as calcium, magnesium, and iron may be important mechanisms of CPA toxicity. In laying hens, it has been demonstrated that CPA perturbs normal egg-shell formation so that hens fed diets containing CPA lay eggs that have thin shells or visibly cracked shells. Approximately 50% of a dose of CPA administered orally or intraperitoneally is distributed to skeletal muscle of rats and chickens within three hours. Other studies have demonstrated that residues of CPA may be transferred into milk and eggs.

Cyclopiazonic acid also has been found to occur in corn, peanuts, cheese, and kodo millet. Because CPA is produced by *A. flavus* and can be found with aflatoxins produced by the same isolate, plus the toxicity of some feeds contaminated by *A. flavus* can be greater than expected from their levels of aflatoxins, the possible role of CPA in diagnosed aflatoxicoses should be explored.

19.2.5 Fescue Alkaloids

Tall fescue is a perennial grass grown on approximately 40 million acres in the United States. In much of its growing range, it is the only cool season perennial forage. Although this forage is able to withstand heavy grazing and environmental stresses, body weight gains and fertility in cattle grazing fescue have been disappointing. Fescue toxicosis in cattle is associated with consumption of tall fescue grass containing the endophytic fungus, *Neotyphodium coenophialum.* A umber of syndromes have been included in fescue toxicosis including "fescue foot," "fat necrosis," "agalactica," "summer slump," and "summer toxicosis"..

Animals grazing tall fescue infected with the endophyte may have decreased weight gains, increased body temperature, rough hair coats, gangrenous necrosis of tissue of feet, tail and ears, and decreased conception rate. Clinical manifestation of the different syndromes is changed by environmental temperature.

During summer months, hyperthermia is a major feature but during winter months fescue foot, which involves gangrene of extremities, is observed in animals grazing tall fescue. Reproductive abnormalities are the major features of horses grazing tall fescue pastures. Various alkaloids isolated from the infected grass suggest that ergot-like alkaloids were involved in this intoxication. Ergovaline is a primary ergot alkaloid found in many samples of infected tall fescue. Endophyte-infected fescue, however, contains many other alkaloids that probably contribute to the overall toxicoses.

Approximately 22% of U.S. cattle graze fescue pastures. If body weight of weaned calves is decreased 25% by effects of the fungus, the annual loss exceeds $500 million.

Decreased calving rate contributes a further $300 million yearly loss. Additional losses stem from fescue foot, a gangrenous loss of the extremities that occurs during colder periods and affects a minority of the herd. Nevertheless, affected animals are lost from the herd. A final condition attributable to fescue is fat necrosis, a herd problem affecting mature animals. Masses of hard, necrotic fat in the abdomen obstruct the flow of ingesta and lead to difficulty in calving. Further research is needed to identify agents that will combat the toxicosis and alter potency of the fungus.

19.2.6 Immunologic Effects

The immune system is a key target of several important mycotoxins. One type of adverse effect occurs when a mycotoxin suppresses one or more functions of the immune system, with the net result being impaired resistance to infectious agents. Mycotoxins might thus predispose livestock to infectious disease, possibly resulting in feed refusal and decreased productivity. Increased infections in food-producing animals might result in more animal-to-human transmission of pathogens such as *Salmonella* and *Listeria.* Veterinary clini cians have long recognized immunosuppression in food-producing animals that have ingested mycotoxin- contaminated grains. Although the entire episode may be primarily a mycotoxin- facilitated disease, such recognition is often overshadowed by the infectious disease itself. Therefore, a descriptive or analytical epidemiological evaluation of the role of mycotoxins in the health of farm animals and humans is often difficult or impossible to accomplish. Only when mycotoxicoses are acute manifestations of disease is there a chance at such an evaluation.

Substantial experimental evidence exists that mycotoxins can be immunotoxic. The difficulty of conducting immunological studies in the field means that mycotoxininduced immune alterations have been examined primarily in experimental animals. Several mycotoxins produced by the *Fusarium* and the *Aspergillus-Penicillium* groups have notable immunotoxic effects.

19.2.7 Aflatoxins

Although they are primarily known as hepatotoxins and hepatocarcinogens, aflatoxins seem to have been involved in outbreaks of infectious disease in domesticated animals. Salmonellosis, a bacterial infection, and candidiasis, a yeast infection, were associated with outbreaks of Turkey X disease caused by aflatoxins in 1960. Outbreaks of salmonellosis occurred in swine in the southeastern United States following the appearance of high concentrations of aflatoxins in the 1977 regional corn crop. Extensive experimental evidence shows that aflatoxins are very capable of lowering the resistance of several animal species to bacterial, fungal, and parasitic infections.

General features of immunosuppression by aflatoxins. Interpretation of results from studies of aflatoxins and immunity requires particular care, because some used mixtures of aflatoxins, while others used purified aflatoxin B1. Differences have been demonstrated in this regard with aflatoxin B1 and its metabolites. In general, evidence from experimental models supports the contention that aflatoxin B_1 particularly suppresses the cell-mediated immune response. Aflatoxin- induced immune modulation has been discussed in several reviews. Because aflatoxin is an economic threat to the poultry industry, its effects on avian immunity have been extensively studied. Cell-mediated responses are particularly sensitive, as reflected by decreased thymus weight and lower peripheral T lymphocyte numbers in chickens fed aflatoxin B1. Graft versus host response is suppressed in chickens given 300 ppm aflatoxin B_1. Delayed hypersensitivity response to dinitrofluorobenzene is decreased in broiler chicks given 1 ppm aflatoxin B_1 feed. Oral administration of aflatoxin B1 to chicks at 0.1 and 0.5 mg/kg body weight lowers peripheral blood lymphocyte proliferation responses to the T cell mitogen concanavalin A (Con A).

Phagocytic functions of macrophages and the reticuloendothelial system are consistently inhibited byaflatoxin B1. In chicks, the percentage of nitroblue tetrazolium-positive cells in spleen tissue is depressed by aflatoxin B_1 (0.3 to 1.0 ppm), indicating depressed macrophage function. Oral administration of aflatoxin B1 (0.35 to 0.7 mg/kg bw) in rats depresses both macrophage number and function. Clearance of circulating colloidal carbon is decreased in chicks given aflatoxin B_1 (0.3 mg/kg feed), suggesting decreased phagocytic status of the reticuloendothelial system. In vitro data also indicate suppression of phagocytic activity in vivo in chickens and rats. Phagocytosis, intracellular killing of *Candida albicans*, and spontaneous superoxide anion (O_2^-) production are suppressed in rat peritoneal macrophages exposed to aflatoxins in vitro. Activation of aflatoxin B1 by mixed function oxidases is apparently required for macrophage toxicity.

Aflatoxins impair the function of more than one cell type of the mononuclear phagocyte system. This activity decrease seems to be related to effects on phagocytic cells (Fig. 4.) but, perhaps more importantly, to a heat-stable substance in the serum necessary for phagocyte activity.

Complement, a serum constituent produced by the liver, plays an important role in several immune reactions. Impairment of this activity means decrease of an integral part of the host's immunologic potential. Serum complement activity is decreased in pigs given feed containing 500 ppm aflatoxin B1; however, complement activity is unaffected in pigs given 300 ppm aflatoxin B1 in feed and in rabbits given 95 ppm aflatoxin B1 in feed. Aflatoxins decrease the hemolytic complement activity in guinea pigs and other species. Modulatory effects of aflatoxin B1 on humoral immunity are not as apparent as effects on cell-mediated immunity, particularly in cross-species comparisons. In swine given up to 500 ppm aflatoxin B1 in feed and inoculated with *Erysipelothrix rhusiopathiae* bacteria, there is no significant difference in antibody titers compared to inoculated swine given uncontaminated feed. The ability of guinea pigs to produce antibodies to *Brucella abortus* is unaffected by aflatoxin (0.045 mg/kg bw)administered orally. In rabbits given aflatoxin (approximately 24 ppm feed), the antibody- forming response to sheep red blood cells (SRBCs), a T cell-dependent antigen, is unchanged compared to animals given aflatoxin B1-free diets. To summarize, aflatoxin B1 suppresses cell-mediated immunity to a greater extent than humoral immunity. Some aspects of innate immunity, especially phagocytic responses, are also inhibited by aflatoxin B1. It is clear that aflatoxins in the low ppm range would be immunomodulatory (Fig. 4).

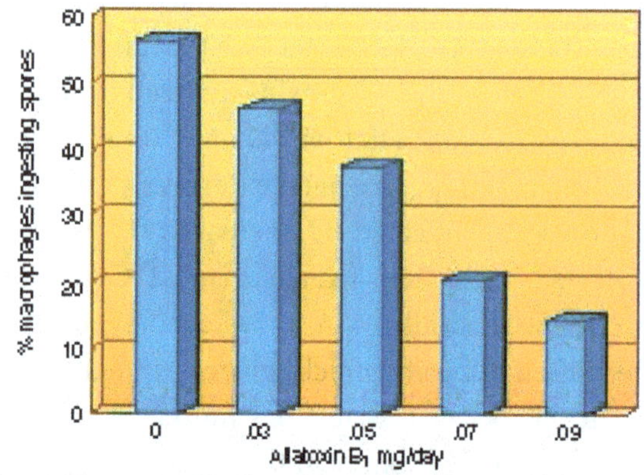

Fig. 4. Decreased phagocytosis of macrophages with increased doses of aflatoxins in rabbits

19.2.8 Trichothecenes

Trichothecenes may be as important as aflatoxins in causing immunosuppression. The capacity of trichothecene mycotoxins to inhibit protein synthesis apparently contributes to their potential to modulate immune function. Acute exposure to trichothecenes results in severe damage to actively dividing cells in tissues such as bone marrow, lymph nodes, spleen, thymus, and intestinal mucosa. General effects on function of immunocompetent cells, host resistance, and immunoglobulin production at lower doses using oral and other routes of exposure have been reported and reviewed extensively. Repeated exposure of experimental animals to trichothecenes results in markedly increased susceptibility to *Candida*, *Cryptococcus*, *Listeria*, *Salmonella*, *Mycobacterium*, and herpes simplex virus type. There is a marked increase in susceptibility to lipopolysaccharide (LPS) when T-2 toxin is coadministered, suggesting that impaired resistance to LPS might be one mechanism for increased susceptibility to gram-negative bacteria.

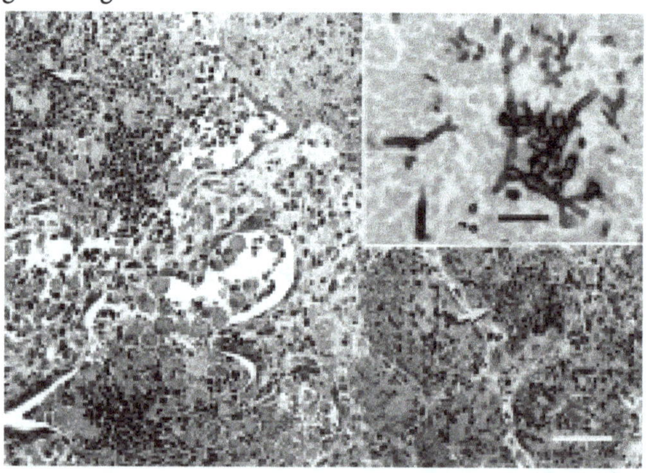

Fig. 5. Photomicrograph of a section of the lung from a rabbit given T-2 toxin orally for 17 days, exposed to aerosols of *Aspergillus fumigatus* conidia on days 7 through 16, and

necropsied on day 17. There are multifocal to confluent necrotic granulomas within alveoli and bronchioles. Notice fungal elements surrounded by radiating clubs (white arrows) within some granulomas. H&E stain; bar = 50 μm. Inset reveals tangled masses of A. fumigatus hyphae in this lung specimen. Gridley stain; bar = 25 μm.

Table 2. Trichothecene effects on immunity

Effect
Effect on cellular immunity and inflammatory responses
Chemotactic migration of neutrophils decreased
Phagocytosis by alveolar macrophages decreased
Skin graft rejection time increased
Mitogen-induced blastogenesis of lymphcytes inhibited
Platelet function inhibited
Cytotoxic to lymphcytes
Increased sensitivity of bacterial endotoxin
Depressed delayed-type hypersensitivity
Effects on Infectious diseases
Resistance to mycobacterial infection in mice decreased
Mortalities of chickens and mice to Salmonella spp. increased
susceptibility to herpes simplex virus in mice increased
Mortality of mice with listeriosis increased (dependent upon time of exposure to toxin)
Infectious response in rabbits with *aspergillosis* increased

Interaction of trichothecenes with infectious disease agents occurs when the resistance mechanisms appear to depend on cellular immunity (Table.4). As reported a decrease in phagocytosis by alveolar macrophages (Table 5) and increased severity of experimental aspergillosis in rabbits treated with T-2 toxin (Fig. 5). Depression of host resistance by trichothecenes seems to involve suppression of several cellular functions, when evaluated using assays such as allograft rejection time, neutrophil migration, macroph age phagocytosis, and delayed-type hypersensitivity.

Determined that short-term, intraperitoneal pre-inoculation treatment with T-2 toxin can be immunostimulatory and actually enhance resistance to *Listeria*, whereas post-inoculation T- 2 toxin treatment is markedly immunosuppressive.

Similarly enhanced resistance to mastitis pathogens by pre-inoculation gavage of T-2 toxin in mice. Dose-dependent decreases or increases in B cell and T cell mitogen responses can be found in lymphocytes obtained from animals exposed to DON, T-2 toxin, orvarious macrocyclic trichothecenes For example, ingestion of DON at 2 ppm for 5 weeks and 5 ppm for one week depresses mitogen responses.

Trichothecene exposure in vitro impairs or, at lower doses, enhances mitogen-induced lymphocyte proliferation.

Trichothecenes can both suppress and stimulate immunoglobulin production. Repeated exposure to trichothecenes impairs murine antibody responses to challenge with sheep red blood cells. Dietary DON causes marked elevation of serum IgA, with concurrent decreases in immunoglobulin M (IgM) and immunoglobulin G (IgG) in the mouse. IgA is polyspecific and reacts with bacterial and self antigens. Concurrent with serum IgA elevation are immunopathologic effects such as increased polymeric IgA and IgA immune complexes, kidney mesangial IgA accumulation, and hematuria. These effects are persistent and highly analogous to the common human glomerulonephritis, IgA nephropathy. IgA nephropathy occurs to a greater extent in males than females and, recently, determined that, in the mouse, males exhibit a predilection for DONinduced IgA nephropathy in terms of threshold toxin dose as well as onset and magnitude of response. Another common feature of the human disease and the murine model is the involvement of polyspecific "natural" IgA that may be associated with immune complex formation and subsequent glomerulonephritis. The suppressive effects of trichothecenes on immune function can be explained by their potent capacity to inhibit protein synthesis; however, their immunostimulatory effects at low doses are problematic.

Superinduction of cytokines has been demonstrated using trichothecenes and other protein synthesis inhibitors. Proposed mechanisms for these effects include interference with synthesis of highturnover proteins that limit transcription or half-life of interleukin messenger RNA (mRNA); analogousmechanisms could be proposed at the level of translation. Dysregulation of IgA production is apparently brought about by macrophage- and T cell-mediated polyclonal differentiation of B cells to IgA secretion at the level of the Peyer's patch.

19.2.3 Ochratoxin A

Ochratoxin A immunomodulation has been reviewed. General indicators of immunosuppression following ochratoxin A ingestion include lymphocytopenia and depletion of lymphoid cells, particularly in the thymus, bursa of Fabricius, and Peyer's patches. Immunosuppression by ochratoxin A has been studied mainly in poultry, with evidence suggesting that this mycotoxin has an effect on immunoglobulins and phagocytic cells. Cellular immune responses in poultry were affected by ochratoxin A more in broilers than in quail or turkeys; species differences are apparent. Air sacculitis caused by *Escherichia coli* was notably increased in turkeys and broilers involved in an outbreak of ochratoxicosis. Interestingly, antibiotic therapy for the air sacculitis was ineffective in birds consuming the ochratoxin A-containing feed but became effective when the diet was changed. This phenomenon could be economically important because of added cost with little or no benefit in treating disease. Although several studies indicate that intraperitoneal injection of ochratoxin A suppresses antibody responses to sheep red blood cells or *Brucella abortus*, depressed T-independent or T-depen dent antigen responses were not seen when ochratoxin A was administered in the diet.

Because phenylalanine has been shown to attenuate ochratoxin A-induced immunosuppression in mice and to decrease mortality in broiler chicks fed ochratoxin A-contaminated diets, dietary factors could account for some of the discrepancies in studies of ochratoxin A immunotoxicity using different administration routes. The complement system is unaffected by dietary ochratoxin A (0.45 mg/ml) in guinea pigs; however, chickens fed ochratoxin A (2 ppm) have significantly decreased complement activity. Perhaps species or other experimental factors are responsible for these differences.

19.2.4 Fumonisins

Fumonisin B1 is the etiologic agent of equine leukoencephalomalacia, porcine pulmonary edema, and is hepatotoxic and carcinogenic to rats. Observed that pigs dosed intravenously with fumonisin over the period of several days (total of 4.6 to 7.9 mg/kg) or fed 48 to 166 ppm of the toxin exhibited membranous material in pulmonary macrophages. Fumonisins B1 and B2 were cytotoxic to turkey lymphocytes with 50% cytotoxic doses ranging between 0.3 to 2 μg/ml. Chicken peritoneal macrophages treated with fumonisin B1 (10 to 100 μg) show morphological alterations, small but significant decreases in cell viability, and decreased phagocytic potential. Tumoricidal factor production in a chicken macrophage cell line is unaffected.

19.2.5 Zearalenone

Although there have been few investigations of potential zearalenone immunotoxicity, other strogens have been shown to decrease host resistance to pathogens, increase tumor susceptibility, and impair cellmediated immunity and inflammatory responses. Zearalenone and its analogues are capable of inhibiting mitogen-stimulated lymphocyte proliferation . Dietary exposure to 10 ppm zearalenone in the mouse for 2 weeks decreases resistance to *Listeria* but has no effects on antibody or delayed-type hypersensitivity responses. The same study indicated that zearalenone and deoxynivalenol do not act synergistically when the same immune functions are monitored. Zearalenone and zearalenol can induce thymic atrophy and macrophage activation.

19.2.6 Patulin

Several studies indicate that patulin inhibits multiple aspects of macrophage function in vitro. found that protein synthesis was inhibited in rat alveolar macrophages exposed to patulin in vitro and that cell membrane function was compromised. In rat and rabbit peritoneal macrophages in vitro, patulin suppressed the oxidative burst. In mouse peritoneal macrophages, patulin significantly decreased O2- production, phagosome-lysosome fusion, and lysosomal enzyme activity. Oral administration of patulin (10 mg/kg for 4 days) decreased the mortality of mice experimentally infected with *Candida albicans* from 80% in control mice to 50% in mice treated with patulin. Patulin administration also resulted in increased circulating neutrophils,which could have contributed to increased resistance.

Neutrophilia in rats administered patulin orally was attributed to GI inflammation.

19.2.7 Citrinin

Citrinin is primarily known as a nephrotoxin; few studies have addressed its potential for immunotoxicity. Administered citrinin (125 to 500 ppm for three weeks) in feed to broiler chicks and found that cellular and humoral immune responses were not affected. In mice given citrinin intraperitoneally (0.12 to 3.0 mg/kg bw), decreased peripheral leukocytes were observed. Antibody response to sheep red blood cells was depressed at the lowest citrinin dose but stimulated at higher doses. Splenic lymphocyte blastogenesis with or without mitogens was stimulated by citrinin administration. From this study, citrinin appeared to be immunostimulatory rather than immunosuppressive; however, it remains to be determined whether ingestion of feed contaminated with low levels of citrinin produces responses similar to intraperitoneal exposure. Wortmannin is a steroid metabolite first isolated from *Penicillium wortmanni* and later found to be producedby other fungi, including *Myrothecium roridum*, *Fusarium oxysporum*, and *F. sambucinum* . This toxin has been used in metabolic studies to increase our understanding of insulin action and glucose homeostasis. The potential for immunosuppression by wortmannin was suggested by the anti-inflammatory activity of structurally related fungal metabolites. 11-desacetoxy- wortmannin inhibits carrageenan-induced rat paw edema and decreases symptoms of arthritis. Exposure of human neutrophils or mouse bone marrow-derived macrophages to wortmannin or its 11-desacetoxy analogue in vitro inhibits induction of the respiratory burst during phagocytosis. Gastric and myocardial hemorrhages are prominent in rats given crude wortmannin in feed or purified wortmannin by gavage. Accompanying the hemorrhagic lesions are necrosis of the spleen, lymph nodes, and gut-associated lymphoid tissue. Rats given crude wortmannin in feed display depletion and mild necrosis of thymus lymphoid tissue, whereas in wortmannin-gavaged rats there is severe necrosis of the thymus cortex accompanied by hemorrhage. There is a possibility that wortmannin contributed to the mycotoxicosis known as alimentary toxic aleukia; however, this has not been proven experimentally.

19.2.8 Fusarochromanone

Fusarochromanone (TDP-1) is a mycotoxin produced by *Fusarium equiseti* that has been associated with tibial dyschondroplasia in poultry. Avian tibial dyschondroplasia is an abnormality of bone growth plates in which there is an accumulation of uncalcified and avascular cartilage in the proximal head of the tibiotarsal bone. This lesion can be induced by fusarachromanone and the toxin has been used by a number of groups to study bone calcification. Several studies indicate that fusarachromanone has some influence on immune function, particularly suppression of humoral immunity. The compound is both inhibitory andstimulatory (at lower levels) in the mitogen-induced lymphocyte proliferation assay. Deacetylation to its metabolite, TDP-2, strongly decreases these immunotoxic effects. Additional studies are needed to verify whether immunotoxicity is similarly evident in vivo.

19.2.9 Gliotoxin

Gliotoxin, a mycotoxin with antimicrobial and immunosuppressive capabilities, is produced by several genera of fungi including the pathogenic fungi, *Aspergillus fumigatus* and *Candida albicans*. This mycotoxin is suspected of being involved in the pathogenesis of these agents and belongs to the class of fungal metabolites called epipolythiodioxopiperazines. The immunosuppressive activity of this compound was first discovered when workers found that gliotoxin secreted by *A. fumigatus* inhibited the phagocytic phenomenon of macrophage adherence to plastic. Since that time other immunosuppressive activities of this mycotoxin have been described, including inhibition of mitogenic stimulation of lymphocytes in vitro, cytotoxicity to avian lymphocytes, and inhibition of allograft rejection.

The possibility of involvement of gliotoxin in the pathogenesis of aspergillosis is strengthened by the fact that it is produced in infected animal tissue. Eichner and coworkers (1988) determined that gliotoxin and its dimethylthioether derivative were detectable in the peritoneal lavages from mice inoculated and infected with *A. fumigatus*. Gliotoxin also has been isolated and identified in naturally infected bovine udder tissue while other known toxins of this agent were apparently absent or were undetectable. Richard and DeBey (1995) demonstrated the presence of gliotoxin in infected tissues of turkey poults experimentally infected with known gliotoxin-producing strains of *A. fumigatus*. Gliotoxin subsequently was found to occur naturally in turkeys infected with *A. fumigatus*. The status of gliotoxin as an agent involved in mycotoxicoses established through ingestion of contaminated feeds is relatively unknown.

19.2.10 Fescue and Ergot Alkaloids

It has been demonstrated that cattle grazing tall fescue pastures infected with endophyte have a suppressed antibody response when immunized with tetanus toxoid. The immunosuppressing effect of the grass could be due to a number of different alkaloids produced by the endophyte. The decrease in immune function in cattle with fescue toxicosis may be due, in part, to the effect of ergot alkaloids on decreasing serum prolactin levels. Prolactin has been reported to function as a regulatory factor in immune responses.

19.3 HEMATOPOIETIC EFFECTS

19.3.1 Aflatoxins

Hemorrhagic anemia syndrome, caused by the consumption of moldy feed by poultry, has been linked to the presence of aflatoxin-producing fungi (Forgacs and Carll 1962). The syndrome is characterized by massive hemorrhagic lesions in the major organs and musculature. A possible hemolytic anemia with bone marrow hyperplasia and an increase in bone marrow nucleic acid occurred in broiler chicks consuming aflatoxin.

Hemoglobin, packed cell volume, and circulating erythrocytes significantly decreased. Aflatoxins also caused significant time increases in whole blood clotting, recalcification, and prothrombin in broiler chicks fed aflatoxins for 3 weeks. However, aflatoxins in the feed (20 ppm) of mature broilers for 4 weeks caused only a slight anemia without increasing erythrocyte fragility. The presented data suggesting that the development of anemia in aflatoxin-treated animals was a secondary effect of severe hypoproteinemia. These effects may be secondary to primary liver damage. Exposure to aflatoxins also may affect hemostasis in developing embryos. Hatched chicks, following embryonic exposure to aflatoxin B1, had significantly decreased cell counts, hematocrits, and haemoglobin concentrations. However, there were no differences in erythrocytes of treated versus control groups.

19.3.2 Trichothecenes

Trichothecenes, e.g., DON, T-2 toxin, cause histologic lesions in experimental animals, including cellular necrosis and karyorrhexis in actively dividing tissues of the intestinal mucosa, bone marrow, spleen, testis, and ovary. Therefore, macrophages, lymphocytes, and erythrocytes may be decreased with prolonged exposure to trichothecenes. Erythrocyte numbers also can be decreased by trichothecene-induced hemolysis. T-2 toxin caused a complete hemolysis of rat erythrocytes following a lag period dependenton the concentration of T-2 toxin. Hemolysis was thought to occur by the formation of free radicals because darkness and free radical scavengers — including vitamin E, mannitol, and histidine — inhibit erythrocyte hemolysis by T- 2 toxin. Trichothecenes may alter hemostasis. T-2 toxin significantly increased prothrombin time but not recalcification time or whole blood clotting time in broiler chickens. However, a trend to prolong recalcification, i.e., formation of a fibrin clot, by T-2 toxin occurred. A single intravenous administration of T-2 toxin in rabbits decreased circulating activity of certain plasma coagulation factors. Similar results were shown in calves given a single intravenous dose of T-2 toxin.

19.4 HEPATOTOXICITY

19.4.1 Aflatoxins

Diseases resulting from consumption of grains contaminated with aflatoxins range from acute liver disease, including necrosis and hemorrhage, to more chronic forms with extensive fibrosis and bile duct proliferation. Clinical manifestations of hepatotoxicity include changes in clinical chemistry values associated with liver function. Jaundice, hemorrhage, and diarrhea, along with decreased performance, may be evident in affected animals with acute disease. Ingested aflatoxins typically are absorbed in the GI tract, where they begin to be metabolically activated or detoxified in the mucosal cells of the intestinal tract and subsequently in the liver (or perhaps the kidney) after hematogenous delivery. Similar pathways likely exist for the other aflatoxins. A major metabolite resulting from cytochrome P450-dependent epoxidation of either aflatoxin B1 or G1 is the -8,9-epoxide. This product interacts with DNA, forming an adduct presumably involved in the carcinogenic activity of aflatoxins. Binding of the epoxide intermediate with proteins could be important in the acute hepatic toxicity of the aflatoxins. Generally, formation of other derivatives of aflatoxin B1 is a detoxification mchanism, as most of these metabolites are less toxic than the parent compound. Notably, epoxides of some of these intermediates of lower toxicity, e.g., aflatoxins M1 and P1, can occur. An important detoxification mechanism is the formation of the glutathione conjugate of aflatoxin B1-8,9- epoxide. The enzyme for this conjugation is known as glutathione S-transferase. Its level of activity in different species may explain their relative susceptibility to aflatoxins.

Other factors, e.g., diet, stress, sex, immune state, drug treatment, may be involved in the susceptibility of a given species at a given time to aflatoxins. The involvement of these factors plus differences in the biotransformation of aflatoxins relative to animal susceptibility has not been fully elucidated. Eliminating aflatoxins from the body must include passage of some of the dosage through the intestinal tract unabsorbed, while biliary excretion (major aflatoxin B1 metabolite is aflatoxin B1- glutathione) and urinary excretion (major aflatoxin B1 metabolites are aflatoxins M1 and P1 and aflatoxin B1-N7-guanine) function as major routes of elimination. Milk is also a route of excretion in lactating animals (major aflatoxin B1 metabolite is aflatoxin M1) and a major route of exposure of the suckling young. Further information on aflatoxin biotransformation can be found.

19.4.2 Ochratoxin

Although ochratoxins are primarily nephrotoxic, they can cause liver damage, particularly at higher doses. In studies with broiler chicks, liver damage was present in concert with nephrotoxicity . Lymphocyte infiltration occurred in the liver along with lymphocytolysis in lymphoid organs. Necrotic changes in periportal cells were observed in rats given an LD50 (20 mg/kg bw) dose of ochratoxin A. Ochratoxin effects on liver seem to be much less pronounced and specific than those of aflatoxins. Interestingly, ochratoxin A apparently prevented fatty degeneration of the liver caused by aflatoxins when the two toxins were given simultaneously to broiler chickens.

19.4.3 Fumonisins

Fumonisins are hepatotoxic but liver disease may be coincident with other manifestations. Even in horses, when the dosage is suficiently high, there is liver disease, along with clinical evidence of elevated system disorders. In rats dosed with fumonisins, the usual manifestation is acute hepatosis and even hepatocellular carcinoma. Fumonisin involvement in diseases of systems, e.g., brain, lungs, kidney, other than the hepatic system is likely based on their mode of action. Fumonisin B1 inhibits a key enzyme in sphingolipid metabolism known as N-acyltransferase. This enzyme is involved in the conversion of sphingosine and sphinganine to ceramide, which is subsequently converted to complex sphingolipid. Disruption of this pathway can produce several outcomes because the basic process is involved in cellular regulation and other important biochemical events.

19.4.4 Sporidesmin

Sporidesmin is a hepatotoxic mycotoxin produced by *Pithomyces chartarum* on certain grasses. When consumed by animals, sporidesmin can cause a photosensitization disease known as facial eczema. This classic mycotoxicosis, which occurs primarily in New Zealand, results from destruction of bile duct epithelial cells, with blocking of the bile ducts causing phylloerythin (a breakdown product of chlorophyll) to accumulate in the circulating blood. Photosensitization occurs when these high energyabsorbing compounds in the peripheral blood circulation absorb energy from sunlight. Photosensitization particularly affects light-skinned or nonwooled areas, thus the name, facial eczema. Sheep, cattle, horses, and deer are known to be affected following consumption of the dead grasses on which the organism thrives.

18.4.5 Rubratoxin

Rubratoxin has been suspected as the cause of hepatotoxic, hemorrhagic disease of cattle and pigs fed moldy corn. Rubratoxin B produces hepatic degeneration, centrilobular necrosis, and hemorrhage of the liver and intestine in experimental animals given the compound. Rubratoxin B is known to be produced by *Penicillium rubrum* and *P. purpurogenum* on corn and both organisms have beenfound in moldy grains that have intoxicated domestic animals. Analytical methods are not sensitive enough to detect rubratoxin in grains, especially at low concentrations. Rubratoxin apparently exerts a synergistic effect with aflatoxin in guinea pigs.

19.4.6 Phomopsins

Lupinosis is a hepatotoxic condition characterized by severe liver damage and icterus in animals ingesting *Lupinus* species infected with the fungus *Phomopsis leptostromiformis*. The mycotoxin phomopsin A is the major fungal metabolite responsible for lupinosis; phomopsin B probably is involved as well. The disease is most common in sheep in Australia, New Zealand, and South Africa but apparently also occurs in the United States. While the disease occurs primarily in sheep grazing *Lupinus* species, it has been found in cattle, horses, and pigs and has been produced experimentally in goats, rabbits, dogs, and mice.

19.5 NEPHROTOXICITY

19.5.1 Ochratoxin A

Ochratoxin A, a mycotoxin produced primarily by *Aspergillus ochraceus* or *Penicillium verrucosum*, is nephrotoxic. Dogs, rats, and swine are known to be affected with kidney problems due to ochratoxin A. A major renal disease in swine known as porcine nephropathy that occurs in certain European countries, particularly Denmark, is associated with the consumption of ochratoxin-contaminated barley. Affected pigs show signs of pain in the kidney area, consume excessive amounts of water, appear depressed, urinate almost continuously, and have decreased feed consumption. The impaired renal function results in glucosuria and proteinuria, with casts evident in the urine. Pathologically, the kidneys are bilaterally affected, showing enlargement and paleness with large amounts of connective tissue evident in the cortex when viewed on a cut surface. Histopathologically, thre is tubular degeneration and atrophy, interstitial fibrosis, and, perhaps, hyalinization of the glomeruli.

Ochratoxin biotransformation is cytochrome P450 dependent in animals and humans and forms metabolic intermediates active in the carcinogenic and other toxic activities of ochratoxin A. The toxic activity of ochratoxin A may relate to one of its structural components, phenylalanine, which likely is involved in inhibition of enzyme reactions where it is known to function. Generally, ochratoxin A is passively absorbed in the GI tract in a non-ionized form and, through enterophepatic circulation, can undergo secretion and reabsorption.

Absorption of this mycotoxin also occurs in the proximal and distal tubules of the kidney. Using the regression equation proposed from the data of Krogh (1976) , approximately 4.6 ng ochratoxin A/g of kidney tissue would occur in swine consuming feed containing 200 ng of ochratoxin A/g. In the blood, which contains higher concentrations of ochratoxin A than other tissues, ochratoxin A is bound to the albumin fraction and thus persists for extended periods. A major mechanism of detoxification is conversion by ruminal and intestinal microflora into the nontoxic ochratoxin alpha, which is then excreted in the urine and feces along with some amounts of the parent compound. Several reviews on ochratoxin metabolism have been published.

19.5.2 Trichothecenes

DON and nivalenol, trichothecenes produced primarily by *Fusarium graminearum*, are associated indirectly with a nephropathy that has been studied in mice. These toxins increase IgA levels in the sera of mice, resulting in mesangial accumulation of this immunoglobulin and a disease similar to that found in the human IgA nephropathy — the most common human glomerulonephritis worldwide.

19.5.3 Fumonisins

Experimental animals fed fumonisin B1 exhibit altered renal histopathology as well as alterations in kidney weight, urine volume, proteinuria, enzymuria, and ion transport. These effects occur in several species and appear to be mediated by altered sphingolipid biosynthesis.

Reproductive Effects

Zearalenone

The major effects of zearalenone are estrogenic and primarily involve the urogenital system. Swine are the most commonly affected animals, although cattle, poultry, and laboratory rodents also are affected. Hyperestrogenism in female swine may be manifested as swelling of the vulva and enlargement of the mammary glands, especially in prepubescent gilts. Dietary concentrations of 1.0 ppm zearalenone or more may produce hyperestrogenism in pigs.

Zearalenone has been associated with feminization in young male swine, including testicular atrophy, swol len prepuce, and mammary gland enlargement; decreased libido may be a variable sequela but mature boars apparently have enough testicular reserve to avoid decreased spermatogenesis. In severe cases, this syndrome may progress to rectal and vaginal prolapse. Other effects related to higher concentrations include anestrus, nymphomania, and pseudopregnancy. High concentrations of zearalenone in cattle diets have been associated with infertility, teat enlargement, and udder secretions, which were most apparent in immature dairy heifers; vent enlargement and enhanced secondary sex characteristics are noted in turkeys and chickens (Fig. 6).

Fig. 6. Enlarged uterus in mouse (left) given an intraperitoneal injection of zearalenone. Compare with uterus of control mouse on the right

High concentrations of zearalenone (50 to 100 ppm) in swine diets have been reported to adversely affect cycling, conception, ovulation, and implantation. Placental membrane and fetal development also may be disrupted, resulting in decreased litter size and diminished viability of neonates. Zearalenone also decreased conception rates in dairy heifers. Zearalenone caused embryonic death, inhibition of fetal development, and decreased numbers of foetuses present in exposed swine. The majority of cases of zearalenone intoxication occur after animals consume cereal grains that have been contaminated by the toxin. It has been demonstrated in New Zealand and Europe, however, that sheep and cattle grazing pastures in which grass is contaminated with zearalenone can subsequently have reproductive problems. The mode of action of zearalenone and its derivatives involves displacement of estradiol from its uterine binding protein, elucidating an estrogenic response. The metabolism of zearalenone in several animal species has demonstrated that the major products of the parent compound are two metabolites, alpha- and betazearalenol, that can be found as the glucuronide conjugate in urine and feces. The major compounds found in plasma and urine samples, treated with beta-glucuronidase, from pigs dosed with zearalenone were zearalenone and alpha- zearalenol. Zearalenone can be converted to a compound known as zeranol (Ralgro(r)) or alpha-zearalanol that is used as an anabolic agent in cattle and sheep.

Ergot Alkaloids

Abortion may be a variable and controversial sequel to ergot ingestion, depending on species affected and alkaloid content of sclerotia. Ergot-contaminated diets of pregnant swine are associated with decreased piglet birth weights and increased stillborn rates; gestation time may be shorter or longer than normal. Ergot also inhibits prolactin secretion in pregnant swine, resulting in diminished udder development and agalactia at farrowing. Severity of effects is directly related to dietary concentrations; feeds containing at least 0.3% ergot sclerotia have definite detrimental effects on overall reproductive performance. Altered reproduction patterns and reproductive failures in swine are associated with ergot. In horses the endophyte of tall fescue is the causative agent for reproductive abnormalities in pregnant mares. Increased gestational length, agalactia, foal and mare mortality, toughened and thickened placentas, weak and dysmature foals, and decreased serum, prolactin, and progesterone levels occur in mares consuming endophyte-infected tall fescue pastures.

Other Mycotoxins

Other mycotoxins also have reproductive effects. Aflatoxin B1 has been associated with bovine abortions and may adversely affect nursing neonates through exposure to metabolites in milk. When administered parenterally to pregnant swine, T-2 toxin elicited embryo-toxic effects, whereas that fed in the diet did not. Clinical signs of T-2 toxin and diacetoxyscirpenol exposure in domestic animals usually are manifested as decreased egg production or rate of weght gain. The T-2 toxin consumption by breeding sows has caused drastically decreased conception rates and weak piglets, with decreased litter sizes in those sows conceiving. Ochratoxin A, rubratoxin B, secalonic acid D, and sterigmatocystin all are reported to affect embryonic survival.

Teratogenic Effects

Experimentally, aflatoxin B1, ochratoxin A, rubratoxin B, T-2 toxin, sterigmatocystin, and zearalenone are teratogenic in at least one mammalian species (Hayes 1981). When administered in combination during organogenesis, ochratoxin A enhanced T-2 toxin-induced teratogenicity. Intraperitoneal administration of ochratoxin A to pregnant mice had a teratogenic effect on surviving pups, increased fetal death, and decreased fetal weight .

19.6. NEUROTOXIC EFFECTS

19.6.1 Trichothecenes

DON and other trichothecenes have two characteristics that suggest neurotoxicity in animals. DON induces vomiting at high acute doses, while, at lower doses, it induces feed refusal. DON is considered one of the least toxic trichothecenes with regard to mortality. However, it has anorexic and emetic potencies that are equivalent to or higher than those reported for more acutely toxic trichothecenes.

For example, although T-2 toxin is at least 10 times more lethal, it has 10-fold less emetic potency than DON. The oral minimum effective dose for DON-induced vomiting in swine is approximately 50 µg/kg whereas observed the oral minimum effective dose for T-2 in pigs to be greater than 500 µg/kg. The threshold for DON-mediated anorexia is similar to that of the much more lethal T-2 toxin. At levels as low as 2 to 3 ppm toxin in the diet, both DON and T-2 depress porcine feed consumption and weight gains. However, effects at this lower concentration are transitory, with animals quickly compensating for any initial losses. At higher levels of contamination (over 8 ppm DON or T-2), pronounced effects on feeding behavior become more long term. While T-2 can produce severe pathological and biochemical alterations at these higher intakes (8 to 20 ppm), direct clinical changes due to DON at the same intakes unrelated to anorexia and resultant weight losses are not apparent. Sensitivity to the anorectic effects of DON differ considerably in the three species most commonly studied, with the order of tolerance being rat > mouse > pig. Neurochemical alterations in laying hens dosed with DON are not apparent, given the high tolerance of poultry to trichothecenes. The serotoninergic system mediates both an anorectic response and an emetic response. DON has been shown to modulate serotonin activity, suggesting a possible link between DON induced emesis and the serotoninergic mechanism. DON-induced vomiting is blocked in swine by pretreatment with selective 5HT type 3 (5HT3) receptor blockers. DON also inhibits small-intestinal motility in rodents, mediated through 5HT3 receptors.

Relatedly, gastric relaxation and/or delayed gastric emptying are important components of both emesis and food intake (Hunt 1980). Even though previous studies demonstrated a central nervous system (CNS) effect, it was not clear whether DON acts directly at the central (brain) level. However, research suggests that at least part of DON's mechanism of action is mediated through action on the peripheral 5HT3 receptors found in the GI tract. It is unclear which substances activate these receptors. DON-induced conditioned taste aversion also has been investigated. Combined taste aversion to a novel taste (saccharin) could be established by concurrent treatment with DON. However this effect was transitory, suggesting that once conditioning was established, the response to pretreatment with DON diminished with time relative to the level of toxin used. The same result has been observed with T-2 toxin. It is found that DON-induced combined taste aversion is mediated by the area postrema (located in the brainstem), which is thought to mediate the emetic action of DON (and T-2 toxin). Tolerance to DON may provide some clues about the toxin's mode of action. At lower dietary DON concentrations, reduction in food intake is transitory in several species, e.g., pigs, mice, lasting only a few days before animals begin to compensate for initial losses. With increased DON levels in feed, animals may not return fully to control intake but the extent of feed refusal diminishes with time. Several lines of evidence suggest that tolerance development occurs with most anorexic compounds that rely on a central serotoninergic mechanism.

19.6.2 Fumonisins

A relationship between moldy corn and sporadic outbreaks of equine leukoencephalomalacia (ELEM), an unique neurotoxic syndrome of Equidae, has been known for many years. ELEM was shown to be caused by fumonisins.

Neurotoxic symptoms — including loss of feed consumption, lameness, ataxia, oral and facial paralysis, head pressing, and recumbency — may begin within days after exposure to fumonisin B1. Seizures may be present and morbidity can occur within hours after the onset of clinical signs.

Focal malacia and liquefication of cerebral white matter with peripheral hemorrhage is the pathognomonic necropsy finding. Microscopically, there are liquifactive necrosis and gliosis. Edema, hemorrhage, and perivascular cuffing of leukocytes are often present in the surrounding neuropil. The exact mechanisms by which fumonisins promote this disease are not known but are apparently related to their ability to inhibit sphingolipid synthesis.

19.6.3 Tremorgens

Fungi represent a rich source of secondary metabolites capable of eliciting a tremorgenic response in animals. Fungal tremorgens have been implicated in the etiology of several tremorgenic syndromes. Some tremorgenic syndromes that occur in pastures involving cattle are referred to as "staggers syndromes." Typically, these are characterized by muscle tremor, uncoordinated movements, a general weakness in the hind legs, and stiff, stilted movements of the forelegs. More severely affected animals cannot stand. These effects are aggravated when the animals are moved or excited. Some deaths can occur; these are indirect, usually from dehydration, pneumonia, or drowning while attempting to drink. The animals generally recover, with no apparent damage, 24 to 48 hours after the toxic feed source is removed. Syndromes include paspalum staggers, ryegrass staggers, corn staggers, and Bermuda grass staggers . With the exception of Bermuda grass staggers, fungal tremorgens have been implicated in the etiology of these diseases. Ryegrass staggers occur following the production of tremorgenic mycotoxins or lolitrems by the ryegrass endophyte, *Neotyphodium lolli*, formerly called *Acremonium lolli*.

Another tremorgenic syndrome in cattle occurs when corn silage from a trench-type silo is improperly unloaded with a front-end loader. This aerates the silage, causing fermentation involving toxigenic strains of *A. fumigatus*. These fungi can produce alkaloids and tremorgens of the verruculogen/fumitremorgen group. The disease is characterized by a geeral deterioration typical of protein deficiency and malnutrition, even though ample pasture and supplemental feed may be available. Other prominent clinical signs are diarrhea, irritability, and abnormal behavior.

One of the most characterized tremorgenic intoxications involves a naturally occurring mycotoxin called penitrem A. This compound has been involved in three episodes of intoxication in dogs consuming either moldy cream cheese, English walnuts, or a hamburger bun (bread). Penitrem A and *P. crustosum* also have been implicated in a human intoxication involving moldy beer . The individual involved was a physician, who became acutely ill approximately four hours after consuming about 30 ml of the contaminated beer. He described the symptoms as a tremor, throbbing frontal headache, feverish feeling, nausea, vomiting, double vision, weakness, and bloody diarrhea. All symptoms disappeared, with no apparent residual effects noted after 30 hours.

The mode of action of fungal tremorgens is through release of neurotransmitters from synaptosomes in the CNS and in peripheral nerves at the neuromuscular junction (Knaus et al. 1994). Although many tremorgens affect the high-conductance, calcium (Ca^{2+})-activated potassium (K^{+}) channels in the release of neurotransmitters, this may not be the mechanism for the tremorgenic activity as some nontremorgenic mycotoxins have similar activity. Tremorgens likely are eliminated quickly by biliary excretion; penitrems can be transformed to more polar compounds by bile.

19.6.4 Ergot

Ergotism is a classic mycotoxicosis caused by products of the plant parasitic fungi, *Claviceps purpurea*, *C. paspalli*, and *C. fusiformis*. These fungi invade the female portion of the host plant, particularly small grains such as rye and barley, and the ovary with a mass of fungal tissue called a sclerotium. These sclerotia (ergots) contain alkaloids produced by the fungus in varying amounts, depending on the host plant and environmental factors. When the sclerotia are ingested, two major signs of disease (gangrene or convulsions and GI disorders) may occur. Additionally, production losses have been noted in animal feeding studies. Lameness and necrosis of ears, tails, and feet have occurred in animals pastured on infected grasses or fed grains containing ergot sclerotia. Nervous signs, predominant in the convulsive form of ergotism, have occurred primarily in animals and consist of ataxia, convulsions, and paralysis.

Ergotism occurs in the United States, particularly in areas where rye, barley, and other susceptible small grains or grasses are grown, i.e., midwestern to northwestern states. Effects of ergot on animal production include decreased weight gains, milk production, reproductive efficiency, and agalactia. Although the sclerotia may contain variable quantities and proportions of the ergot alkaloids, the tolerances established in the United States are based on the percentage by weight (less than 0.3%) of sclerotia in graded grains.

19.6.5 Slaframine

Slobbers or salivary syndrome, characterized by excessive salivation in horses and cattle, has been caused by consumption of red clover hay infected with *Rhizoctonia leguminicola*. Other clinical signs of toxicity include diarrhea, piloerection, increased lacrimation, feed refusal, and respiratory failure in very severe cases. Outbreaks of salivary syndrome have led to decreased cultivation of red clover in the Midwest. Slaframine, the compound responsible for salivary syndrome, was isolated and identified from twostrains of *Rhizoctonia* .

Fig. 7. Black sclerotia (ergots) caused by Claviceps purpurea infection in barley

Slaframine is a parasympathomimetic metabolite. Its active ketoimine liver metabolite has an affinity for muscarinic receptors of the nervous system. Slaframine has a particularly high affinity for receptors in the GI tract. Chicks and cattle exposed to this compound showed an increase in salivation, pancreatic flow, and growth hormone release. Thus, slaframine may be an important therapeutic agent in treating digestive disorders.

19.7 CARCINOGENESIS

19.7.1 Aflatoxins

Induction of neoplasms (abnormal growth of tissue) by aflatoxins has been extensively studied (Eaton and Groupman 1994). Aflatoxin B1 —as well as the secondary aflatoxin B1 metabolites, aflatoxin M1 (present in milk of cows ingesting high levels of aflatoxin-contaminated feed), aflatoxicol, and aflatoxin G1 — have caused hepatic, renal, and colonic neoplasms in rats (0.2 to 0.12 μg/day) and hepatic neoplasms in trout (0.4 ppb for 9 months), ducks (30 ppb for 2 to 4 weeks), ferrets (0.3 ppm) , mice (20 μg/week for 76 weeks) and monkeys. Nasal neoplasms were induced in sheep fed 1.75 ppm aflatoxin in the diet for 3-1/2 years. Pulmonary neoplasms were produced in mice injected with 0.2 μg of aflatoxin B1/day for 4 weeks. In in vitro mutagenicity tests, aflatoxin B1 activated by rat liver microsomal preparations has caused point mutations and frame shift mutations in host cell DNA. Aflatoxin B1 also chemically (covalently) binds to DNA in vitro. This tight (covalent) binding of aflatoxin B1 to DNA causes structural alterations in the DNA. Similar covalent interactions between chemicals and DNA and the resultant structural alterations have caused genomic mutations in *in vitro* and *in vivo* systems.

19.7.2 Sterigmatocystin

Sterigmatocystin, a precursor of aflatoxin, has lower toxicity. However, its carcinogenic potential, which experimentally is about one tenth that of aflatoxin B1, is not trivial. Sterigmatocystin has caused liver cell carcinoma in laboratory rats. It was also mutagenic by several *in vitro* tests, e.g., Ames test, Rec assay, and *Bacillus subtilis* assay. Tumorigenicity of sterigmatocystin was considerably less than aflatoxin B1 to rats (0.1 to 0.01). Sterigmatocystin covalently bonds to DNA at approximately 20 to 30% of the level observed with aflatoxin B1.

19.7.3 Ochratoxin A

Although mutagenicity of ochratoxin A has not been established during several in vitro assays (e.g., Ames test, Rec assay), tumorigenesis/carcinogenesis was reported in laboratory animals. Renal and hepatic neoplasms have been induced in laboratory mice fed 40 ppm ochratoxin A for 20 months. These results were confirmed in a study performed under the direction of the National Institute of Environmental Health Sciences as a function of the National Toxicology Program. It. is reported a 40% increase in urinary tract cancer in eastern Europeans with Balkan endemic nephropathy, a condition now suspected by many to result from ingestion of local grains contaminated with ochratoxin.

19.7.4 Fumonisins

Fumonisins with a free primary amino group have been shown to be liver cancer promoters in rats. It is found hepatocarcinomas in male rats fed 50 ppm fumonisin B1 (≥90% purity) for 26 months. The National Toxicology Program has recently completed two-year carcinogenicity studies of fumonisin B1 in rats and mice.

19.8 DERMAL TOXICITY

19.8.1 Trichothecenes

Several trichothecene mycotoxins are skin irritants. Skin tests have been used to determine qualitatively the biologic activity of this class of compounds.As determined the minimum effective dose for dermal toxicity of 18 trichothecenes and compared their structure-activity relationships. T-2 toxin, the most toxic compound tested, caused erythema on the shaved backs of guinea pigs at a dose of 5 ng. T-2 toxin and macrocyclic trichothecenes caused an edema ap proximately 24 hours after application. Trichothecenes, e.g., nivalenol, DON, exhibited very low dermal toxicity. The cutaneous penetration rate and the metabolism of diacetoxyscirpenol, verrucarin A, and T-2 toxin varied in human and guinea pig skin, depending on structure, dosage, and carrier. In vivo cutaneous absorption of T-2 toxin in guinea pigs was compared with in vitro cutaneous penetration in static and flow through diffusion cells. Penetration data from static studies may underestimate the rate and amount of absorption through intact skin. No T-2 toxin was present in plasma or urine of guinea pigs in the in vivo study; however, metabolites of this parent compound were found in urine.

Limitations of Systems Approach for Classification of Mycotoxins Because studies have advanced from large-dose, complete-animal responses to small-dose, subcellular/molecular responses, the systems approach for classification of mycotoxins becomes increasingly difficult. Also, as we learn more about responses in different species, the variation or multiplicity in systems affected become evident. This multiplicity of effects is readily exemplified by the trichothecene mycotoxins, which impact productivity, liver, kidney, hematopoetic system, CNS, and immune system. Similar problems are found with fumonisins, which affect brain, kidney, liver, and lung. In addition to the multiplicity of effects, the study of animal mycotoxicoses has become increasingly complex because of the interaction of systems affected, the basic metabolic pathways affected, and the physiopathologic nature of the intoxications.

19.8.2 Synergistic Effects

The scientific literature is replete with information on the effects of individual mycotoxins in various livestock species. Multiple mycotoxin contamination is an area that recently has become a greater concern to the livestock industry. The concern arises from the fact that concentrations of individual mycotoxins associated with poor livestock performance and/or disease syndromes in commercial operations usually are lower than those reported to cause toxic effects in controlled laboratory studies. Additionally, in some of these reports the feed contained more than one mycotoxin. For example, aflatoxin and fumonisin B1, and vomitoxin and zearalenone commonly occur together in the same grain. Many fungal species also are capable of simultaneously producing several mycotoxins.

Therefore, an individual grain may be naturally contaminated with more than one mycotoxin, or the incorporation of numerous grain sources, which are each contaminated with a different mycotoxin(s), into a single feed may result in a diet that contains a number of different mycotoxins.

Poor livestock performance and/or disease syndromes, reported in commercial operations, may be due to synergistic interactions between multiple mycotoxins. When mycotoxins are fed in combination, interactive effects can be classified as additive, less than additive, synergistic, potentiative, or antagonistic. Additive effects occur when the combined effects of two mycotoxins are equal to the sum of the effect of each toxin given alone (example: 2 + 4 = 6). Less than additive effects occur when the combined effects of two mycotoxins are less than the sum of the effect of each toxin given alone (example: 2 + 4 = 5). Synergistic effects occur when the combined effects of two mycotoxins are much greater than the individual effects of each toxin alone (example: 2 + 2 = 10). Potentiative effects occur when one mycotoxin does not cause a toxic effect on a certain organ or system, but when fed with another mycotoxin it makes the latter much more toxic (example: 0 + 2 = 10). Antagonistic effects occur when one mycotoxin interferes with the effects of another (example: 4 + 0 = 1).

Contains a summary of mycotoxin interaction studies published in refereed journals. In all cases, the interactive effects of only two mycotoxins were studied. For purposes of brevity, and due to the uncertainty of the biological and economic significance of statistical changes in other physiological response variables, the summary was restricted to production effects. The majority of these studies was conducted with poultry with a few studies with pigs and one with sheep. Results of the review indicated that additive or less than additive effects (26/33; 78%) were the predominant interactions observed, with synergistic interactions occurring in 19% (6/33) of the studies, and antagonistic interactions occurring in the remaining study (1/33; 3%). Aflatoxin was one of the two myc otoxins involved in five of the six instances of synergism.

These should be interpreted with some caution. Many of the studies involved acute or subacute levels of mycotoxins, whereas syndromes reported in the field occur with much lower concentrations of mycotoxins. It is plausible that with these high mycotoxin doses the occurrence of interactive effects may not be detectable because one mycotoxin singly has already produced a maximum deleterious response. Furthermore, the studies cited do not consider the more subtle changes that may be occurring within the immune system. Multiple mycotoxin-induced immunosuppressions may make livestock species more susceptible to disease agents present in commercial production systems. Also, only pairs of mycotoxins were evaluated in these studies, whereas under commercial conditions contaminated feeds may contain two or more mycotoxins. Finally, studies were conducted under laboratory conditions where animals generally are not exposed to many of the environmental stressors (heat, ammonia, disease, etc.) that occur under commercial conditions.

Future research in the area of mycotoxin interactions should include combinations of two or more mycotoxins at concentrations that commonly occur in the field. Two such studies where chicks and poults were fed a combination of six toxins that included 1 mg deoxynivalenol/kg, 5 mg moniliformin/kg, 5 mg fumonisin B1/kg, 100 mg aflatoxin/kg, 1 mg zearalenone/ kg, and 0.5 mg ochratoxin A/kg in the diet, were re- cently reported. Results indicated that the combination of mycotoxins at these concentrations depressed chick performance but did not affect poult performance. However, several serum biochemical values were altered in poults fed the combination of mycotoxins. To be determined is if the presence of all six mycotoxins was required to cause the effects observed. Additionally, response variables such as performance, serum chemistry, organ weights, hematology, and histopathology, immune function also should be evaluated in future studies to determine if a compromised immune system may be contributing to previously reported disease syndromes. Both humoral and cell-mediated immunity should be evaluated and the evaluation also should include the effects of mycotoxins on vaccine efficacy and the ability of animals to respond to a disease challenge. The effects of heat stress, a common stressor in livestock operations worldwide, on immune function of livestock fed multiple mycotoxins also should be evaluated.

Mycotoxin interactions indicated that additive or less than additive effects were the predominant interactions observed. Synergistic interactions were only observed in six of the 33 reported studies, and in five of the six studies aflatoxin was one of the two toxins evaluated. These data suggest that mycotoxin synergism may be less of a concern than had been suggested by field outbreaks where multiple mycotoxins had been implicated as possible intoxicants.

19.9 DIAGNOSIS OF MYCOTOXICOSES OF ANIMALS

Many individuals in the field of veterinary medicine long ago recognized that mycotoxins were involved in the diseases of animals. The recognition that fungi or perhaps their toxins were involved in sporadic outbreaks of disease in humans, however, met with only limited means to diagnose these diseases in humans. The difficulties of diagnosing mycotoxicoses in humans likely relate to the variety of foods consumed causing unrecognized intoxications because the overall concentration of mycotoxins in the diet was quite low. Most human intoxications were reported in localized groups of individuals where there was consumption of a primary staple in the diet. When contamination of that dietary staple occurred, there was an outbreak of disease and the cause could be easily traced to the toxin in a single staple in the diet. Therefore, criteria for diagnosis of mycotoxicoses in humans is based on those established in the veterinary literature and will not be addressed specifically in this report.

Diagnosis of mycotoxicoses in animals often is attempted long after the incriminating feedstuff is consumed and other possible causes of disease have been eliminated. This presents the diagnostician with the difficult task of trying to determine whether a mycotoxin(s) is the culprit. An accurate diagnosis depends on several factors, not the least of which is availability of an adequate sample of the feed involved in the intoxication. However, the disease manifestation in the animal must be considered first, because it can often give the analytical chemist insight into the mycotoxin(s) for which to look. Examining a feed sample for any mycotoxin could represent a major expense and lead to misdiagnosis, as there are hundreds of mycotoxins and only some are likely important animal disease agents. The impact of mycotoxins on economic aspects such as animal health and productivity is difficult to assess because of the following factors associated with their occurrence, consumption, and disease-causing potential: (1) chemical diversity, (2) agronomic practices, (3) climate, (4) harvest and storage methods, (5) food processing, (6) dietary diversity, (7) exposure routes, (8) age and species susceptibility, (9) indiscriminate disease manifestation, (10) varied production by fungal isolates, (11) co-occurrence with other toxins (interactions), and (12) overshadowing by secondary disease.

Because mycotoxins generally have a specific target organ in their disease-producing capacity, only a single body system is often considered to be involved. However, a true understanding of disease, and specifically mycotoxicoses, includes the awareness that most mycotoxins affect more than one body system. This is important for making an accurate diagnosis of mycotoxicoses.

The single target organ is valid for a general discussion of mycotoxin effects. This "systems approach" can help reach an accurate diagnosis, as categories of mycotoxins may be suspected when it is recognized that an individual body system is affected. Mycotoxins exhibit the following primary disease effects: hepatotoxic, GI toxic, hematotoxic, nephrotoxic, reproductive toxic, neurotoxic, immunotoxic, endocrine toxic, teratogenic, mutagenic, and carcinogenic. Most data about mycotoxin effects on animals have come from well-defined toxicological laboratory studies. Thus, considerable information about a single purified mycotoxin on a specific animal under wellcontrolled experimental conditions exists.

The characteristics of these studies often include the following:

- Pure toxin is used.
- Calculated dosages are used.
- Dosages are administered at the same time each day.
- Sufficient numbers of animals are given similar doses to establish significance.
- Dosages usually are given as a separate entity, to ensure the dose was administered.
- If the dosage is given with food, food intake is measured.
- Healthy animals are used.
- Animals are kept in a controlled environment.
- Animals are fed and watered adequately.
- Animals have ample room and are often housed individually.
- The light/dark cycle is usually controlled.
- Animals are kept under constant surveillance or observed frequently.
- Careful measurement of effects/changes is conducted.

However, results of these controlled studies can be far from the "real-life" or natural situation shown in Fig. 8. In this situation, one can see all of the aspects of an animal's life that may affect the outcome of an intoxication. A diagnosis may be enhanced by investigating fungi in the feed. In some cases, the mycotoxin-producing fungus may no longer be viable, although the product of its growth and metabolism remains in the specific commodities incorporated in the feed mix. Therefore, a fungus responsible for producing a specific mycotoxin may be absent. Also, it is possible to isolate a toxigenic fungus and find no respective mycotoxin in a sample as not all strains of a species produce toxins. Environmental conditions, e.g., the appropriate substrate, may not have been met for toxin production.

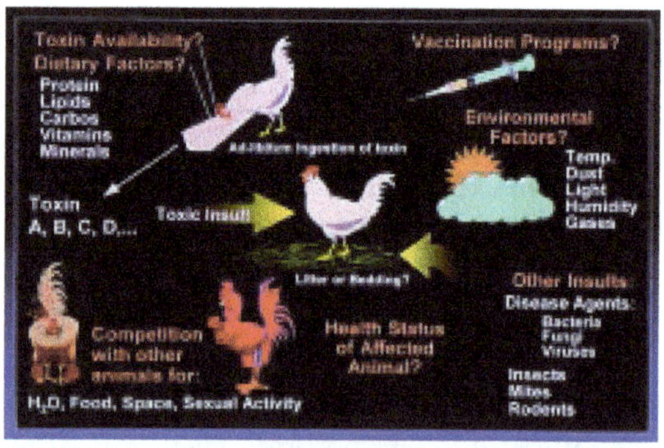

Fig. 8. Cartoon depicting the potential for interactive factors involved in the exposure of animals to mycotoxins.

Accurate diagnosis involves several criteria but one of the most important is obtaining a representative sample of the feed involved in the toxicosis. Thus, a mycotoxicosis should be considered as a possibility early in the manifestation of the respective disease and a feed sample should be preserved. It may be necessary to obtain a sample from the trough or bunk. Far too often, a mycotoxicosis is considered only after other potential causes for the disease have been eliminated; then it is too late to obtain an adequate sample that could be linked to the toxicosis.

19.91 Diagnostic Criteria

Major criteria for making a thorough mycotoxicosis diagnosis include the following:

All Aspects of Disease Manifestation

- Observed changes in animal condition and behavior, e.g., coat quality, salivation, eye movements, gait.
- Measured changes in animal condition and behavior, e.g., temperature, respiration rate.
- Nonintervention, with measurement of urine using chemistry and microbiology and of feces using chemistry.
- Blood chemistry, including immune status and microbiology.

19.9.2 Postmortem Examination

- Gross visible observations, including position at death.
- Histopathological observations of tissues selected during postmortem examination.
- Chemistry of tissues and fluids collected during postmortem examination, e.g., urine, blood, rumen or stomach content.

19.9.3 Examination of Feed

By carefully observing the animal during the intoxication and following the established criteria, one can get a good idea of which mycotoxin(s) could have produced the effects seen. Once an accurate diagnosis has been made, it is important to report the case in the literature, so that others can learn from the experience.

The article should include observations plus the analytical procedure used in the analysis of the feed sample involved in the intoxication to make the diagnosis of the mycotoxicosis. When the observed and measured changes in the animal fit the description of the mycotoxicosis from the scientific literature, and the amount and type of mycotoxin found in the feed involved in the intoxication are commensurate with the mycotoxicosis, then the best effort has been made to achieve an accurate diagnosis.

20

RISK ASSESSMENT AND REGULATIONS FOR MYCOTOXINS

20.1 INTRODUCTION

The control of mycotoxins in foods and feeds is a constantly evolving process. The occurrence of mycotoxins is influenced by certain environmental factors; hence, the extent of contamination of a particular commodity with a particular toxin is unpredictable and will vary with geographic location, agricultural practices, and the susceptibility of commodities to fungal invasion during preharvest, storage, and/or processing periods. Mycotoxins may exhibit various adverse toxicological manifestations in humans and animals. In many cases, the effects of mycotoxins on human health, at the levels at which humans are sometimes exposed, are not known. In highly developed countries, acute toxic effects of mycotoxins are rarely observed because advances in processing technology and quality control programs, such as GAP and GMP prevent heavily contaminated food from entering the food supply. From a public health perspective, there is concern regarding the health effects of long-term exposure to low levels of mycotoxins in foods. Because the occurrence of these toxins in foods and feeds is not entirely avoidable, small amounts of these contaminants may be legally permitted, provided the amounts involved are not considered injurious to human and animal health. The legal basis for regulating toxin or poisonous substances in foods in the United States is the Federal Food, Drug and Cosmetic Act, which is enforced by the FDA. In one section of this Act, a food is considered adulterated if it bears or contains any poisonous or deleterious substance that may render it harmful to health. By enforcing this statute, the FDA can prohibit the entry of, and remove from interstate commerce, any food or feed that is so adulterated. One strategy used by the United States to minimize mycotoxins in the food supply is to establish regulatory limits for certain mycotoxins in various commodities. Once these limits are established, the food industry is monitored routinely to determine if it is adhering to established regulatory limits. The worldwide occurrence of mycotoxins in foods and feeds has been recognized by the FAO, and the World Health Organization (WHO) for many years. Many countries worldwide have established food laws that impose limits on the concentrations of specific mycotoxins in foods. Analytical data obtained from food monitoring programs are coupled with available toxicological data and are used for making science based risk assessments that can serve as a basis for establishing a particular regulatory level. The objective of this chapter is to give an overview of the importance of the use of science-based risk assessments coupled with the establishment of regulatory limits as an ideal approach to minimize the levels of mycotoxin contamination in the food supply and thereby protect the health of consumers.

Mycotoxins are considered unavoidable contaminants in foods in that agronomic technology has not yet advanced to the point where preharvest infection of susceptible crops can be completely eliminated. Mycotoxins may exhibit various adverse toxicological manifestations in humans and animals. They can enter food directly or indirectly during the production, transportation, processing, or storage of food. Small amounts of these contaminants may be legally permitted in foods and feeds because they are not entirely eliminated by good agricultural practices (GAP) and good manufacturing practices (GMP), provided the amounts involved are not injurious to human and animal health. Many countries have established regulatory limits or food laws that restrict the concentrations of mycotoxins that are permitted in certain foods and feeds; at least 77 countries involved in international trade have now established regulatory limits for some of the mycotoxins. Factors that should play a role in deciding appropriate regulatory limits for a mycotoxin in a particular country include the availability of analytical methods, toxicological and analytical survey data for use in preparing risk assessments, and knowledge concerning the occurrence and distribution of the mycotoxin in various commodities. It is also important that each country recognizes the need to maintain an adequate supply of food at reasonable cost to consumers in their country, and to be aware of legislation in other countries with which trade contacts exist.

20.2 RISK ASSESSMENT

Many mycotoxins have potent acute and chronic biological activities as determined by animal studies, *in vitro* bioassays, and human epidemiological studies. These substances are present in the food supply on an intermittent or chronic basis. Regulatory agencies regulate unavoidable contaminants of foods, i.e., those that cannot be avoided by good manufacturing practices, which usually is the case with mycotoxins. In the United States, the FDA regulates mycotoxins under Section 402 or 406 of the Food, Drug and Cosmetic Act (FDCA). The FDA balances the benefits of food availability against the risk that a contaminant in that food might pose to the public health. Quantitative risk assessment for carcinogenic contaminants and safety assessment for noncarcinogenic contaminants are tools used by the FDA to estimate the upper limit of risk to public health posed by the contaminant. Safety/risk assessment requires determining the severity and scope of adverse effects of a hazard (toxicity assessment) and also the degree of exposure of an individual or population (exposure assessment).

The FDA's current approach to toxicity assessment can be traced that introduced the use of 10-fold safety factors, later known as uncertainty factors. Such factors are used to extrapolate data on the effects of specific chemicals on laboratory animals to humans, to derive an acceptable or tolerable daily intake (ADI/TDI); in particular, the no-observed-adverse-effect level (NOAEL) is divided by the safety/uncertainty factors (multiples of 10). One process of cancer risk estimation, generally used for foodborne substances, employs a simple, direct method. A point on the dose-response curve (tumor incidence versus dose) for a chemical is chosen below which the data no longer appear to be reliable (generally in the range of a 1% to 10% tumor incidence).

A straight line is drawn from the upper confidence limit on risk at that point to the origin. This provides the slope of the line used to provide upper bound estimates of cancer risk at low doses where risk ≥slope X dose. For this procedure, the target tissue dose for the active carcinogenic agent is desirable; without this information, the whole dose administered to an animal in a bioassay is used. In the absence of information to the contrary, dose scaling across species is based on body weight to the 3/4 power (arrived at by an interagency committee) or dose is expressed as concentration in the diet. The FDA recognizes that food consumption and body weight of animals influence survival, tumor incidence in some tissues, and toxicity. Carcinogenic potency in animals tends to overestimate carcinogenic potency in humans for about 20 chemicals where data are available for making comparisons.

20.3 SAFETY ASSESSMENT FOR DEOXYNIVALENOL

DON is a natural metabolite produced by various members of the genus *Fusarium*. These fungi grow on various grains and are ubiquitous. The FDA evaluated a number of hazards of exposure to DON which will be described as under.

20.3.1 Acute Toxicity

Swine appear to be the species that is most sensitive to vomitoxin, with vomiting, diarrhea, muscular weakness, tremors and coma observed after 24 hours following dietary doses as low as 2 mg/kg. In acute studies in mice, LD50 values were calculated to be 78 mg/kg bw by gavage and 49 mg/kg bw intraperitoneally.In these animals, lesions were observed in the GI tract, kidney, heart and lymphoid tissue. In rats fed 10 mg/kg bw vomitoxin, no signs of toxicity, e.g., feed refusal, emesis, unusual behavior, were observed. Chickens seem to be less sensitive than swine. In laying hens, concentrations of up to 5 mg/kg vomitoxin in the diet decreased egg weights and shell thickness.

Dairy cattle are apparently able to tolerate relatively high levels of vomitoxin; nonlactating cows can consume 1 mg vomitoxin/kg bw/day without any serious side effects.

20.3.2 Teratogenicity and Reproductive Toxicity

In swine-feeding studies using 3.5 mg/kg bw, treated gilts had a lower feed consumption and growth rate than did controls; increasing levels of vomitoxin yielded a linear trend toward lower fetal weight and length. Mice embryos are fairly sensitive to vomitoxin. Doses of 2.5, 5, 10, and 15 mg/kg bw were associated with a specific lethal effect on the mouse embryo with or without maternal vaginal bleeding but not with any apparent maternal toxicity. Multiple teratogenic effects occurred at doses of 2.5 and 5 mg/kg bw, while skeletal malformations were present in the 1, 2.5, and 5 mg/kg bw groups in a dose- related manner. Resorptions at the 10 and 15 mg/kg bw dose were 100% and 80%. The mean number of live fetuses/litter at these doses showed no statistically significant decrease, although resorptions were increased. In rats, dietary concentrations of vomitoxin of up to 20 ppm caused neither maternal effects nor convincing evidence of fetal effects. In rabbits, increased resorptions were observed at doses of 1.8 or 2 mg/kg bw and fetal weight was decreased at maternal doses of 1 and 1.6 mg/kg bw. All doses caused maternal decreases in body weight and feed intake.

20.3.3 Effects on the Immune System

Chronic exposure to DON has been shown to impart decreased humoral immunity, cell-mediated immunity, and host resistance. An increase in serum IgA can be demonstrated in mice fed levels as low as 2 ppm DON in some experiments, with the greatest effect shown at 25 ppm. Vomitoxin induced elevations in the percentages of IgA+ cells, total T cells, and CD4+ cells in mice, suggesting that it induced regulation of IgA production originating in the Peyer's patches. Increases in serum IgA and the pIgA to mIgA ratio have been associated with IgA nephropathy in mice.

The following discussion shows how toxicology data can be used to estimate a tolerable daily intake. In actual practice, the FDA has employed other data in addition to these mice teratology data. The teratogenicity data in mice were selected as being most appropriate for modeling a quantitative relationship between DON and fetal resorption in mice; this model was used to estimate a maximum likelihood estimate of a dose that leads to a 5% increase in fetal resorption. This dose may also be called a benchmark dose and is equivalent to a NOAEL. A dose that results in a 5% increase in fetal resorption of 1.15 mg/kg/day was obtained.

Evidence indicates that minor renal dysfunction associated with dietary exposure to vomitoxin has a dose-response function similar to that for reproductive hazard. However, insuffcient evidence exists to estimate the magnitude, severity, or dose-response relationship of the hazard that may result.

By applying a factor of 10 to extrapolate across species and a factor of 10 to account for individuals who may be more sensitive, a TDI of 12 g/kg/day can be estimated. For a 60 kg adult, this corresponds to an exposure of 0.7 mg/kg/day. This endpoint is based on an effect that may occur from subchronic exposure (weeks to months). Therefore, a short-term exposure to this concentration from a single crop may be as hazardous as a long-term exposure.

20.4 QUANTITATIVE RISK ASSESSMENT FOR A CARCINOGENIC MYCOTOXIN AND AFLATOXIN

An excellent example of using animal studies, human epidemiology studies, and in vitro metabolism data to perform a quantitative risk assessment for the carcinogenic mycotoxin class of aflatoxins was done by the Joint FAO/WHO Expert Committee on Food Additives (JECFA) in 1997 and in 2001 (United Nations 2001). This subcommittee of internationally recognized experts was charged with reviewing the experimental evidence concerning aflatoxin carcinogenicity, evaluating potencies, linking potencies to intake estimates, and discussing the impact of hypothetical aflatoxin standards on sample populations and their overall risks. A number of factors influence the risk of primary liver cancer, most notably, carriage of hepatitis B virus as determined by the presence in serum of the hepatitis B surface antigen. Aflatoxin potency seems to be significantly enhanced in individuals with simultaneous hepatitis B infection. This interaction makes it difficult to interpret the epidemiological studies, e.g., from Shanghai, Thailand, China, Gambia, Taiwan, in the context of aflatoxin as an independent risk factor.

In addition, identification of hepatitis C virus is an important recent advance in understanding etiology of liver cancer. A limited number of studies with inconclusive results have investigated interactions between hepatitis C infection, aflatoxin, and liver cancer.

An estimated 50 to 100% of liver cancer cases are associated with persistent infection with hepatitis B and/or C. A review of aflatoxin carcinogenic potencies estimated from laboratory animal studies and human epidemiological research yielded potency estimates ranging from 0.01 (cancers/year per 100,000 people per ng aflatoxin/kg bw per day) for humans without hepatitis B virus infection to 1.0 in the Fischer rat (one of the most-sensitive animal species). Observations concerning the interaction of hepatitis B and aflatoxin suggested two separate aflatoxin potencies in populations where chronic hepatitis infections were common versus those where such infections are rare. To evaluate population risk from intake of aflatoxin, potencies were estimated from the positive epidemiological studies for both hepatitis B-negative and hepatitis B-positive individuals. Potency values of 0.3 cancers/year per 100,000 population per ng aflatoxin B1/kg bw per day (with an uncertainty range of 0.05 to 0.5) in hepatitis B-positive individuals and of 0.01 cancers/year per 100,000 population per ng aflatoxin B1/kg bw per day (with an uncertainty range of 0.002 to 0.03) in hepatitis B-negative individuals were calculated. These estimates were subject to potential biases for several reasons: (1) only studies showing a positive association between aflatoxin and liver cancer were used, as opposed to considering all studies, leading to overestimation of aflatoxin potency; (2) relating current levels of intake to current levels of liver cancer (presumably with a long induction period) ignored historical levels (likely to have been higher); (3) the earliest studies systematically underestimated hepatitis B prevalence in cases of liver cancer by a factor of up to 30%, due to limitations in the methodology used to detect hepatitis B, which also led to overestimation of the relative potency of any other factor, including aflatoxins; and (4) histological confirmation of liver cancer cases was limited in most epidemiological studies, so nonprimary liver cancer cases could have been included, underestimating or overestimating aflatoxin potency.

Thus, estimated potencies may be considered overestimates of aflatoxin potency, leading to the hypothesis that humans may be less sensitive to aflatoxin than species tested in laboratory experiments. Thus, the fraction of the incidence of liver cancer in populations attributable to aflatoxin intake was derived by combining aflatoxin potency estimates (risk per unit dose) and estimates of aflatoxin intake (dose per person). Many of the data on aflatoxin contamination levels around the world were derived from non-random samples that appeared to be biased upward, since monitoring studies often focus on commodity lots that are thought to be contaminated. Hence, these contamination levels must be used with caution when inferring patterns of importance in setting aflatoxin standards. Several important conclusions relevant to the impact of aflatoxin exposure on public health and conservation of important food crops have been reached. First, aflatoxin potency in hepatitis B-positive individuals is substantially higher than in hepatitis Bnegative individuals. Thus, decreasing aflatoxin intake in populations with many hepatitis B-positive individuals will have a greater impact on decreasing liver cancer rates than decreasing intake in populations with few such individuals. Second, aflatoxin M1, the hydroxylated metabolite of B1, may be considered to have a potency approximately one order of magnitude less than B1. Third, vaccination against hepatitis B will decrease the prevalence of carriers; the JECFA analysis suggests that this would decrease the potency of aflatoxin in vaccinated populations and consequently decrease liver cancer risks. Fourth, analysis of the application of hypothetical aflatoxin standards (10 g/kg or 20 g/ kg in food) to model populations indicates that (1)populations with a low prevalence of hepatitis B-positive individuals and/or with a low mean aflatoxin intake (less than 1 g/kg bw/day) are unlikely to exhibit detectable differences in population risks for a standard in the range of the hypothetical cases, i.e., western European and U.S. populations and (2) populations with a high prevalence of hepatitis B-positive individuals and high mean aflatoxin intake would benefit from decreased aflatoxin intake.

20.5 BIOMARKERS FOR SELECTED MYCOTOXINS

The exposure risk in humans to selected mycotoxins can be more precisely determined if there are biomarkers that can be used. These biomarkers can be selective for certain mycotoxins and epidemiological studies can assess the validity of their use. If the biomarkers prove to be valid, the exposure of individuals or populations of individuals can be determined with a high degree of accuracy.

Presently, only sufficient work has been done with a few mycotoxins to determine the possibility of using biomarkers to assess exposure. The mycotoxins haing biomarkers of interest for further study are aflatoxins, fumonisins, and ochratoxin. Several biomarkers have been developed for the aflatoxins, especially aflatoxin B1, that can potentially be used in risk assessments for human exposure.

These included the use of aflatoxin-albumin adducts in the serum, aflatoxin N7-guanine adducts in the urine, aflatoxin M1 (a metabolite of aflatoxin B1) in the urine, and mutational patterns in the P53 tumor suppressor gene. The report by the Joint FAO/WHO Expert Committee on Food Additives suggested, however, that sensitive markers for exposure to hepatitis B and/or C viruses likely were more importantly related to liver cancer assessment in humans. Regardless, in determining the exposure to aflatoxins, the biomarkers listed above could be used in the exposure assessments.

One of the earliest known indications of the exposure of humans and other animals species to fumonisin B1 is an increase in the sphinganine/sphingosine (Sa/So) ratio in urine or serum. This has been proposed as a biomarker for exposure to fumonisins. The basis for this change is that fumonisins disrupt lipid metabolism, especially sphingolipids, through the inhibition of ceramide synthase leading to the increase in concentration of the sphingoid bases. In human volunteers, there was an increase in the urinary Sa/So ratio during a one month period of consumption of corn containing fumonisin B1 ranging from 0.08 to 41.1 mg/ kg. The study results suggested that males are more sensitive to fumonisin B1 disruption of sphingolipid metabolism than females. Notably, fumonisins are not genotoxic and no DNA adducts are formed. Therefore, no biomarkers are available based on DNA modification.

While obtaining evidence for the involvement of ochratoxin A in human disease, especially cancer, remainssomewhat problematic, there have been confirmations of human exposure based on the presence of ochratoxin A in blood samples of both apparently healthy individuals and of patients with nephropathy in various countries (Fink-Gremmels et al. 1995a; Frolich et al. 1991; Hald 1991). Its presence in humanmilk was also found (Micco et al. 1991). Interestingly, although ochratoxin A has been found in human blood and breast milk, the major routes of exposure remain unidentified and data from foodsources only partially explain the high exposure rates found in humans. Little information is available regarding use of absolute biomarkers for exposure especially when considering the routes of ingestion versus that of, perhaps, inhalation. Studies in rats show that ochratoxin binds to albumin in the serum, is transported throughout the circulatory system, and is released and reabsorbed in various tissues especially renal tubular epithelium. Thus, the major biomarker for exposure to ochratoxin A is identification of this compound in milk, serum, or perhaps urine of humans.

20.6 REGULATORY CONTROL PROGRAMS FOR MYCOTOXINS

In the United States, the primary mission of the FDA is to protect the public health. The FDA is responsible for enforcing several statutes; one of the most important is the FDCA. This statute and its amendments provide FDA authority to regulate poisonous or deleterious substances, such as mycotoxins, in human food and animal feed. The FDCA also prohibits introduction of adulterated food products into interstate commerce. Under the FDCA, the producer is responsible for ensuring that food products are safe, wholesome, and truthfully labeled.

Under Section 402(a)(1) of the FDCA, "A food shall be deemed to be adulterated if it bears or contains any poisonous or deleterious substance that may render it injurious to health; but in case the substance is not an added substance such food shall not be considered adulterated under this clause if the quantity of such substance in such food does not ordinarily render it injurious to health." Mycotoxins — specifically, aflatoxins — are considered added poisonous and deleterious substances because their presence in human food and animal feeds can be, in part, avoided by good manufacturing or agronomic practices. Therefore, strategies used by the FDA to minimize mycotoxin contamination in the U.S. food supply include (1) establishing standards or guidelines, e.g., action levels or advisory levels; (2) monitoring foods to determine compliance with established standards and guidelines; (3) taking appropriate enforcement action, e.g., seizure of adulterated domestic foods, detention of imports; (4) working with other federal agencies such as USDA, the states, and foreign countries; and (5) guiding and encouraging industry to address contamination problems by developing and implementing HACCP programs.

One approach that the FDA uses to control mycotoxin contamination in human food and animal feed is establishing action levels. For example, the FDA established an action level of 20 ppb (ng/g) total aflatoxin in most human foods. However, as a result of a decision by the U.S. Court of Appeals in a case involvingaflatoxin contamination, the FDA, in its 1988 notice, clarified the status of action levels. Specifically, action levels are not binding on the courts, the public (including food processors), or the agency. Action levels are used as a guide by field staff to determine when enforcement actions should be taken. Situations may occur where enforcement action should be considered at levels below an action level or a decision not to consider enforcement action where an action level is exceeded. Another approach used by the FDA to minimize human and animal exposure to mycotoxins is to issue advisory levels directed at state and federal officials as well as trade associations. FDA advisory levels are not enforceable levels. An advisory level for a particular contaminant in human food or animal feed is a guideline below which, based on available toxicological information, a public health hazard is not anticipated.

While enforcement is not the fundamental purpose of an advisory level, the FDA reserves the right to take appropriate enforcement actions when circumstances warrant such actions. Further, the FDA can establish tolerances or regulatory limits for added poisonous or deleterious substances in human foods through formal rulemaking and notice-and-comment rulemaking processes, respectively. However, the FDA has not established any tolerances or regulatory limits for the control of mycotoxin contamination in human foods and animal feeds. FDA standards, e.g., action levels, for control of mycotoxin contamination in human food and animal feeds are based, in part, on practical considerations relating to the capabilities of sampling procedures, analytical methodology, agronomic practices, and availability of technological procedures to minimize human and animal exposure to these contaminants. In addition, FDA mycotoxin standards weigh heavily on the unavoidability of the toxins and therefore do not represent permissible levels of contamination where it is avoidable. Furthermore, blending human food or animals feeds containing a contaminant at levels that would be considered adulterated with unadulterated food or feeds is not permitted; the result of blending would be considered adulterated, regardless of contamination level.

20.6.1 Aflatoxins

Mycotoxins, particularly aflatoxins, have been regulated under Section 402(a)(1) of the FDCA since 1965. In addition to aflatoxins, major mycotoxins that occur naturally on foods and feeds include DON, fumonisins, ochratoxins, zearalenone, and patulin. To establish the need for regulatory control programs for these and other toxins, background exposure data and toxicological data must be obtained and carefully evaluated. More studies have been reported on aflatoxins than any of the other mycotoxins that have been identified since 1960. The motivating force was a series of findings that aflatoxins were potent liver toxins, carcinogens in susceptible animal species, and contaminants in a variety of human foods. Aflatoxins (B1, B2, G1, G2, M1) are the only mycotoxins for which action levels have been established; all othersare advisory levels. The current action levels for aflatoxins in foods and feeds are shown in Table 1. These levels represent the best guidance available on levels that FDA considers to be of regulatory interest and will thus enhance the safety of the food supply when implemented and adhered to by the food and feed industries. However, the FDA expects, based on surveillance data, that aflatoxin levels typically are well below action levels due to good agronomic and manufacturing practices. These levels are reviewed periodically as more knowledge regarding analytical, toxicological, and technological procedures become available. Experience and knowledge obtained through creating an effective control program for aflatoxins has been used to address regulatory considerations involving other mycotoxins.

20.6.2 Deoxynivalenol

FDA advisory levels were first issued in 1982 for DON in wheat and wheat products. This action was taken because spring wheat crops in some parts of the United States were heavily infected with head blight or pink scab; unusually high levels of DON were also reported. DON is a trichothecene mycotoxin produced by several molds of the genus

Fusarium, especially *F. graminearum*, a common contaminant of several grains. When the advisory levels were issued, there was great concern about the safety of this wheat, which might be used for human food as well as animal feed. Toxicological data on DON were extremelylimited; therefore, it was difficult for FDA to estimate its potential public health hazard. The agency believed it was important to provide guidance to assist state and local government to deal with the problem; hence, advisory levels for DON in wheat and wheat products were developed. Products that contained the toxin at or below these levels were not believed to present a public health hazard. The FDA has reviewed scientific data that became available after 1982, including reports of DON-associated acute GI illness in humans in China in 1984–1985 and in India in 1987. DON's precise role in those outbreaks is not certain but the data suggest an association between human exposure to DON-contaminated wheat and occurrence of the illnesses..

Table 1: U.S. Food and Drug Administration action levels for total aflatoxins in food and feed

Commodity	Concentration (ng/g)
Corn for immature animals and dairy cattle swine, and mature poultry	20
Corn and peanut products for finishing beef cattle	300
All other feedstuffs	20
All products, except milk, designated for humans	20
Corn and peanut products for breeding beef cattle,	100
Corn and peanut products for finishing swine	200
Cottonseed meal (as a feed ingredient)	300

In 1993, based on new scientific information, updated advisory levels were announced. The advisory levels (Table 2) amended those for DON found in milled wheat products intended forhuman consumption, e.g., flour, bran, and in wheat and other grains used for animal feed. An advisory level was not set for raw wheat intended for milling because normal manufacturing practices and additional technology available to millers can substantially decrease DON levels in the finished wheat product. These advisory levels were sent to state agriculture directors; state feed control officials; and food, feed, and grain trade associations.

20.6.3 Fumonisins

The FDA has issued a guidance document to the food industry concerning fumonisins; however, no action level has been established to date. The fumonisins (B1 and B2 are the major ones) are natural byproducts of the fungi *Fusarium verticillioides* (syn., *moniliforme*) and *F. proliferatum*, common contaminants of corn. The finding of a high incidence of fumonisins in corn-based products for human consumption, epidemiological investigations indicating a possible association of *F. verticillioides* and its metabolic products with esophageal cancer, the association of *F. verticillioides* and feminising B1 with ELEM, and animal studies demonstrating the carcinogenicity of feminising B1 in rats, have heightened concerns about the public health impact of these contaminants. Surveys of corn, corn products, and other foods and feeds have been conducted by the FDA, with a number of chronic toxicological studies underway. Results suggest that fumonisins do not accumulate in edible tissues of food animals and no detectable levels have been found in milk from cows dosed with high levels of fumonisins. This information, along with additional data from ongoing toxicological studies, will be used to assess human health risks and determine the need for regulatory programs. Based on the existing health hazard data, the FDA believes that exposure to fumonisins should be decreased o the extent practicable. This can be accomplished through issuance of guidance levels or implementation by industry of an appropriate voluntary monitoring and control program. The FDA has issued a guidance document to the food industry that contained proposed maximum feminising levels that FDA considers adequate to protect human and animal health and that are achievable in human foods and animal feeds with the use of good agricultural and manufacturing practices. These levels will be in effect until a final document is published. The recommended levels for total fumonisins (FB1, FB2, and FB3) in human food products and animal feeds are shown in Table 3.

Table 2: U.S. Food and Drug Administration advisory levels for deoxynivalenol in wheat/derived products, other grainsa

Product	Concentration (g/g)
All finished wheat products, e.g. flour, bran and germ,	1
that may be consumed by humans All grains and grain by-products destined for ruminating	10
beef and feedlot cattle older than 4 months and for chickens; these ingredients should not exceed 50% of the diet of cattle or chickens All grains and grain by-products destined for swine; these	5
ingredients should not exceed 20% of the diet All grains and grain by-products for all other animals; these ingredients should not exceed 40% of the diet	5

20.6.4 Patulin

Patulin is most frequently found as a natural contaminant of apples and apple products. It has been isolated almost exclusively from apples and apple products contaminated with *Penicillium expansum*. The FDA believes that patulin levels in apple products, particularly apple juice (major source of patulin in the human diet) can be controlled by processors principally by removing spoiled and visibly damaged apples from the product stream used for the production of apple juice. The FDA recently published a document titled "Draft Guidance Document of FDA Components and Industry on Apple Juice, Apple Juice Concentrates, and Apple Juice Products — Adulteration with Patulin" (U.S. Food and

Drug Administration 2000b). This document and its accompanying Draft Compliance Policy Guide supports and establishes an action level of 50 ìg/kg for patulin in apple juice, apple juice concentrates, and apple juice products

20.6.5 Other Mycotoxins

Other mycotoxins of concern to the FDA are ochratoxin A, zearalenone, penicillic acid, sterigmatocystin, citrinin, cyclopiazonic acid, and the *Alternaria* toxins. No regulatory standards have been initiated for these toxins; results from exposure data and/or risk assessments have indicated that regulatory standards are not warranted at this time. The FDA continuously follows the development of newer data regarding these mycotoxins, thereby constantly evaluating the need to set regulatory standards.

20.7 WORLDWIDE REGULATIONS FOR MYCOTOXINS

20.7.1 Introduction

Worldwide food legislation safeguards the health of consumers and the economic interests of producers and traders. Food laws often impose limits on the concentrations of specific contaminants, e.g., mycotoxins, in foods. Overviews of regulations for mycotoxins and their rationales have been published . This section will focus on factors that may influence the establishment of regulations for mycotoxins and summarizes current regulations that exist worldwide for mycotoxins.

The need for legislation imposing limits on mycotoxin concentration in foods and feeds is generally recognized by the industrialized world. Virtually all countries with fully developed market economies have regulations. In contrast, many developing countries where subsistence farming is significant have not established regulations for mycotoxins, as shown in Fig. 8.1. How the limits are chosen and for which commodities depends on several factors, such as (1) availability of toxicological data, (2) availability of data on mycotoxin occurrence in various commodities, (3) homogeneity of the concentration in a lot, (4) availability of analytical methods, (5) legislation in other countries with which trade contacts exist, and (6) the need for sufficient food supply.

20.7.2 Toxicological Data

Measures are primarily taken on the basis of known toxic effects. For the mycotoxins currently considered most significant (aflatoxins, ochratoxin A, patulin, some trichothecenes including deoxynivalenol, fumonisins, zearalenone), the Joint Expert Committee on Food Additives (JECFA) of the World Health Organization and the Food and Agriculture Organization, has recently evaluated the hazard. JECFA is a scientific advisory body. It provides a mechanism for assessing the toxicity of additives, veterinary drug residues, and contaminants. Safety evaluation of contaminants incorporates various steps in a formal health risk assessment approach. The qualitative indication that a contaminant can cause adverse effects on health (hazard identification) is dex

Alimentarius) purposes, factors other than hazard assessment play a role; these will be discussed below.

Table 3: Proposed guidance levels for fumonisin B1 in human foods and animal feed

	Total fumonisins ($FB_1 + FB_2 + FB_3$)
Human food product	
Degermed dry milled corn products	2 µg/g
Whole/Partially degermed dry milled corn product	4 µg/g
Dry milled corn bran	4 µg/g
Cleaned corn intended for masa production	4 µg/g
Cleaned cron intended for poporn	3 µg/g
Corn and corn byproducts for animals	
Equids and rabbits	5 µg/g > 20% of diet
Swine and catfish	20 µg/g >50% of diet
Breeding ruminants, poultry, milk, dairy cattle and laying hens	30 µg/g > 50% of diet
ruminants >3 months before slaughter and mink for pelt production	60 µg/g > 50% of diet
Poultry raised for slaughter	100 µg/g > 50% of diet
All other species or classes of livestock and pet animals	10 µg/g > 50% of diet

20.7.3 Occurrence of Mycotoxins in Commodities

In addition to hazard assessment, another factor is necessary to make risk assessment possible: exposure assessment. Data on mycotoxin occurrence in various commodities are needed to enable exposure assessment. They also allow the effects of regulations to be estimated on the availability of the foods and feeds concerned.

Risk assessment is the product of hazard assessment and exposure assessment. Among the factors it involves are (1) hazard assessment based on animal toxicity and possibly epidemiological data, (2) exposure assessment based on data concerning actual concentrations in commodities, (3) extrapolation of hazard assessment from hih-exposure animal data to low-exposure human data, and (4) comparison of the product of hazard assessment and exposure assessment with accepted risk.

20.7.4 Distribution of Mycotoxins over Commodities

The homogeneity of mycotoxin concentration in products is a key factor in carrying out proper sampling. Distribution can be very nonhomogeneous, as with aflatoxins in peanuts . While the number of contaminated peanut kernels in a lot is usually very low, the contamination level within a kernel can be very high. Similar results are known for corn. If insufficient care is taken to achieve a representative sample, the mycotoxin concentration in an inspected lot may be easily estimated incorrectly. Consumption of peanuts may lead to an accidental high single dose of aflatoxins, instead of chronic intake at a relatively low level. A similar situation may occur with pistachios and figs and other commodities.

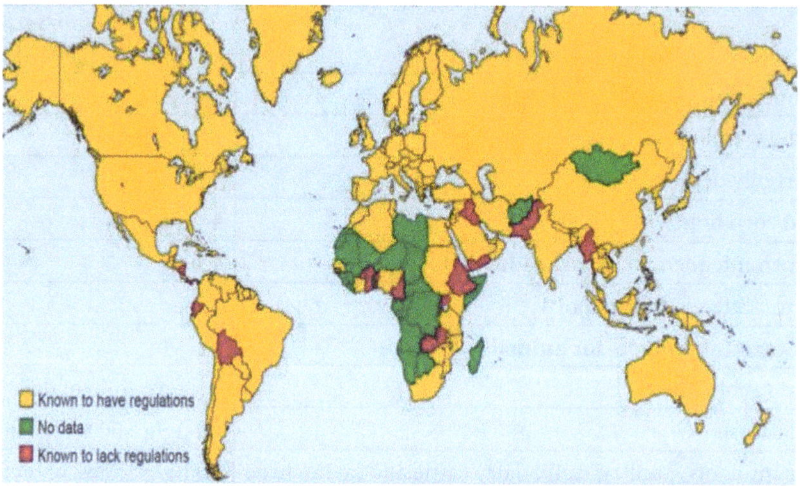

Fig. 1. Counties known to regulate mycotoxins in food and feed (yellow), those where it is unknown whether regulations exist (green), and nations known to have no specific regulations (red)

20.7.5 Methods of Analysis

Legislation calls for control, so reliable, accurate analytical methods must be available. Tolerance levels that do not have a reasonable expectation of being measured may waste resources and condemn products that are perfectly fit for consumption . Besides accuracy and reliability, simplicity is required, as it will influence the amount of data that will be generated and the practicality of the ultimate measures taken. Data reliability can be improved by using methods that fulfill performance criteria (as demonstrated in collaborative studies). Application of AQA procedures is recommended, including the use of reference materials. Certified reference materials should be used when a high degree of comparison and accuracy is required. Improvements in AQA and use of reference materials are likely to emerge for mycotoxin control in foods and feeds.

20.7.6 Trade Contacts

Preferably, regulations should be harmonized with hose in other countries with which trade contacts exist. Unnecessarily strict regulative actions make it difficult for importing countries to obtain supplies of essential commodities such as food grains and animal feedstuffs. Exporting countries may have difficulty finding markets for their products. For example, stringent regulations for aflatoxin Bl in animal feedstuffs in the European Union made it difficult for some developing countries to export feed ingredients to their European trading partners.

20.7.7 Food Supply

Regulatory philosophy should not jeopardize availability of basic commodities at reasonable prices.Especially in the developing countries, where food supplies are already limited, drastic legal measures my cause food shortages and excessive prices.

Weighing the factors in the decision-making process of establishing mycotoxin tolerances is not trivial. Common sense is the major factor for reaching a decision. Public health officials are confronted with a complex problem: mycotoxins, particularly aflatoxins, should be excluded from food as much as possible. Since the substances are present in foods as natural contaminants, however, human exposure cannot be completely prevented, so exposure of the population to some level of mycotoxins has to be tolerated. Despite the dilemma, mycotoxin regulations have been established in the past decades in many countries.

20.8 CURRENT REGULATIONS

Schuller and colleagues (1983) produced the first overview of worldwide mycotoxin legislation that was updated by van Egmond (1987). Another update has been published as FAO Food and Nutrition Paper 64. Table.4 summarizes the maximum tolerated levels for mycotoxins in foodstuffs, dairy products, and animal feedstuffs, as they existed on October 1, 1996. It shows that tolerated levels for mycotoxins vary widely between countries, making harmonization of regulations highly desirable. Since 1996, the situation on mycotoxin regulations in foodstuffs and feedstuffs around the world has changed significantly. At the time of publishing this CAST report, an international inquiry was ongoing, conducted by H. P. van Egmond on behalf of the Food and Agriculture Organization, to produce an updated document on the mycotoxin regulations. This document is scheduled for publication as an FAO Food and Nutrition Paper in 2003, and it will cover the mycotoxin regulations as they existed worldwide in 2002 in more than 100 countries.

The document also will describe the results of regional attempts to harmonize existing regulations. A group of countries in which new (harmonized) regulations for mycotoxins in foodstuffs have recently come into force is the European Union. As a spin-off of the ongoing inquiry, the new EU regulations for mycotoxins in food. The new EU regulations overrule the corresponding existing national regulations in the EU countries, as presented in Table 4 under the various country headings. These countries retain the right, however, to keep national regulations for those mycotoxin/matrix combinations not covered in the harmonized EU regulations.. Most new information comes from South America. Seventy-seven countries regulate mycotoxins, an increase of 38% over the 1987. Thirteen countries are known to have no specific regulations, while no data are available for about 50 countries, most of them in Africa.

Most mycotoxin regulations concern aflatoxins and in fact all countries with mycotoxin regulations at least have tolerances for aflatoxins in foods and/or animal feedstuffs. Less frequently, specific regulations exist for other mycotoxins, i.e., patulin, ochratoxin A, DON, diacetoxyscirpenol, zearalenone, T-2 toxin, chetomin, stachybotryotoxin, phomopsin, fumonisins. Much work is needed to arrive at consensus tolerances for some of these toxins that may significantly impact human and animal health, e.g., fumonisins, ochratoxin A, some trichothecenes, zearalenone.

At the time of publication of this report, many new limits and regulations for mycotoxins have come into force or are in development; therefore, the information contained herein is rapidly changing. The number of countries known to have specific mycotoxin regulations has increased, whereas more mycotoxins are regulated in more commodities and tolerance limits tend to decrease. At the same time, regulations have been harmonized or are in some stage of harmonization. They have become more diverse and detailedwith newer requirements regarding official analytical methodology and sampling procedures. Therefore, FAO has decided to produce an update of Food and Nutrition paper No. 64 that is scheduled for publication in 2003.

20.9 MAXIMUM TOLERATED LEVELS OF MYCOTOXINS IN FOODSTUFFS, DAIRY PRODUCTS, AND ANIMAL FEED-STUFFS

To reiterate, 77 countries regulated mycotoxins in 1996, 13 had no regulations, and no data were available for 50. Compared to 1987, maximum tolerated levels for aflatoxins had not changed dramatically and continued to differ greatly between countries, requiring harmonization to remove the extreme variability of tolerated levels. A new update was in preparation in 2002. This update, scheduled for publication by the FAO in 2003, will show increased and more up-to-date information on mycotoxin regulations in foodstuffs and feedstuffs. It also will show the first results of regional harmonization of regulations for foodstuffs, e.g., in the European Union.

Table 4: Maximum tolerated levels of mycotoxins in foodstuffs, dairy products, and animal feedstuffs

Mycotoxins	Abbreviations used here	CAS Registry Number
aflatoxin B_1	afla B_1	1162-65-8
aflatoxin B_2		7220-81-7
aflatoxin G_1		1165-39-5
aflatoxin G_2		7241-98-7
aflatoxins $B_1B_2G_1G_2$	afla $B_1B_2G_1G_2$	
aflatoxin M_1	afla M_1	6795-23-9
aflatoxin M_2	afla M_2	6885-57-0
agaric acid		666-99-9

Mycotoxins	Abbreviations used here	CAS Registry Number
deoxynivalenol	DON	51481-10-8
diacetoxyscirpenol	DAS	2270-40-8
fumonisin B_1		116355-83-0
fumonisin B_2		116355-84-1
fumonisin B_3		136379-59-4
HT-2 toxin		26934-87-2
ochratoxin A		303-47-9
patulin		149-29-1
phomopsin A		64925-80-0
sterigmatocystin		10048-13-2
T-2 toxin		21259-20-1
zearalenone		17924-92-4

RELATED TERMINOLOGY

Aflatoxicosis: Mycotoxicosis caused by ingestion of aflatoxins in contaminated foods and feeds.

Aflatoxin B_1: Potent hepatotoxic, hepatocarcinogenic, mutagenic and teratogenic mycotoxins produced by *Aspergillus flavus* and *parasiticus*. Formed during growth on a wide range of crops, including peanuts, corn and other cereals, and oilseeds. Metabolized to aflatoxin M1 and aflatoxin Q1.

Aflatoxin B_2: Moderately potent hepatotoxic, hepatocarcinogenic, mutagenic and teratogenic mycotoxins produced by *Aspergillus flavus* and *parasiticus*. Dihydroxy derivatives of aflatoxin B1. Formed during growth on the same commodities as aflatoxin B1 (including peanuts, corn and other cereals, and oilseeds), but in smaller amounts. Metabolized to aflatoxin M_2, and excreted in milk in this form.

Aflatoxin B_3: Toxic mycotoxins produced by older cultures of *Aspergillus parasiticus* and *flavus*. Alternative name for parasiticol.

Aflatoxin D_1: Carboxylated product of aflatoxin B1, produced by the reaction between aflatoxin B1 and heated ammonium hydroxide. Possesses lower toxicity than aflatoxin B1.

Aflatoxin G_1: Potent carcinogenic and genotoxic mycotoxins produced by *Aspergillus parasiticus* and other *Aspergillus* species. Formed during growth on a wide range of crops, including peanuts, corn and other cereals, and oilseeds. Possess toxicity and structure similar to those of aflatoxin B_1.

Aflatoxin G_2: Mildly carcinogenic and genotoxic mycotoxins produced by *Aspergillus parasiticus* and other *Aspergillus* species. Occur in a wide range of foods, including nuts, seeds, beans, spices and fruits. Dihydroxy derivative of aflatoxin G_1, with lower toxicity.

Aflatoxin M_1: The toxic, 4-hydroxy derivative of aflatoxin B_1, found in the livers, kidneys, blood, faeces, urine and milk of mammals that have consumed aflatoxin B_1 contaminated feeds or foods. Subsequently occurs in dairy products, particularly cheese, and human milk. Produced in small quantities by *Aspergillus flavus* and *parasiticus,* and can occur in corn, nuts and soybeans. Associated with liver damage and cancer. Possesses lower toxicity than aflatoxin B1. Aflatoxin M1 can be degraded by UV radiation.

Aflatoxin M_2: The toxic, 4-hydroxy derivative of aflatoxin B2, found in the livers, kidneys, blood, faeces, urine and milk of mammals that have consumed aflatoxin B2 contaminated feeds or foods. Also occurs in human milk. Aflatoxin M2 is considerably less toxic than aflatoxin M1. Produced in small quantities by *Aspergillus flavus* and *parasiticus.*

Aflatoxin Q_1: The 3-hydroxy derivative and major metabolite of aflatoxin B1 in humans, rats and primates. Considerably less toxic than aflatoxin B1.

Aflatoxins: Mycotoxins produced by certain strains of *Aspergillus,* most notably *flavus* and *A. parasiticus.* Formed during growth of these fungi on commodities such as cereals (e.g. corn), nuts (e.g. peanuts) and oilseeds (e.g. soybeans). Contamination can take place both pre- and post-harvest. Host crops are particularly susceptible to infection following prolonged exposure to high humidities or damage during drought conditions. Once ingested, aflatoxins are metabolized by the liver to a reactive intermediate, aflatoxin M1. Hepatotoxic and hepatocarcinogenic in humans and animals, and can result in aflatoxicosis.

Akakabi-byo: Red-mold disease of cereal grains caused by trichothecenes. (See also kokuten-byo.)

Antibiosis: An association between microorganisms in which one is adversely affected; a type of amensalism.

Antibody screen: test to make sure that a potential blood recipient has not produced antibodies to antigens other than the ABO and Rh antigens

Antibody: Y-shaped glycoprotein molecule produced by B cells that binds to specific epitopes on an antigen

Antibody-dependent cell-mediated cytotoxicity (ADCC): mechanism by which large pathogens are marked for destruction by specific antibodies and then killed by secretion of cytotoxins by natural killer cells, macrophages, or eosinophils

Antigen (also, immunogen): a molecule that stimulates an adaptive immune response

Antigenic drift: form of slight antigenic variation that occurs because of point mutations in the genes that encode surface proteins

Antigenic shift: form of major antigenic variation that occurs because of gene reassortment

Antigenic variation: changing of surface antigens (carbohydrates or proteins) such that they are no longer recognized by the host's immune system

Antigenic: able to stimulate an adaptive immune response

Antimetabolites: compounds that are competitive inhibitors for bacterial metabolic enzymes

Antimicrobial drugs: chemical compounds, including naturally produced drugs, semisynthetic derivatives, and synthetic compounds, that target specific microbial structures and enzymes, killing specific microbes or inhibiting their growth

Antimicrobial peptides (AMPs): class of nonspecific, cell-derived chemical mediators with broad-spectrum antimicrobial properties

Ascospore: A sexual (meiotic) spore; one of usually eight haploid spores characteristically formed inside a specialized cell in fungi of the Ascomycotina.

Buyer's risk: The type of mistake when "bad" commodity lots (i.e., mycotoxin concentration greater than the guidelines) test as good and are accepted by an inspection program; also called a "false negative." (See also seller's risk.)

Calvin-Benson cycle: most common CO_2 fixation pathway in most photoautotrophs; involves light-independent reactions of photosynthesis that occur in the cytoplasm of photosynthetic bacteria and in the stroma of eukaryotic chloroplasts

***Campylobacter jejuni* gastroenteritis:** gastroenteritis caused by *C. jejuni*; generally mild but sometimes with serious complications

Candidiasis: fungal infection caused by *Candida* spp., especially *C. albicans*; can affect various regions of the body, e.g., skin (cutaneous candidiasis), oral cavity (oral thrush), or vagina (yeast infection) clinical signs include muscle tremor, uncoordinated movements, general weakness in the hind legs, and stiff, stilted movements of the forelegs.

Colistin: membrane-active polymyxin that was historically used for bowel decontamination but now used for systemic infections with drug-resistant pathogens

Colitis: inflammation of the large intestine

Collagenase: enzyme that digests collagen, the dominant protein in connective tissue

Conidiuma(conidiospore): An asexual spore produced by many fungi, especially ascomycetes, often borne on specialized erect hyphae.

Direct contact transmission: movement of a pathogen between hosts by physical contact or transfer in droplets at a distance less than one meter

Direct Coombs' test: assay that looks for antibodies *in vivo* against red blood cells caused by various types of infections, drug reactions, and autoimmune disorders

Direct ELISA: enzyme-linked immunoabsorbent assay in which the antigens are immobilized in the well of a microtiter plate; only a single antibody is used in the test

Direct fluorescent antibody (DFA) test: FA technique in which the labeled antibody binds to the target antigen

Direct hemagglutination assay: test that determines the titer of certain bacteria and viruses that causes clumping of red blood cells

Direct microscopic cell count: counting of cells using a calibrated slide under a light microscope

Direct repair (light repair or photoreactivation): light-dependent mechanism for repairing pyrimidine dimers involving the enzyme photolyase

Dry milling: A dry process that separates grain components into various particle sizes, resulting in fractions such as grits, germ, meal, and flour. (See also wet milling.)

Early splits: A condition in pistachios where the hull splits before the nut is mature.

Ergot: A hardened mass of fungus tissue; the sclerotium of Claviceps spp.

Fescue foot: The gangrenous loss of extremities in cattle that occurs during colder periods.

Fescue toxicosis: Ergot-like poisoning on tall fescue pastures.

Flag leaf: The innermost leaf surrounding an immature inflorescence.

Fluorescence microscope: a microscope that uses natural fluorochromes or fluorescent stains to increase contrast

Fluorescence-activated cell sorter (FACS): technique for using a flow cytometer to physically separate cells into two populations based on fluorescence intensity

Fluorescent antibody (FA) techniques: suite of assays that use a fluorescently labeled antibody to bind to and so make an antigen easy to visualize

Fluorescent enzyme immunoassay (FEIA): EIA in which the substrate is a fluorogen that becomes fluorescent following reaction with the enzyme

Fluorescent: the ability of certain materials to absorb energy and then immediately release that energy in the form of light

Fluorochromes: chromophores that fluoresce (absorb and then emit light)

Fluorogen: nonfluorescent molecule that becomes fluorescent on enzyme or laser activation

Head blight: Wheat and barley head scab.

Honeydew: A sweet liquid suspension of asexual spores transferred by insects from floret to floret.

Inoculum: Fungal propagules that initiate disease in plants or animals.

Kokuten-byo: Black-spot disease of cereal grains, caused by trichothecenes. (See also akakabi-byo.)

Kwashiorkor: A disease linked to consumption of aflatoxin-containing food. Attributes include hypoalbuminemia, fatty liver, and immunosuppression.

LD_{50}: The single dose of a substance required to cause 50% mortality in a population.

Macroconidia: Long, multicelled spores produced by fungi.

Microconidia: One-celled, asexual spores.

Mycotoxicosis: A disease resulting from exposure to a fungal toxin.

Mycotoxicosis: Disease of humans and animals resulting from the ingestion of mycotoxins in foods or feeds.

Mycotoxin: A compound produced by a fungus that is toxic to humans and other animals.

Nixtamalization: The process used to make masa flour from corn.

Pericarp: The wall of a fruit, derived from the maturing ovary wall.

Perithecia: Fruiting bodies containing the sexual spores of fungi.

Phylloerythin: A breakdown product of chlorophyll.

Replacement parasite: A parasite that replaces the developing plant structures with fungal structures, e.g., ergots or sclerotia.

Reye's syndrome: A disease linked to consumption of aflatoxincontaining food. Attributes include acute encephalopathy with fatty degeneration of viscera.

Saprophyte: A plant, fungus, or bacterium that obtains its nutrition from dead organic materials.

Sclerotium: The resting body of certain fungi, which can survive long periods of adverse conditions to produce either a mycelium or fruiting bodies and serve as important sources of inoculum.

Secondary metabolite/metabolism: Compounds not essential for the growth of the producing organism in culture, but may be important in the ecology of the organism in nature.

Seller's risk: The type of mistake when "good" commodity lots (i.e., mycotoxin concentration less than or equal to the guidelines) test as "bad" and are rejected by an inspection program; also called a "false positive." (See also buyer's risk.)

Shoshin kakke: Acute cardiac beriberi caused by the fungal extract citreoviridin.

St. Anthony's fire: Description of ergotism in the Middle Ages in humans whose early symptoms are swollen limbs with burning sensations.

Staggers: Disease of cattle caused by tremorgenic mycotoxins.

Stroma: Proliferating fungal structure; a solid mass of plectenchyma (a form of "tissue" commonly found among the higher fungi composed of a mass of interwoven hyphae) that may bear perithecia on its surface or embedded within the tissue.

Tombstones: Chalky white seeds present in head scab or blight; premature bleaching caused by Fusarium infection.

Toxins: Poisonous substances, especially those that are produced by one living organism, and are poisonous to other living organisms. Aflatoxins. Polyketides derived from acetate as a starter unit.

Water activity (aW): The ratio of the vapor pressure of the product to that of pure water.

Wet milling: (See also dry milling.) A wet process that separates grain components into its various constituents such as starch, gluten feed, and gluten meal.

REFERENCES

Abbas, H. K., Reddy, K. R. N., Salleh, B., Saad, B., Abel, C. A., and Shier, W. T. (2010). An overview of mycotoxin contamination in foods and its implications for human health. *Toxin Rev.* 26–3 ,29. doi: 15569541003598553/10.3109

Abdel-Hadi, A., Carter, D., and Magan, N. (2010). Temporal monitoring of the nor-1 (aflD) gene of *Aspergillus flavus* in relation to aflatoxin B1 production during storage of peanuts under different environmental conditions. *J. Appl. Microbiol.* ,109 1922–1914. doi: 10.1111/j.2672.2010.04820-1365.x

Akande, K. E., Abubakar, M. M., Adegbola, T. A., and Bogoro, S. E. (2006). Nutritional and health implications of mycotoxin in animal feed. *Pak. J. Nutr.* 403–398 ,5. doi: 10.3923/pjn.2006.398.403

Alkhayyat, F., and Yu, J. H. (2014). Upstream regulation of mycotoxin biosynthesis. *Adv. Appl. Microbiol.* 278–251 ,86. doi: 10.1016/B6-9.00005-800262-12-0-978

Alonso, V. A., Monge, M. P., Dalcero, A. M., Keller, K. M., Rosa, C. A., Cavaglieri, L. R., et al. (2009). Contribution of raw materials on dairy cattle feedstuff aflatoxin contamination in central Argentina. *Rev. Bras. Med. Vet.* 99–92 ,31.

Alshannaq, A., and Yae-Hiuk, Y. (2017). Occurence, toxicity and analysis of major mycotoxins in food. *Int. J. Environ. Res. Public Health* 14:632. doi: 10.3390/ijerph14060632

Amaike, S. A., and Keller, N. P. (2011). *Aspergillus flavus. Annu. Rev. Phytopathol.* ,49 133–107. doi: 10.1146/annurev-phyto095221-072910-

Amaike, S., and Keller, N. P. (2011). *Aspergillus flavus. Annu. Rev. Phytopathol.* –107 ,49 133. doi: 10.1146/annurev-phyto095221-072910-

Amare, M. G., and Keller, N. P. (2014). Molecular mechanisms of *Aspergillus flavus* secondary metabolism and development. *Fungal Genet. Biol.* 18–11 ,66. doi: 10.1016/j.fgb.2014.02.008

Asemoloye, M. D., Jonathan, S. G., Saddaf, R., Habiba, Z., Okoawo, E. E., and Bello, T. S. (2017). "Incidence and chemical Implications of aflatoxin in street-vended foods," in *Aflatoxin-Control, Analysis, Detection and Health Risks*, ed. L. Abdulra'Uf, (London: IntechOpen Limited), 153–176. doi: 10.5772/intechopen.68478

Ashworth, L. J. Jr., McMeans, J. L., and Brown, C. M. (1969). Infection of cotton by *Aspergillus flavus*: epidemiology of the disease. *J. Stored Prod. Res.* 202–193 ,5. doi: 474-0022/10.1016X(2-90033(69

Assaf, J. C., Atoui, A., Khoury, A. E., Chokr, A., and Louka, N. (2018). A comparative study of procedures for binding of aflatoxin M1 to *Lactobacillus rhamnosus* GG. *Braz. J. Microbiol.* 127–120 ,49. doi: 10.1016/j.bjm.2017.05.003

Assaf, J. C., Nahle, S., Chokr, A., Louka, N., and El Khoury, A. (2019). Assorted methods for decontamination of aflatoxin M1 in milk using microbial adsorbents. *Toxins* 11:E304. doi: 10.3390/toxins11060304

Battacone, G., Nudda, A., Cannas, A., Cappio Borlino, A., Bomboi, G., and Pulina, G. (2003). Excretion of aflatoxin M1 in milk of dairy ewes treated with different doses of aflatoxin B1. *J. Dairy Sci.* 2675–2667 ,86. doi: 10.3168/jds.S-73862(03)0302-0022 4

Battacone, G., Nudda, A., Palomba, M., Mazzette, A., and Pulina, G. (2009). The transfer of aflatoxin M1 in milk of ewes fed diet naturally contaminated by aflatoxins and effect of inclusion of dried yeast culture in the diet. *J. Dairy Sci.* 5004–4997 ,92. doi: 10.3168/jds.1684-2008

Battilani, P., Rossi, V., Giorni, P., Pietri, A., Gualla, A., Van Der Fels-Klerx, H. J., et al. (2012). *Modelling, Predicting and Mapping the Emergence of Aflatoxins in Cereals in the EU Due to Climate Change.* Parma: EFSA.

Bertuzzi, T., Rastelli, S., Mulazzi, A., and Pietri, A. (2012). Evaluation and improvement of extraction methods for the analysis of aflatoxins B1, B2, G1 and G2 from naturally contaminated maize. *Food Anal. Method* 519–512 ,5. doi: 10.1007/s-011-12161 5-9274

Bhatnagar, D., Ehrlich, K. C., and Cleveland, T. E. (2003). Molecular genetic analysis and regulation of aflatoxin biosynthesis. *Appl. Microbiol. Biotechnol.* 93–83 ,61. doi: 10.1007/s-1199-002-00253x

Bhatt, R. V., and Krishnamachari, K. A. V. R. (1978). Food toxins and disease outbreaks in India. *Arogya J. Health Sci.* 100–92 ,4.

Borbély, M., Sipos, P., Peles, F., and Györi, Z. (2010). Mycotoxin contamination in cereals. *J. Agroaliment. Process. Technol.* 98–96 ,16.

Borreani, G., and Tabacco, E. (2010). The relationship of silage temperature with the microbiological status of the face of corn silage bunkers. *J. Dairy Sci.* 2629–2620 ,93. doi: 10.3168/jds.2919-2009

Bovo, F., Corassin, C. H., and Rosim, R. E. (2015). Efficiency of lactic acid bacteria strains for decontamination of aflatoxin M1 in phosphate buffer saline solution and skimmed milk. *Food Bioprocess. Technol.* 2234–2230 ,6. doi: 10.1007/s-011-11947 9-0770

Brown, D. W., Adams, T. H., and Keller, N. P. (1996). *Aspergillus* has distinct fatty acid synthases for primary and secondary metabolism. *Proc. Natl. Acad. Sci. U.S.A.* ,19 14877–14873. doi: 10.1073/pnas.93.25.14873

Bryden, W. L. (2012). Mycotoxin contamination of the feed supply chain: implications for animal productivity and feed security. *Anim. Feed Sci. Technol.* 158–134 ,173. doi: 10.1016/j.anifeedsci.2011.12.014

Bueno, D. J., Casale, C. H., Pizzolitto, R. P., Salvano, M. A., and Oliver, G. (2007). Physical adsorption of aflatoxin B1 by lactic acid bacteria and *Saccharomyces cerevisiae*: a theoretical model. *J. Food Prot.* 2154–2148 ,70. doi: 028-0362/10.4315x70.9.2148-

Caloni, F., and Cortinovis, C. (2011). Toxicological effects of aflatoxins in horses. *Vet. J.* 273–270 ,188. doi: 10.1016/j.tvjl.2010.06.002

CAST, 2003. Council for Agricultural Science and Technology. Mycotoxins: Risks in plant, animal and human system. Ames, Iowa, USA.

Castells, M., Ramos, A. J., Sanchis, V., and Marín, S. (2007). Distribution of total aflatoxins in milled fractions of hulled rice. *J. Agric. Food Chem.* 2764–2760 ,55. doi: 10.1021/ jf063252d

Chulze, S. (2010). Strategies to reduce mycotoxin levels in maize during storage: a review. *Food Addit. Contam. Part A Chem. Anal. Control Expo. Risk. Assess.* ,27 657–651. doi: 19440040903573032/10.1080

Cleveland, T. E. (2006). "The use of crop proteomics and fungal genomics inelucidating fungus–crop interactions," in *Proceedings of the Myco-Globe Conference*, Bari, 32.

Cole, R. J., and Cox, E. H. (1987). *Handbook of Toxic Fungal Metabolites.* New York, NY: Academic Press.

Corassin, C. H., Bovo, F., Rosim, R. E., and Oliveira, C. A. F. (2013). Efficiency of *Saccharomyces cerevisiae* and lactic acid bacteria strains to bind aflatoxin M1 in UHT skim milk. *Food Cont.* 83–80 ,31. doi: 10.1016/j.foodcont.2012.09.033

Cornea, C. P., Ciuca, M., Voaides, C., Gagiu, V., and Pop, A. (2011). Incidence of fungal contamination in a Romanian bakery: a molecular approach. *Rom. Biotechnol. Lett.* 5871–5863 ,16.

Costa-Fernandez, J. M., and Sanz-Medel, A. (2000). Room temperature phosphorescence biochemical sensors. *Quim. Anal.* 204–189 ,19.

Cotty, P. J., and Jaime-Garcia, R. (2007). Influences of climate on aflatoxin producing fungi and aflatoxin contamination. *Int. J. Food Microbiol.* 115–109 ,119. doi: 10.1016/j.ijfoodmicro.2007.07.060

Council for Agricultural Science and Technology [CAST], (ed.) (2003). "Mycotoxicoses of animals," in *Mycotoxins: Risks in Plant, Animal, and Human Systems, Task Force Report No. 139*, (Ames, IA: CAST), 58–79.

Cserhati, M., Kriszt, B., Krifaton, C., Szoboszlay, S., Hahn, J., Toth, S., et al. (2013). Mycotoxin-degradation profile of *Rhodococcus* strains. *Int. J. Food Microbiol.* ,166 185–176. doi: 10.1016/j.ijfoodmicro.2013.06.002

Cui, X., Muhammad, I., Li, R., Jin, H., Guo, Z., Yang, Y., et al. (2017). Development of a UPLC-FLD method for detection of aflatoxin B1 and M1 in animal tissue to study the effect of curcumin on mycotoxin clearance rates. *Front. Pharmacol.* 8:650. doi: 10.3389/fphar.2017.00650

Diaz, G. J., Murcia, H. W., and Cepeda, S. M. (2010). Cytochrome P450 enzymes involved in the metabolism of aflatoxin B1 in chickens and quail. *Poult. Sci.* 2469–2461 ,89. doi: 10.3382/ps.00864-2010

Diaz-Rios L and Jaffe S. 2008. Standards, Competitiveness, and Africa's groundnut Exports to Europe: Barrier, Catalyst, or Distraction? Agriculture & Rural Development Department. Discussion Paper 39. The International Bank for Reconstruction and Development/the World Bank.11.

Ehrlich, K. C. (2014). Non-aflatoxigenic *Aspergillus flavus* to prevent aflatoxin contamination in crops: advantages and limitations. *Front. Microbiol.* 5:50. doi: 10.3389/fmicb.2014.00050

Fink-Gremmels, J. (2008b). The role of mycotoxins in the health and performance of dairy cows. *Vet. J.* 92–84 ,176. doi: 10.1016/j.tvjl.2007.12.034

Frisvad, J. C., Smedsgaard, J., Samson, R. A., Larsen, T. O., and Thrane, U. (2007). Fumonisin B2 Production by *Aspergillus niger*. *J. Agric. Food Chem.* 9732–9727 ,55. doi: 10.1021/jf0718906

FSANZ, 2017. Food Standards Australia New Zealand, https://www. mpi.govt.nz/law-policy/legal-overviews/food-safety/australianew-zealand-co-operation/food-standards-australia-newzealand-fsanz/

Gil-Serna, J., Vázquez, C., and Patiño, B. (2019). Genetic regulation of aflatoxin, ochratoxin A, trichothecene, and fumonisin biosynthesis: a review. *Int. Microbiol.* doi: 10.1007/s10123-019-00084-2 [Epub ahead of print].

Giovati, L., Magliani, W., Ciociola, T., Santinoli, C., Conti, S., and Polonelli, L. (2015). AFM1 in milk: physical, biological, and prophylactic methods to mitigate contamination. *Toxins* 4349–4330 ,7. doi: 10.3390/toxins7104330

Giray, B., Girgin, G., Engin, A. B., Aydın, S., and Sahin, G. (2007). Aflatoxin levels in wheat samples consumed in some regions of Turkey. *Food Control* 29–23 ,18. doi: 10.1016/j.foodcont.2005.08.002

Glamočić, D., Polovinski Horvatović, M., Jajić, I., Krstović, S., and Guljaš, D. (2019). Occurrence of aflatoxin B1, ochratoxin A and zearalenone in maize silage in the region of Vojvodina, Serbia. *Acta Vet.* 115–106 ,69. doi: 10.2478/acve0007-2019-

Gnonlonfin, G. J., Hell, K., Adjovi, Y., Fandohan, P., Koudande, D. O., Mensah, G. A., et al. (2013). A review on aflatoxin contamination and its implications in the developing world: a sub-Saharan African perspective. *Crit. Rev. Food Sci. Nutr.* 365–349 ,53. doi: 10408398.2010.535718/10.1080

Gomah, N. H., Ragab, W. S., and Bullerman, L. B. (2010). Inhibition of fungal growth and aflatoxin B1 production by some *Lactobacillus* strains. *Assiut. J. Agric. Sci.* 40, 27–36. doi: 10.1007/s12602-017-9350-2

González Pereyra, M. L., Dogi, C., Torres Lisa, A., Wittouck, P., Ortíz, M., Escobar, F., et al. (2014). Genotoxicity and cytotoxicity evaluation of probiotic *Saccharomyces cerevisiae* RC016: a -60day subchronic oral toxicity study in rats. *J. Appl. Microbiol.* 833–824 ,117. doi: 10.1111/jam.12552

Grenier, B., and Applegate, T. (2013). Modulation of intestinal functions following mycotoxin ingestion: meta-analysis of published experiments in animals. *Toxins* ,5 430–396. doi: 10.3390/toxins5020396

Griffin, G. J., and Garren, K. H. (1976). Colonization of rye green manure and peanut fruit debris by *Aspergillus flavus* and *Aspergillus niger* group in field soils. *Appl. Environ. Microbiol.* 32–28 ,32.

Guan, S., Ji, C., Zhou, T., Li, J., Ma, Q., and Niu, T. (2008). Aflatoxin B1 degradation by *Stenotrophomonas maltophilia* and other microbes selected using coumarin medium. *Int. J. Mol. Sci.* 1503–1489 ,9. doi: 10.3390/ijms9081489

Hadavi, E., Feizi, H., and Gheibi, B. (2017). Aflatoxin-contaminated nut separation by applied machinery and processing stages in fresh pistachio processing plant. *Front. Microbiol.* 8:2404. doi: 10.3389/fmicb.2017.02404

Hameed, H. G. (1993). *Extrusion and Chemical Treatments for Destruction of Aflatoxin in Naturally-Contaminated Corn.* Ph.D. thesis, University of Arizona, Tucson, AZ.

Harris KL and Carl JL. 1976. Postharvest Grain Loss Assessment Methods - A Manual of Methods for the Evaluation of Postharvest Losses. American Association of Cereals Chemists.

Hashemi, M., Gholampour Azizi, I., Rezai, Z., and Rouhi, S. (2012). Mycological survey and total aflatoxin analyze in silage from Qaemshahr City (Northern Iran). *J. Chem. Health Risks* 56–51 ,2. doi: 10.22034/jchr.2012.543995

Hassan, Z. U., Khan, M. Z., Khan, A., Javed, I., and Hussain, Z. (2012). Effects of individual and combined administration of ochratoxin A and aflatoxin B1 in tissues and eggs of White Leghorn breeder hens. *J. Sci. Food Agric.* 1544–1540 ,92. doi: 10.1002/jsfa.4740

Hawkins, L. K., Windham, G. L., and Williams, W. P. (2005). Effect of different postharvest drying temperatures on *Aspergillus flavus* survival and aflatoxin content in five maize hybrids. *J. Food Prot.* 1524–1521 ,68. doi: 028-0362/10.4315X68.7.1521-

Heathcote, J.G. and Hibbert, J.R. 1978. Aflatoxins : Chemical and biological aspect. Elsevier, New York. pp.173-186.

Hedayati, M. T., Pasqualotto, A. C., Warn, P. A., Bowyer, P., and Denning, D. W. (2007). *Aspergillus flavus*: human pathogen, allergen and micotoxin producer. *Microbiology* 1692–1677 ,153. doi: 10.1099/mic.0-007641/0.2007

Helferich, W. G., Garrett, W. N., Hsieh, D. P., and Baldwin, R. L. (1986). Feedlot performance and tissue residues of cattle consuming diets containing aflatoxins. *J. Anim. Sci.* 696–691 ,62. doi: 10.2527/jas1986.623691x

Herzallah, S. M. (2013). Eggs and flesh of laying hens fed aflatoxin B1 contaminated diet. *Am. J. Agric. Biol. Sci.* 161–156 ,8. doi: 10.3844/ajabssp.2013.156.161

Herzallah, S. M., Al-Ameiri, N., Al-Dmoor, H., Mosoud, S., and Shawabkeh, K. (2014). Meat and organs quality of broiler chickens fed diet contaminated with B1 Aflatoxin. *Glob. Vet.* 380–376 ,12.

IARC (2002). *Monographs on the Evaluation of Carcinogenic Risks to Humans*, Vol. 82. Lyon: International Agency for Cancer Research.

Idris, Y. M. A., Mariod, A. A., Elnour, I. A., and Mohamed, A. A. (2010). Determination of aflatoxin levels in Sudanese edible oils. *Food Chem. Toxicol.* 2541–2539 ,48. doi: 10.1016/j.fct.2010.05.021

Inan, F., Pala, M., and Doymaz, I. (2007). Use of ozone in detoxification of aflatoxin B1 in red pepper. *J. Stored Prod. Res.* 429–425 ,43. doi: 10.1016/j.jspr.2006.11.004

Intanoo, M., Kongkeitkajorn, M. B., Pattarajinda, V., Bernard, J. K., Callaway, T. R., Suriyasathaporn, W., et al. (2018). Isolation and screening of aflatoxin-detoxifying yeast and bacteria from ruminal fluids to reduce aflatoxin B1 contamination in dairy cattle feed. *J. Appl. Microbiol.* 125, 1603–1613. doi: 10.1111/jam.14060

Iqbal, S. Z., Mustafa, H. G., Asi, M. R., and Jinap, S. (2014). Variation in vitamin E level and aflatoxins contamination in different rice varieties. *J. Cereal Sci.* 355–352 ,60. doi: 10.1016/j.jcs.2014.05.012

Isikber, A. A., and Athanassiou, C. G. (2015). The use of ozone gas for the control of insects and micro-organisms in stored products. *J. Stored Prod. Res.* 64B, 145–139. doi: 10.1016/j.jspr.2014.06.006

Ismail, A., Gonçalves, B. L., de Neeff, D. V., Ponzilacqua, B., Coppa, C., Hintzsche, H., et al. (2018). Aflatoxin in foodstuffs: occurrence and recent advances in decontamination. *Food Res. Int.* 85–74 ,113. doi: 10.1016/j.foodres.2018.06.067

Jacobsen, B. J. (2008). "Predicting the Incidence of Mycotoxins," in *Proceeding of 26th Annual International Symposium on Man and His Environment in Health and Disease: Molds & Mycotoxins, Hidden Connections for Chronic Diseases,* (Dallas, TX: American Environmental Health Foundation), 13–17.

Jaimez, J., Fente, C. A., Vazquez, B. I., Franco, C. M., Cepeda, A., Mahuzier, G., et al. (2000). Application of the assay of aflatoxins by liquid chromatography with fluorescence detection in food analysis. *J. Chromatogr. A* 10–1 ,882. doi: 10.1016/S0-00212(00)9673-0021

Klein, P. J., Buckner, R., Kelly, J., and Coulombe, R. A. Jr. (2000). Biochemical basis for the extreme sensitivity of turkeys to aflatoxin B1. *Toxicol. Appl. Pharmacol.* 52–45 ,165. doi: 10.1006/taap.2000.8926

Klein, P. J., Van Vleet, T. R., Hall, J. O., and Coulombe, R. A. Jr. (2002). Biochemical factors underlying the age-related sensitivity of turkeys to aflatoxin B1. *Comp. Biochem. Phys. C* 201–193 ,132. doi: 10.1016/S0-00065(02)0456-1532

Kordasht, H. K., Moosavy, M.-H., Hasanzadeh, M., Soleymani, J., and Mokhtarzadeh, A. (2019). Determination of aflatoxin M1 using aptamer based biosensor on the surface of dendritic fibrous nano-silica functionalized by amine groups. *Anal. Methods* 3919–3910 ,11. doi: 10.1039/C9AY01185D

Kos, J., Hajnal, E. J., Jajić, I., Krstović, S., Mastilović, J., Šarić, B., et al. (2016). Comparison of ELISA, HPLC-FLD and HPLC-MS/MS methods for determination of aflatoxin M1 in natural contaminated milk samples. *Acta Chim. Sloven.* 756–747 ,63. doi: 10.17344/acsi.2016.2451

Krishnamachari, K. A. V. R., Bhat, R. V., Nagarajan, V., and Tilak, T. B. G. (1975). Hepatitis due to aflatoxicosis-An outbreak in western India. *Lancet* 1063–1061 ,1. doi: 10.1016/S2-91829(75)6736-0140

Lai, X. W., Sun, D. L., Ruan, C. Q., Zhang, H., and Liu, C. L. (2014). Rapid analysis of aflatoxins B1, B2, and ochratoxin A in rice samples using dispersive liquid-liquid microextraction combined with HPLC. *J. Sep. Sci.* 98–92 ,37. doi: 10.1002/jssc.201300970

Lancaster, M. C., Jenkins, F. P., and Philip, J. M. C. (1961). Toxicity associated with certain samples of ground nuts. *Nature* 1096–1095 ,12. doi: 1921095/10.1038a0

Larsson, P., Busk, L., and Tjälve, H. (1994). Hepatic and extrahepatic bioactivation and

GSH conjugation of aflatoxin B1 in sheep. *Carcinogenesis* 955–947 ,15. doi: 10.1093/carcin/15.5.947

Le Lay, C., Coton, E., Le Blay, G., Chobert, J. M., Haertlé, T., Choiset, Y., et al. (2016). Identification and quantification of antifungal compounds produced by lactic acid bacteria and propionibacteria. *Int. J. Food Microbiol.* 85–79 ,239. doi: 10.1016/j.ijfoodmicro.2016.06.020

Lee, J., Her, J.-Y., and Lee, K.-G. (2015). Reduction of aflatoxins (B1, B2, G1, and G2) in soybean-based model systems. *Food Chem.* 51–45 ,189. doi: 10.1016/j.foodchem.2015.02.013

Lüthy, J., Zweifel, U., and Schlatter, C. H. (1980). Metabolism and tissue distribution of [14C] aflatoxin B1 in pigs. *Food Cosmet. Toxicol.* 256–253 ,18. doi: -0015/10.1016 0-90103(80)6264

M. S., Yu, J., Nierman, W. C., Kim, H. S., Pritchard, B., Jacobus, C. A., et al. (2006). The aflatoxin pathway regulator AflR induces gene transcription inside and outside of the aflatoxin biosynthetic cluster. *FEMS Microbiol. Lett.* 279–275 ,255. doi: 10.1111/j.6968.2005.00084-1574.x

Ma, Z. X., Amaro, F. X., Romero, J. J., Pereira, O. G., Jeong, K. C., and Adesogan, A. T. (2017). The capacity of silage inoculant bacteria to bind aflatoxin B1 in vitro and in artificially contaminated corn silage. *J. Dairy Sci.* 7210–7198 ,100. doi: 10.3168/jds.12370-2016

Macia-Vicente, J. G., Jansson, H. B., Talbot, N. J., and Lopez-Llorca, L. V. (2009). Real-time PCR quantification and live-cell imaging of endophytic colonization of barley (Hordeumvulgare) roots by *Fusarium equiseti* and Pochonia chlamydosporia. *New Phytol.* 228–213 ,182. doi: 10.1111/j.8137.2008.02743-1469.x

MacLachlan, D. J. (2011). Estimating the transfer of contaminants in animal feedstuffs to livestock tissues, milk and eggs: a review. *Anim. Prod. Sci.* 1078–1067 ,51. doi: 10.1071/AN11112

Madden, U. A., and Stahr, H. M. (1995). Retention and distribution of aflatoxin in tissues of chicks fed aflatoxin-contaminated poultry rations amended with soil. *Vet. Hum. Toxicol.* 29–24 ,37.

Magan, N., Aldred, D., Hope, R., and Mitchell, D. (2010). Environmental factors and interactions with mycoflora of grain and grapes: effects on growth and deoxynivalenol and ochratoxin production by *Fusarium culmorum* and *Aspergillus carbonarius*. *Toxins* 366–353 ,2. doi: 10.3390/toxins2030353

Magan, N., and Aldred, D. (2007). Post-harvest control strategies: minimizing mycotoxins in the food chain. *Int. J. Food Microbiol.* 139–131 ,119. doi: 10.1016/j.ijfoodmicro.2007.07.034

Magnoli, C., Astoreca, A., Ponsone, L., Fernandez-Juri, M. G., Chiacchiera, S., and

Dalcero, A. (2006). Ochratoxin A and the occurrence of ochratoxin A-producing black aspergilla in stored peanut seeds from Cordoba, Argentina. *J. Sci. Food Agric.* 2373–2369 ,86. doi: 10.1002/jsfa.2625

Mahanti, N., Bhatnagar, D., Cary, J. W., Joubran, J., and Linz, J. E. (1996). Structure and function of fas-1A, a gene encoding a putative fatty acid synthetase directly involved in aflatoxin biosynthesis in *Aspergillus parasiticus*. *Appl. Environ. Microbiol.* ,62 195–191.

Maldonado-Siman, E., Bai, L., Ramírez-Valverde, R., Gong, S., and Rodríguez-de Lara, R. (2014). Comparison of implementing HACCP systems of exporter Mexican and Chinese meat enterprises. *Food Control* 115–109 ,38. doi: 10.1016/j.foodcont.2013.10.017

Nakai, V. K., Rocha, L. O., Goncalez, E., Foneseca, H., Ortega, E. M., and Correa, B. (2008). Distribution of fungi and aflatoxins in stored peanut variety. *Food Chemistry* ,106 285–190. doi: 10.1016/j.foodchem.2007.05.087

Nielsen, H. B., Sonne, A. M., Grunert, K. G., Banati, D., Pallak-Toth, A., Lakner, Z., et al. (2009). Consumer perception of the use of high-pressure processing and pulsed electric field technologies in food production. *Appetite* 126–115 ,52. doi: 10.1016/j.appet.2008.09.010

Ogunade, I. M., Arriola, K. G., Jiang, Y., Driver, J. P., Staples, C. R., and Adesogan, A. T. (2014). Effects of 3 sequestering agents on milk aflatoxin M1 concentration and the performance and immune status of dairy cows fed diets artificially contaminated with aflatoxin B1. *J. Dairy Sci.* 6273–6263 ,9. doi: 10.3168/jds.10905-2016

Ogunade, I. M., Martinez-Tuppia, C., Queiroz, O. C. M., Jiang, Y., Drouin, P., Wu, F., et al. (2018). Silage review: mycotoxins in silage: occurrence, effects, prevention, and mitigation. *J. Dairy Sci.* 4059–4034 ,101. doi: 10.3168/jds.13788-2017

Oguz, H., and Parlat, S. (2005). Effects of dietary mannanoligosaccharide on performance of Japanese quail affected by aflatoxicosis. *S. Afr. J. Anim. Sci.* 148–144 ,34. doi: 10.4314/sajas.v34i3.3957

Ouakhssase, A., Chahid, A., Choubbane, H., Aitmazirt, A., and Addi, E. A. (2019). Optimization and validation of a liquid chromatography/tandem mass spectrometry (LC-MS/MS) method for the determination of aflatoxins in maize. *Heliyon* 5:e01565. doi: 10.1016/j.heliyon.2019.e01565

Oyelami, O. A., Maxwell, S. M., Adelusola, K. A., Aladekoma, T. A., and Oyelese, A. O. (1995). Aflatoxins in the autopsy brain tissue of children in Nigeria. *Mycopathology* ,132 38–35. doi: 10.1007/BF01138602

Palumbo, J. D., Baker, J. L., and Mahoney, N. E. (2006). Isolation of bacterial antagonists of *Aspergillus flavus* from almonds. *Microb. Ecol.* 52–45 ,52. doi: 10.1007/s-00248 -9096-006y

Pantaya, D., Morgavi, D. P., Silberberg, M., Martin, C., Wiryawan, K. G., and Boudra, H. (2014). Low pH enhances rumen absorption of aflatoxin B1 and ochratoxin A in sheep. *Glob. Vet.* 232–227 ,13. doi: 10.5829/idosi.gv.2014.13.02.84118

Paterson, R. R. M., and Lima, N. (2011). Further mycotoxin effects from climate change. *Food Res. Int.* 2566–2555 ,44. doi: 10.1016/j.foodres.2011.05.038

Paterson, R., and Lima, N. (2010). How will climate change affect mycotoxins in food? *Food Res. Int.* 1914–1902 ,43. doi: 10.1016/j.foodres.2009.07.010

Paterson, R., and Lima, N. (2011). Further mycotoxin effects from climate change. *Food Res. Int.* 2566–2555 ,44. doi: 10.1016/j.foodres.2011.05.038

Plaizier, J. C., Danesh Mesgaran, M., Derakhshani, H., Golder, H., Khafipour, E., Kleen, J. L., et al. (2018). Review: enhancing gastrointestinal health in dairy cows. *Animal.* 12, s399–s418. doi: 10.1017/S1751731118001921

Poppy, G. D., Rabiee, A. R., Lean, I. J., Sanchez, W. K., Dorton, K. L., and Morley, P. S. (2012). A meta-analysis of the effects of feeding yeast culture produced by anaerobic fermentation of *Saccharomyces cerevisiae* on milk production of lactating dairy cows. *J. Dairy Sci.* 6041–6027 ,95. doi: 10.3168/jds.5577-2012

Pusztahelyi, T., Holb, I. J., and Pócsi, I. (2015). Secondary metabolites in fungus-plant interactions. *Front. Plant Sci.* 6:573. doi: 10.3389/fpls.2015.00573

Qian, G., Tang, L., Guo, X., Wang, F., Massey, M. E., Su, J., et al. (2014). Aflatoxin B1 modulates the expression of phenotypic markers and cytokines by splenic lymphocytes of male F344 rats. *J. Appl. Toxicol.* 249–241 ,34. doi: 10.1002/jat.2866

Raju, M.V.L.N. and G. Devegowda, 2002. Esterified glucomannan in broiler chicken diets contaminated with Aflatoxin, Ochratoxin and T-2 toxin: evaluation of its binding ability (in vitro) and efficacy as immunomodulatory. Asian – Aust. J. Anim. Sci. 15(7):1051-1056.

Ramos, A.J. and E. Hernandez. 1997. Prevention of aflatoxicosis in farm animals by means of hydrated sodium calcium aluminosilicate addition to feedstuffs: a review. Anim. Feed Sci. Technol. 65:197-206.

Rawal, S., Kim, J. E., and Coulombe, R. A. Jr. (2010). Aflatoxin B1 in poultry: toxicology, metabolism and prevention. *Res. Vet. Sci.* 331–325 ,89. doi: 10.1016/j.rvsc.2010.04.011

Reddy, B. N., and Raghavender, C. N. (2007). Outbreaks of aflatoxicoses in India. *Afr. J. Food Agric. Nutr. Dev.* 15–1 ,7.

Reddy, K. R. N. (2010). An overview of mycotoxin contamination in foods and its implications for human health. *Toxin Rev.* 26–3 ,29. doi: 15569541003598553/10.3109

Robert, H., Payros, D., Pinto, P., Théodorou, V., Mercier-Bonin, M., and Oswald, I. P. (2017). Impact of mycotoxins on the intestine: are mucus and microbiota new targets? *J. Toxicol. Environ. Health B Crit. Rev.* 275–249 ,20. doi: 10937404.2017.1326071/10.1080

Romani, L. (2004). Immunity to fungal infections. *Nat. Rev. Immunol.* 23–1 ,4. doi: 10.1038/nri1255

Rooney-Latham, S., Janousek, C. N., Eskalen, A., and Gubler, W. D. (2008). First report of *Aspergillus carbonarius* causing sour rot of table grapes (*Vitisvinifera*) in California. *Plant Dis.* 92:651. doi: 10.1094/PDIS0651-4-92-A

Saalia, F. K., and Phillips, R. D. (2011a). Degradation of aflatoxins by extrusion cooking: effects on nutritional quality of extrudates. *LWT Food Sci. Technol.* 1501–1496 ,44. doi: 10.1016/j.lwt.2011.01.021

Sen, Y., Onal-Ulusoy, B., and Mutlu, M. (2019). Detoxification of hazelnuts by different cold plasmas and gamma irradiation treatments. *Innov. Food Sci. Emerg. Technol.* ,54 259–252. doi: 10.1016/j.ifset.2019.05.002

Shcherbakova, L., Statsyuk, N., Mikityuk, O., Nazarova, T., and Dzhavakhiya, V. (2015). Aflatoxin B1 degradation by metabolites of *Phoma glomerata* PG41 isolated from natural substrate colonized by aflatoxigenic *Aspergillus flavus*. *Jundishapur J. Microbiol.* 8:e24324. doi: 10.5812/jjm.24324

Sheppard, G. S. (2008). Determination of mycotoxins in human foods. *Chem. Soc. Rev.* ,37 2477–2468. doi: 10.1039/b713084h

Shi, H., Ileleji, K., Stroshine, R. L., Keener, K., and Jensen, J. L. (2017). Reduction of aflatoxin in corn by high voltage atmospheric cold plasma. *Food Bioprocess Technol.* 10:1042. doi: 10.1007/s8-1873-017-11947

Shi, H., Stroshine, R. L., and Ileleji, K. (2014). "Aflatoxin reduction in corn by cleaning and sorting," in *Proceedings of the 2014 ASABE Annual International Meeting Paper*, St. Joseph, MI. doi: 10.1007/s11947-017-1873-8

Shim, W.-B., Mun, H., Joung, H.-A., Ofori, J. A., Chung, D.-H., and Kim, M.-G. (2014). Chemiluminescence competitive aptamer assay for the detection of aflatoxin B1 in corn samples. *Food control* 35–30 ,36. doi: 10.1016/j.foodcont.2013.07.042

Siciliano, I., Spadaro, D., Prelle, A., Vallauri, D., Cavallero, M. C., Garibaldi, A., et al. (2016). Use of cold atmospheric plasma to detoxify hazelnuts from aflatoxins. *Toxins* 8:125. doi: 10.3390/toxins8050125

Sineque, A. R., Macuamule, C. L., and Dos Anjos, F. R. (2017). Aflatoxin B1 contamination in chicken livers and gizzards from industrial and small abattoirs, measured by ELISA technique in Maputo, Mozambique. *Int. J. Environ. Res. Public Health* 14:951. doi: 10.3390/ijerph14090951

Sirhan, A. Y., Tan, G. H., and Wong, R. C. S. (2013). Determination of aflatoxins in food using liquid chromatography coupled with electrospray ionization quadrupole time of flight mass spectrometry (LC-ESI-QTOF-MS/MS). *Food Control* 44–35 ,31. doi: 10.1016/j.foodcont.2012.09.016

Sirot, V., Fremy, J. M., and Leblanc, J. C. (2013). Dietary exposure to mycotoxins and health risk assessment in the second French total diet study. *Food Chem. Toxicol.* ,52 11–1. doi: 10.1016/j.fct.2012.10.036

Siwela, A. H., Siwela, M., Matindi, G., Dube, S., and Nziramasanga, N. (2005). Decontamination of aflatoxin-contaminated maize by dehulling. *J. Sci. Food Agric.* 2538–2535 ,85. doi: 10.1002/jsfa.2288

Śliżewska, K., and Smulikowska, S. (2011). Detoxification of aflatoxin B1 and change in microflora pattern by probiotic in vitro fermentation of broiler feed. *J. Anim. Feed Sci.* 309–300 ,20. doi: 10.22358/jafs/2011/66187

Squire, R. A. (1981). Ranking animal carcinogens: a proposed regulatory approach. *Science* 880–877 ,214. doi: 10.1126/science.7302565

Stadler, D., Sulyok, M., Schuhmacher, R., Berthiller, F., and Krska, R. (2018). The contribution of lot-to-lot variation to the measurement uncertainty of an LC-MS-based multi-mycotoxin assay. *Anal. Bioanal. Chem.* 4418–4409 ,410. doi: 10.1007/s5-1096-018-00216

Udomkun, P., Wiredu, A. N., Nagle, M., Müller, J., Vanlauwe, B., and Bandyopadhyay, R. (2017). Innovative technologies to manage aflatoxins in foods and feeds and the profitability of application - A review. *Food Control* 138–127 ,6. doi: 10.1016/j.foodcont.2017.01.008

Udovicki, B., Audenaert, K., De Saeger, S., and Rajkovic, A. (2018). Overview on the mycotoxins incidence in Serbia in the period 2004–2016. *Toxins* 10:279. doi: 10.3390/toxins10070279

Upadhaya, S. D., Sung, H. G., Lee, C. H., Lee, S. Y., Kim, S. W., Kyung, J., et al. (2009). Comparative study on the aflatoxin B1 degradation ability of rumen fluid from Holstein steers and Korean native goats. *J. Vet. Sci.* 34–29 ,10. doi: 10.4142/jvs.2009.10.1.29

Usman, M., Javed, M. R., Mehmood, M. A., Huma, T., and Ijaz, A. (2019). Isolation of aflatoxigenic *Aspergillus flavus* from animal-feed and exploration of the genetic basis of aflatoxin biosynthesis. *Pak. Vet. J.* (in press). doi: 10.29261/pakvetj/2019.078

van Asselt, E. D., van der Fels-Klerx, H. J., Marvin, H. J. P., van Bokhorst-van, H., and Nierop Groot, M. (2017). Overview of food safety hazards in the European dairy supply chain. *Compr. Rev. Food Sci. Food Saf.* 75–59 ,16. doi: -1541/10.1111 4337.12245

Varga, J., Baranyi, N., Chandrasekaran, M., Vágvölgyi, C. S., and Kocsubé, S. (2015). Mycotoxin producers in the *Aspergillus* genus: an update. *Acta Biol. Szeged* ,59 167–151.

Variane, A. C. F., Santos, F. C., Castro, F. F., Barbosa-Tessmann, I. P., Santos, G. T., and Santos Pozza, M. S. (2018). The occurrence of aflatoxigenic *Aspergillus* spp. in dairy cattle feed in Southern Brazil. *Braz. J. Microbiol.* 928–919 ,49. doi: 10.1016/j.bjm.2018.05.005

Wang, H., Li, W., Muhammad, I., Sun, X., Cui, X., and Cheng, P. (2018). Biochemical basis for the age-related sensitivity of broilers to aflatoxin B1. *Toxicol. Mech. Methods* ,28 368–361. doi: 15376516.2018.1428258/10.1080

Wang, W., Heitschmidt, G. W., Ni, X., Windham, W. R., Hawkins, S., and Chu, X. (2014). Identification of aflatoxin B1 on maize kernel surfaces using hyperspectral imaging. *Food Control* 86–78 ,42. doi: 10.1016/j.foodcont.2014.01.038

Wang, Y., Zhang, H., Yan, H., Yin, C., Liu, Y., Xu, Q., et al. (2018). Effective biodegradation of aflatoxin B1 using the *Bacillus licheniformis* (BL010) strain. *Toxins* 10:497. doi: 10.3390/toxins10120497

Weaver, M. A., and Abbas, H. K. (2019). Field displacement of aflatoxigenic *Aspergillus flavus* strains through repeated biological control applications. *Front. Microbiol.* 10:1788. doi: 10.3389/fmicb.2019.01788

Westlake, K., Mackie, R. I., and Dutton, M. F. (1989). In vitro metabolism of mycotoxins by bacterial, protozoal and ovine ruminal fluid preparations. *Anim. Feed Sci. Tech.* 25, 169–178. doi: 10.1016/0377-8401(89)90117-X

Wu, F., Bhatnagar, D., Bui-Klimke, T., Carbone, I., Hellmich, R., Munkvold, G., et al. (2011). Climate changes impacts on mycotoxin risks in US maize. *World Mycotoxin J.* 93–79 ,4. doi: 10.3920/WMJ2010.1246

Wyllie, T.D. and Morchause,L.G. 1978. Mycotoxin Fungi, Mycotoxins, Mycotoxicoses-An Encyclopedic Handbook.Vols.1,2, and 3.Marcel Dekker, Inc. New york.

Xiao, M. W., Bai, X. L., Liu, Y. M., Yang, L., and Liao, X. (2018). Simultaneous determination of trace aflatoxin B1 and ochratoxin A by aptamer-based microchip capillary electrophoresis in food samples. *J. Chromatogr. A* 228–222 ,1569. doi: 10.1016/j.chroma.2018.07.051

Xu, F., Ren, K., Yang, Y., Gio, J., Ma, G., Liu, Y., et al. (2015). Food safety special issue: immunoassay of chemical contaminants in milk: a review. *J. Integr. Agric.* ,14 2295–2282. doi: 10.1016/S2-61121(15)3119-2095

Yan, M., Li, A., Liang, G., and Ligang, L. (2017). Recent advances in mycotoxin determination for food monitoring via microchip. *Toxins* 9:324. doi: 10.3390/toxins9100324

Yilmaz, S., Kaya, E., Karaca, A., and Karatas, B. (2018). Aflatoxin B1 induced renal and cardiac damage in rats: protective effect of lycopene. *Res. Vet. Sci.* 275–268 ,119. doi: 10.1016/j.rvsc.2018.07.007

Yu, J. (2012). Current understanding on aflatoxin biosynthesis and future perspective in reducing aflatoxin contamination. *Toxins* 1057–1024,4. doi: 10.3390/toxins4111024

Yunus, A. W., Razzazi-Fazeli, E., and Bohm, J. (2011). Aflatoxin B1 in affecting broiler's performance, immunity, and gastrointestinal tract: a review of history and contemporary issues. *Toxins* 590–566 ,3. doi: 10.3390/toxins3060566

Zheng, N., Zhang, H., Li, S., Wang, J., Liu, J., Ren, H., et al. (2018). Lactoferrin inhibits aflatoxin B1- and aflatoxin M1-induced cytotoxicity and DNA damage in Caco-2, HEK, Hep-G2, and SK-N-SH cells. *Toxicon* 85–77 ,150. doi: 10.1016/j.toxicon.2018.04.017

Zhu, F. (2018). Effect of ozone treatment on the quality of grain products. *Food Chem.* ,264 366–358. doi: 10.1016/j.foodchem.2018.05.047

Zhu, W., Wei, Z., Xu, N., Yang, F., Yoon, I., Chung, Y., et al. (2017). Effects of *Saccharomyces cerevisiae* fermentation products on performance and rumen fermentation and microbiota in dairy cows fed a diet containing low quality forage. *J. Anim. Sci. Biotechnol.* 8:36. doi: 10.1186/s3-0167-017-40104

Zielińska, K. J., and Fabiszewska, A. U. (2018). Improvement of the quality of maize grain silage by a synergistic action of selected lactobacilli strains. *World J. Microbiol. Biotechnol.* 9–1 ,34. doi: 10.1007/s9-2400-017-11274

Zulkifli, N. A., and Zakaria, L. (2017). Morphological and molecular diversity of *Aspergillus* from corn grain used as livestock feed. *HAYATI J. Biosci.* 34–26 ,24. doi: 10.1016/j.hjb.2017.05.002

INDEX

www.ingramcontent.com/pod-product-compliance
Lightning Source LLC
LaVergne TN
LVHW062220200926
848484LV00006B/1443
9781787150362